Spectroscopy of Biological Molecules

Spectroscopy of Biological Molecules

Edited by

R.E. Hester and R.B. Girling
Chemistry Department, University of York

THE ROYAL SOCIETY OF CHEMISTRY
1841-1991

Proceedings of the Fourth European Conference on the
Spectroscopy of Biological Molecules, York, England,
1st–6th September 1991

Special Publication No. 94

ISBN 0-85186-437-6

A catalogue record of this book is available from the British
Library

© The Royal Society of Chemistry 1991

Published by The Royal Society of Chemistry,
Thomas Graham House, Science Park, Cambridge CB4 4WF

Printed and bound in Great Britain by Bookcraft (Bath) Ltd.

PREFACE

This volume contains articles prepared by the invited lecturers and those making poster presentations at the Fourth European Conference on the Spectroscopy of Biological Molecules, held at the University of York, England, on 1-6 September, 1991. Previous meeting places for this Conference have been Reims, France (1985), Freiburg, Germany (1987), and Rimini, Italy (1989). These have become truly international meetings and the present volume contains articles from many countries outside as well as within Europe.

The scientific emphasis remains on vibrational spectroscopy, mainly infrared and Raman, as applied to the study of structure and dynamics in biological and related model systems. However, a significant number of the papers are devoted to applications of other methods, such as NMR, CD, MCD, X-ray absorption and diffraction, optical absorption and fluorescence, and molecular graphics simulations and other theoretical methods. The aim has been to achieve a well balanced, critically comparative review of recent progress in this field of biomolecular structure, bonding, and dynamics, based on applications of the full range of spectroscopic methods. Within the context of this biophysical spectroscopy approach, an emphasis has been given to currently "hot" topics such as protein folding, biomolecular interactions and dynamics, supramolecular organization, and ultrafast photobiological processes.

As is seen from the following list of contents, the book is organized in Sections I to VIII. These correspond to the lecture sessions at the Conference, with poster papers being assigned to the section most relevant to their scientific content. Four pages are associated with most of the lectures, two with posters. A balance between established world authorities and younger scientists with exciting new results to report was sought for the lectures, and many other distinguished leaders in the field are included as authors in the poster papers.

We are grateful to all those authors who submitted their camera-ready manuscripts on time and in the standard format prescribed by our publisher. We have made relatively minor corrections and additions to some papers for clarity and legibility but have been obliged to restrain ourselves to only very light editing in the interests of meeting tight publication schedules. Our special thanks go to our secretaries Shirley Leedham and Jean Scott for their hard work in making many somewhat dubious manuscripts presentable and in preparing the list of contents and author index.

These certainly are exciting times to be involved in research into the structure, bonding, and dynamics of biological molecules. The relation between these properties and the biological function of such molecules as proteins, lipids, and nucleic acids is fundamental. This book shows much of what can now be achieved by skilled practitioners using the powerful spectroscopic "tools of the trade" and we hope that it will serve to inform and thus to inspire others to still greater achievements. Its minimal purpose is to serve as a handbook for conference participants but we believe that its influence can be greater and its impact stronger.

Finally, we wish to acknowledge the co-sponsorship of this meeting by the International Union of Pure and Applied Chemistry (IUPAC) and the following learned societies: The Association of British Spectroscopists, The Biochemical Society, The Federation of European Chemical Societies (FECS), The Institute of Biology, The Institute of Physics, The International Union of Biochemistry (IUB), The Royal Society of Chemistry (RSC), The Society of Chemical Industry (SCI). We are grateful for financial support received from the Commission of the European Communities, Glaxo Group Research Ltd., Imperial Chemical Industries plc, The International Council of Scientific Unions, Pfizer Ltd., The Royal Society, and SmithKline Beecham plc.

Ronald E Hester   Chemistry Department
Reuben B Girling   University of York
                                                    York YO1 5DD
Editors   England

CONTENTS

Page

I

THEORY

PROTEIN STRUCTURE ANALYSIS

MEMBRANE PROTEINS

MOLECULAR GRAPHICS AND MODELLING

R.E. Hubbard

Department of Chemistry
University of York
York
YO1 5DD

INTRODUCTION

During the early 1980s molecular graphics systems were only
available in the more progressive drug design groups or in a few
large protein crystallographic groups, mainly because of the cost of
the computing and graphics systems required. Over the past six
years however, the combination of a rapid decrease in computer
costs and the development of new software and techniques has
ensured that molecular graphics and modelling have become
increasingly available tools in all areas of the physical sciences. In
particular, computer graphics systems have become almost
indispensible for the study of macromolecular structures and
interactions.

Currently (mid-1991) it is possible for the price of a standard
uv/vis spectrometer to acquire a molecular graphics system which is
capable of displaying the structures of macromolecules with real
time rotation, depth cueing and colour. Such a system also provides
the disk storage and computer power to effectively interrogate and
analyse the structure, to perform moderately complex calculations
on the molecule, and to compare features of the molecule against
databases of other structures.

My brief is to give a summary of the current status of molecular
graphics and modelling with particular emphasis on the study of
the structure of biological molecules. In my presentation I will use
examples of recent modelling exercises at York to illustrate the
range and power of the techniques available as applied to real
problems. Here I will describe the techniques.

REPRESENTATION OF STRUCTURE

The visualisation of the structure of a macromolecule is a non-trivial
task. Mechanical models of molecules containing many thousands
of atoms are cumbersome, difficult to construct, usually inaccurate

and prone to gravity. Many of the scientists who have acquired molecular graphics systems over the past few years, have done so mainly for the ability to effectively manage and display these structures.

The standard technique for representing structures is to draw just the bonds between the atoms to produce a stick representation of the molecule. As most of the graphics systems available contain hardware optimised for drawing vectors very rapidly it is possible to view molecules drawn in this way with real time rotation, depth cueing, clipping and in stereo. Such dynamic viewing is crucial to gain the three dimensional perception of the molecule essential for full appreciation of the structure. In addition, colour can be used to represent various properties of parts of the molecule, such as hydrophobicity, charge distribution, hydrogen bonding potential, subunit or domain structure, binding sites etc.

The representation of the surface of a molecule is important as it is through the surface that most molecules perform their function. A variety of techniques are available which represent the surface as dots drawn typically at the van der Waals radius of each atom, or representing the surface accessibility to various sized probe molecules. These dot surfaces (as very short vectors) are also manipulable in real time on most graphics systems. Very recently there have been great improvements in the performance of raster graphics hardware such that a large number of shaded polygons can be drawn rapidly. This means that cartoon-like representations of protein secondary structure as cylinders and ribbons are now possible as real time displays. However for most macromolecules, a shaded CPK representation of a molecule will remain a static representation for a few more years on all but the most powerful graphic systems.

INTERROGATION AND COMPARISON OF STRUCTURE

As a molecular graphics system holds the coordinates for the molecule under study, it can rapidly report on the distance and angles between specified atoms, display hydrogen bonding patterns and rapidly present other geometrical information, such as Ramachandran diagrams for proteins.

The past few years have seen an explosion in the number of macromolecular structures determined by X-ray crystallography, and increasingly by multi-dimensional nmr. Appropriately organised computer databases are currently under development which allow the scientist to explore the wealth of structural information contained in this data. For example it is now possible to rapidly extract related protein molecules, to overlap the structures on a variety of criteria and to analyse the origin of structural similarity and conformational change.

MODELLING MOLECULAR INTERACTIONS

The early days of molecular modelling saw intense development of techniques for drug design. Because there were few crystal structures available for proteins of medicinal interest, much of this work focussed on analysing the structure of pharmacologically active molecules to explore the relationship between structure and activity. The revolution in molecular biology of the 1980s and the speed of modern structure determination methods has stimulated a more proactive strategy. Once a target protein has been identified, it can be cloned, over expressed and if all goes well, structure(s) determined, often with a variety of drugs/inhibitors bound.

The availability of these structures has revitalised development of techniques for designing molecules to fit particular binding sites. These are based on mapping the functional group requirements of a binding site, followed by generation from either templates or structural databases of molecules that have the required chemistry and shape. The aim is to suggest new compounds for modification and design, using the molecular graphics system to rationalise the results of experiment.

STRUCTURE PREDICTION

The prediction of protein structure from sequence alone remains a holy grail of molecular modelling. Despite substantial efforts, our ability to predict the secondary structure of a protein on the basis of sequence alone remains little better than random. Much greater success has been achieved by homology modelling where the central theme is that if a group of proteins have similar sequence then they probably have similar structure.

Given the sequence of a protein of interest, the crucial step is to establish whether it has a similar sequence to other proteins whose structure is known. If there is reasonable homology, particular in the major regions of secondary structure or at particular amino acids crucial for activity, then a model for the unknown sequence can be built. Typically the conserved core of the structure is taken and the loops constructed, usually from fragment databases. It is in the loops connecting conserved regions of secondary structure where most of the sequence and structural variability is seen in families of structures. Finally the side chain conformations can be built for the variable amino acids, again drawing on database and other geometric information.

PROPERTY PREDICTION

It is here that molecular modelling embraces the macromolecular calculations introduced by Martin Karplus. Great efforts are under way to develop and apply molecular dynamics calculations, free energy simulations, exotic quantum mechanics/molecular mechanics, and detailed electrostatic calculations to understand the details of macromolecular structures extending from the structures provided by X-rays and nmr. Molecular graphics systems can provide an important interface to these calculations and a crucial tool for analysing their results.

SATISFYING EXPERIMENTAL CONSTRAINTS

One of the most successful recent developments in computational techniques is the application of simulated annealing/molecular dynamics techniques to generate and refine structures from X-ray crystallographic and nmr data. Here, the x-ray gradients or the NOE distances are added to the standard geometric terms describing the molecule to provide additional constraints on the structure as it explores conformational space.

This type of approach can be extended to other experimental information. Data from a variety of spectroscopic measurements can be converted into constraints that the structure must satisfy (for example, certain amino acids on the surface or buried). The increasing speed of computers, the wealth of information available from other structures and the sophistication of the algorithms mean that this area of satisfying experimental constraints will be an important area of development.

CONCLUDING REMARKS

This is an exciting time to be involved in the study of macro-molecular structure. Many of the crucial technologies and methods are maturing so that some of the central secrets of molecular biology will be unlocked over the next few years. Molecular graphics systems will provide both a window and an invaluable tool in displaying, analysing and modelling these structures.

QUANTITATIVE DETERMINATION OF THE SECONDARY STRUCTURE OF SOLUBLE AND MEMBRANE PROTEINS FROM FACTOR ANALYSIS OF INFRARED SPECTRA

D. C. Lee[*#], P. I. Haris[+], D. Chapman[+] and R. C. Mitchell[#]

[#]SmithKline Beecham Pharmaceuticals
The Frythe
Welwyn AL6 9AR, UK

[+]Royal Free Hospital School of Medicine
Rowland Hill St.
London NW3 2PF, UK

INTRODUCTION

The structure of proteins is now accessible by a variety of techniques. Of these, X-ray crystallography currently provides the most detailed descriptions but has the disadvantage of being restricted to those proteins that can be persuaded to form crystals of adequate size and order. Significant advances are being made by two-dimensional NMR spectroscopy but these are limited to proteins of relatively low molecular weight. CD spectroscopy can provide both qualitative and quantitative assessments of overall secondary structure. However, β-turn and β-sheet structures are difficult to distinguish and the technique is less easily applied to membrane proteins. The development of both qualitative assignments and, in recent years, quantitative methods makes vibrational spectroscopy a powerful technique for determining overall secondary structure in both soluble and membrane proteins. This approach is being applied to conformational changes associated with function and the interactions with drug molecules.

Qualitative analysis of the secondary structure of proteins has developed from early measurements on the position of the amide I band maximum in the infrared spectra of ^2H_2O and H_2O solutions.[1,2] Further advances came with the application of deconvolution and derivative techniques for the visualisation of the several overlapping components of the amide bands.[3-5] Neither method offers complete 'resolution' of the amide I components assigned to each secondary structure element. This, together with the influence of band width on relative intensities, means that quantitative estimations should not be obtained directly from these spectra. A favoured approach has been to use deconvolution to identify the overlapping components followed by curve-fitting of these components to either the deconvolved[6] or original[7] spectra. Good agreement between infrared values for the percentages of α-helix and β-sheet and the original estimations from X-ray data has been obtained.[6] However, there are a number of problems with this approach. Deconvolution procedures should be used conservatively since over-deconvolution produces negative side-lobes and reveals noise and water-vapour. It is difficult to apply curve-fitting procedures reproducibly between laboratories. Finally, each amide I component cannot always be assigned with confidence so that summing the components according to structure may not be accurate. As an alternative we have investigated the use of factor analysis of the infrared spectra of proteins after subtraction of the appropriate water/buffer spectrum.[8,9] A method based on factor analysis followed by correlation of the factor loadings with structural composition was used. This gave Standard Errors of Prediction (SEP) of 3.9%, 8.3% and 6.6% for α-helix, β-sheet and turn content respectively, of 17 soluble proteins (calibration set).[9] A similar method based on partial least squares (PLS) analysis of the infrared spectra of H_2O solutions of 13 proteins has been reported.[10] We now discuss the application of the partial least squares method to our calibration set of 17 proteins and apply this method to the spectra of membrane proteins.

METHOD

The mathematical basis of factor analysis is described elsewhere.[11] A calibration set was constructed from the infrared spectra of 17 proteins whose secondary structure has been determined by analysis[12] of original X-ray data. Factor analysis generates a series of abstract spectra or eigenspectra which may be combined to generate the original data within experimental error. The contribution (loading) of each factor to each spectrum in the calibration set is determined and correlations obtained between the factor loadings and the structural composition. Those factors which show high correlation are retained in the regression equations. Analysis of an unknown proceeds by determination of the factor loadings followed by substitution in the relevant regression equation for each structure. The method is validated by prediction of each protein using a calibration set regenerated from the spectra of the remaining proteins. The PLS method used in this paper differs from our earlier method mainly in the calculation of the factors. A separate factor analysis is performed for each structure and is weighted according to structural composition. The effect of this is to include more concentration information in the first few factors and as a consequence fewer factors are required for the prediction of each component. The PLS method we use is that of Haaland and Thomas[13] as supplied by Galactic Industries Inc.

The spectra of the soluble proteins used to generate the calibration set were as described earlier[9] and pretreatment was identical. Essentially, pretreatment involves correction for variation in baseline position at $1700 cm^{-1}$ and normalisation of amide I band area to account for variations in pathlength when working with $6\mu m$ pathlength cells. The spectra of the various membrane proteins were recorded after purification of each protein in its endogenous lipid environment by established methods. Difference spectra were generated by interactive subtraction of the appropriate buffer spectrum. Prior to PLS analysis the spectra were pretreated as above.

RESULTS

The accuracy of the PLS method was investigated by the validation procedure described above. The method was applied to the amide I region, $1700-1600 cm^{-1}$. Figure 1 presents plots of predicted versus actual composition for α-helix, β-sheet and turn structures. The SEP values were calculated as for our earlier method[9] by comparison with the composition according to X-ray data.[12] The following values were obtained:4.8% for α-helix (5 factors), 6.4% for β-sheet (1 factor) and 3.3% for turn structures (7 factors). Thus, compared to our earlier method, the accuracies for β-sheet and turns are improved at the expense of a small reduction in accuracy for α-helix. Using the amide I/II region ($1700-1500 cm^{-1}$) the SEP values were 7.1% for α-helix (5 factors), 6.2% for β-sheet (2 factors) and 4.3% for turns (4 factors). The reduction in accuracy for helical and turn structures may occur because of interference from the absorption of amino acid side-chains in the $1600-1500 cm^{-1}$ region.

The predictions for the membrane proteins were made using the $1700-1600 cm^{-1}$ region and a calibration set containing all 17 soluble proteins. The results are presented in Table 1.

Table 1   Secondary Structure Prediction for Membrane Proteins using PLS

Membrane Protein	% α-helix	% β-sheet	% turn
Ca^{2+}-ATPase	65.1	15.7	6.8
H^+/K^+-ATPase	47.6	33.0	21.5
Photosystem II Reaction Centre	60.0	17.6	22.2
Bacteriorhodopsin	71.5	5.8	-11.2

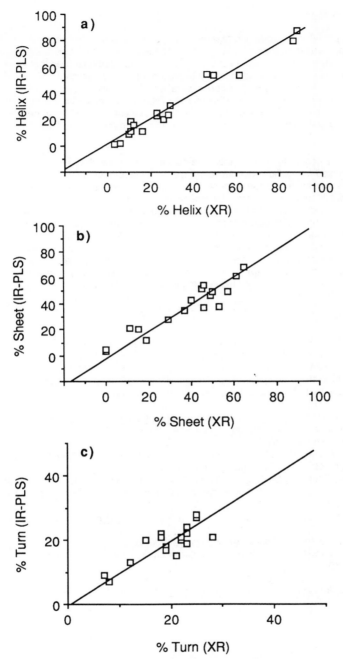

Figure 1   Comparison of secondary structure predictions by partial least squares analysis of infrared spectra (IR-PLS) with estimations[12] from X-ray data (XR) for 17 soluble proteins.

CONCLUSION

This method provides accurate and reproducible measurements of the secondary structure of proteins from their infrared spectra. The accuracies are comparable to those obtained from CD and Raman spectra. We have also demonstrated that, unlike CD which can be affected by scattering artefacts, the method is applicable to membrane proteins. Our approach offers several advantages over the deconvolution/curve-fitting method. Deconvolution is not required and pre-treatment of the data is minimised. Assignment of the several amide I components is not necessary. Additionally, no curve-fitting, a procedure which is difficult to apply reproducibly, is required.

The results obtained for the membrane proteins are consistent with previous estimations based on predictions from sequence data, electron diffraction data or qualitative interpretation of infrared spectra. For example, the prediction for Ca^{2+}-ATPase is consistent with earlier measurements by Raman, infrared and CD spectroscopies and a prediction based on amino-acid sequence. In addition, the helix estimation for bacteriorhodopsin is in close agreement with the available diffraction data and sequence prediction. The appearance of a negative value is probably a reflection of the spectral properties of bacteriorhodopsin lying outside the range of properties defined by the calibration set. Nevertheless this approach, which is free from artefacts caused by the membrane environment and may readily be applied to membrane proteins in their native form, shows high promise for accurate quantitative descriptions of secondary structure.

REFERENCES

1. H. Susi, 'Structure and Stability of Biological Molecules', S.N. Timasheff and G.D. Fasman eds., Marcell Dekker, New York, 1969. pp. 573-633.
2. J.L. Koenig and D.L. Tabb, 'Analytical Applications of FTIR to Molecular and Biological Systems', J.R. Durig ed., Reidel, Holland, 1980. pp. 241-255.
3. J.K. Kauppinen, D.J. Moffat, H.H. Mantsch and D.G. Cameron, Appl. Spectrosc., 1981, 35, 271.
4. H. Susi and D.M. Byler, Biochem. Biophys. Res. Commun., 1983, 115, 391.
5. D.C. Lee and D. Chapman, Bioscience Rep., 1986, 6, 235.
6. D.M. Byler and H. Susi, Biopolymers, 1986, 25, 469.
7. P.W. Yang, H.H. Mantsch, J.L.R. Arrondo, I. Saint-Girons, Y. Guillou, G.N. Cohen and O. Barzu, Biochemistry, 1987, 26, 2706.
8. D.C. Lee, P.I. Haris, D. Chapman and R.C. Mitchell, 'Spectroscopy of Biological Molecules - State of the Art', A. Bertoluzza, C. Fagnano and P. Monti eds., Societa Editrice Esculapio, Bologna, 1989. pp. 57-58.
9. D.C. Lee, P.I. Haris, D. Chapman and R.C. Mitchell, Biochemistry, 1990, 29, 9185.
10. F. Dousseau and M. Pezolet, Biochemistry, 1990, 29, 8771.
11. E.R. Malinowski and D.G. Howery, 'Factor Analysis in Chemistry', Wiley, New York, 1980.
12. M. Levitt and J. Greer, J. Mol. Biol., 1977, 114, 181.
13. D.M. Haaland and E.V. Thomas, Anal. Chem., 1988, 60, 1193.

UV RAMAN STUDIES OF BIOLOGICAL STRUCTURE AND DYNAMICS AND ENERGY TRANSFER

Sanford A. Asher

The development of UV resonance Raman spectroscopy as a probe of biological structure and function presents opportunities to study both average ground state macro-molecular structure as well as molecular dynamics (1). In addition, the Raman excitation profiles and the enhanced vibrational modes detail information on the molecular electronic excited states. These excited states may be biologically relevant for biomolecular function since they may contribute to transition state structure and be involved in electron and energy transfer pathways.

We will review our work in this area and touch on aspects of these issues. For proteins we will examine enhancement of the amide modes and correlate the extent of amide mode enhancement with geometry changes in the excited state (2,3). The strong enhancement of the amide II and amide III modes of polylysine in the random coil, alpha helix and beta sheet conformations demonstrate that the geometry and or the bonding of the amide CO-N linkage is distorted along the C-N bond (Figure 1). The facile photochemical cis-trans photoisomerization and the observed enhancement of the putative amide V overtone vibration indicates that the peptide π^* excited state is twisted compared to the ground state (2). In addition, the excited state appears to be conjugated over the peptide backbond. This result may have great significance for electron transfer rates in peptides as well as for peptide conformation geometry constraints.

We will also review our examination of the conjugation between the peripheral vinyl groups of the heme prosthetic group of hemoglobins and myoglobin. We earlier demonstrated the lack of conjugation of the vinyl groups with the heme pi orbitals; the vinyl stretches are selectively enhanced by the isolated vinyl $\pi \rightarrow \pi^*$ transitions (4). The vinyl group resonance Raman depolarization ratio for excitation in the vinyl group $\pi \rightarrow \pi^*$ is 0.33 for both heme and a monovinyl heme derivative. This indicates that the vinyl C=C motion is uncorrelated between the two heme vinyl groups. The observed lack of vinyl-heme group conjugation suggests that control of the vinyl group orientation is not likely to be a useful route for the protein to modulate the heme ligand affinities or oxidation potentials.

We have utilized the pulsed laser sources typically used in Raman spectral measurements to examine T_1 relaxation rates in proteins (5,6). The technique, which is called Raman Saturation Spectroscopy, uses the Raman intensities to monitor the concentration of ground state species during the ca 10 nsec laser pulse. Absorption removes ground state population while relaxation repopulates the ground state. The non linear variation of the Raman intensity with increases in the incident intensity, can be used to determine the T_1 relaxation rate (Figures 2 and 3). This T_1 relaxation rate depends upon the radiative and non radiative relaxation rates which are influenced by near neighbor groups which are involved in electron transfer and Forster energy transfer. We will discuss the use of saturation Raman spectroscopy to examine electronic communication between the tyrosines and tryptophans at the hemoglobin subunit interfaces with the heme rings.

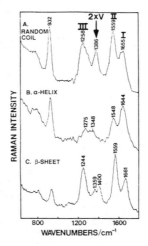

Figure 1. Raman spectra of PLL (0.17 mM) in water at 218-nm excitation, with $NaClO_4$ (0.2 M). The 932-cm^{-1} band derives from ClO_4^- stretching: (A) random coil form, pH = 4.0, 25°C; (B) α-helix form, pH = 11.0, 25 °C; (C) β-sheet form, pH = 11.3, 52 °C.

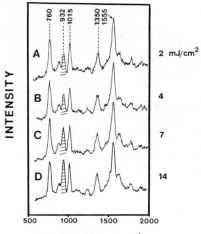

Figure 2. Resonance Raman spectra of TRP (pH 6.5) excited at 225 nm with pulse energy fluxes below 15 mJ/cm^2. Pulse energy fluxes listed in units of mJ/cm^2. Internal standard band shaded.

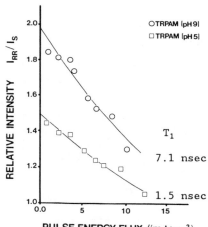

Figure 3. Raman saturation of the 1555-cm^{-1} band of TRPAM (pH 9) (open circles) and TRPAM (pH 5) (open squares) at 225-nm excitation. 0.5 mM analyte, 1 M perchlorate. Solid curves derive from a quantitative model for the saturation. See text for details.

13

REFERENCES

1. S.A. Asher, <u>Annual Review Physical Chemistry</u>, 1988, <u>39</u>, 537.
2. S. Song, S.A. Asher, S. Krimm, and K.D. Shaw, <u>J. Am. Chem. Soc.</u>, 1991, <u>113</u>, 1155.
3. S. Song and S. A. Asher, <u>J. Am. Chem. Soc.</u>, 1989, <u>111</u>, 4295.
4. V.L. DeVito and S.A. Asher, <u>J. Am. Chem. Soc.</u>, 1989, <u>111</u>, 9143.
5. J. Teraoka, P.A. Harmon, and S.A. Asher, <u>J. Am. Chem. Soc.</u>, 1990, <u>112</u>, 2892.
6. P.A. Harmon, J. Teraoka, and S.A. Asher, <u>J. Am. Chem. Soc.</u>, 1990, <u>112</u>, 8789.

THE STRUCTURE AND DYNAMICS OF M13 PROCOAT PROTEIN MEMBRANE INSERTION

Horst Vogel, Mark Soekarjo, Eric Thiaudiere and Andreas Kuhn[+]

Ecole Polytechnique Fédérale de Lausanne, Switzerland,
[+]University of Karlsruhe, FRG

The aim of our research is to determine which structural and dynamical features of a newly synthesized integral membrane protein allows it to insert and (partly) translocate across a lipid bilayer. A combination of molecular biological and biophysical methods have been applied to reveal the kinetics of the folding of the M13 procoat protein at a molecular level.

The coat protein of coliphage M13 has attracted increasing attention as a simple model to study the protein insertion into biological membranes [1]: (i) M13 coat protein is synthesized in E. *coli* cells as a precursor procoat, with a 23-residue leader sequence at its N-terminus (Fig. 1). (ii) After synthesis, procoat binds to the inner surface of the plasma membrane. (iii) In the presence of a transmembrane electrical potential, procoat is inserted across the bilayer. (iv) In this configuration procoat is cleaved by leader peptidase yielding the transmembrane coat protein and leader peptide.

$^{-23}$MKKSLVLKASVAVATLVPMLSFA^{-1}
$^{+1}$AEGDDPAKAA FNSLQASATEYIGYAWAMVV VIVGATIGIK LFKKFTSKAS^{50}

FIGURE 1: Amino acid sequence of M13 procoat protein

At present the molecular structure of procoat at the different stages is unknown as is the molecular mechanism by which the polypeptide binds to and inserts into the lipid bilayer. Because the protein is soluble at low concentrations in water, the whole assembly pathway can be reconstituted with purified M13 procoat protein.

The following basic questions are addressed: (i) Which conformational changes does the procoat protein undergo during the membrane insertion process? (ii) What is the structural flexibility of the protein in aqueous solution, in its membrane bound form and its transbilayer form? (iii) What is the mechanism whereby the protein binds to lipid membranes and inserts across the bilayer in the presence of an electrical potential?

A procedure has been developed in our laboratory allowing the isolation of milligram quantities of the M13 procoat protein, both the wild type form as well as several mutant proteins [2] with defects in membrane binding and/or transbilayer insertion. The structure of the procoat protein(s) (conformation, orientation and state of aggregation) at various stages of membrane insertion has been determined by using spectroscopic techniques such as circular dichroism, infrared and Raman spectroscopy.

Furthermore, the intrinsic fluorescence of the procoat protein due to a single tryptophan residue was used to study the thermodynamics and kinetics of procoat protein binding to and insertion into lipid bilayers (steady state titration, stopped-flow and temperature jump experiments detecting changes of the intrinsic fluorescence). The internal protein mobility was investigated by time-resolved fluorescence measurements. The question of how the different procoat proteins affect the properties of the lipid bilayer have been addressed by spectroscopic (fluorescence, infrared) and electrical conductance measurements on planar lipid bilayers.

The results of all these experiments were taken as a proper basis for molecular dynamics calculations. Such calculations yielded a reasonable description of the protein structural fluctuations at the various stages of association with lipid membranes. For a summary of the biophysical methods used here see refs. 3-5 .

References
1.   W. Wickner (1988) Biochemistry 27, 1081-1086.
2.   A. Kuhn, W. Wickner and G. Kreil (1986) Nature 322, 335-339.
3.   H. Vogel and F. Jähnig (1986) Biophys. J. 50, 573-582.
4.   H. Vogel, et al (1988) Proc. Natl. Acad. Sci. USA 85, 5067-5071.
5.   H. Vogel, et al (1989) in "Proceedings of the 3rd European Conference on Spectroscopy of Biological Molecules" pp. 219 - 222.

DEVELOPMENT OF SELF-CONSISTENT MOLECULAR MECHANICS ENERGY PARAMETERS FOR PEPTIDES FROM VIBRATIONAL SPECTROSCOPY

K. Palmö, L-O. Pietilä and S. Krimm
Biophysics Research Division and Department of Physics,
University of Michigan, Ann Arbor, Michigan 48109, USA

The rapid development of computers in recent years has brought increasingly complex organic compounds into the range of high level *ab initio* calculations. These calculations provide a valuable source of information that is well worth utilizing in the construction of molecular mechanics (MM) potential energy functions.

We have previously reported on a method by which scaled *ab initio* force fields and structures can be transformed into MM force constants and reference geometry parameters, provided that reasonable parameters for the nonbonded interactions are known [1]. We call the resulting MM force field a Spectroscopically Determined Force Field (SDFF), since it retains the *ab initio* vibrational frequencies. When testing the SDFF procedure on N-methylacetamide and some alanine dipeptides, a major difficulty was to find parameters of the van der Waals interactions that were reasonably consistent with the *ab initio* results. It became evident that adjustment of these parameters was needed as a complement to the original method. In this work the SDFF procedure has therefore been further developed to allow for optimization of van der Waals parameters using the criterion that, after being transformed, certain force constants should have values close to one another. The optimization is done simultaneously with the calculation of the MM valence force field and reference geometry. In order to take full advantage of the procedure, *ab initio* results for several conformations of a group of related molecules should be available.

The method is being applied to a series of alanine dipeptides [2,3] with the purpose of constructing a complete molecular mechanics energy function capable of accurately reproducing vibrational frequencies as well as structures. Of the *ab initio* conformations we have chosen to start with the β_2-, α_R-, α_L-, and α'- structures, because they do not contain any hydrogen bonds. In the transformation we used van der Waals parameters and atomic charges from refs. [4] and [5], respectively. The van der Waals parameters of ref. [4], which

Table I   Initial and final one-atom Lennard-Jones 12-6 parameters.
Units: Energy (kcal/mol), distance (Å)

Atom type	Initial		Final	
	a	b	a	b
C	653.0	19.8	653.0	19.8
O	118.6	8.7	118.6	8.7
N	300.6	13.0	300.6	13.0
H (on C)	57.6	3.2	57.6	3.2
H (on N)	57.6	3.2	17.7	3.2

are used by the CHEM-X program, are simple and involve only one atom type each for hydrogen, carbon, nitrogen, and oxygen. However, their compatibility with the *ab initio* force fields and structures improved considerably when a new value was optimized for the repulsive parameter of hydrogen attached to nitrogen. The initial and final van der Waals parameters are shown in Table I.

The transformation produces a complete general valence force field, which is impractical to use in MM calculations. However, the number of force constants is greatly reduced if the cross terms that do not significantly affect the frequencies are omitted. If necessary, the reduced force field and the reference geometry can of course be reoptimized to the *ab initio* frequencies and structures, or to experimental data, using the normal CFF-procedure [6,7].

The SDFF technique offers, among other things, a convenient method to detect unphysical features in a model for nonquadratic interactions, provided that the *ab initio* results are reliable. Thus, the compensation of such features by unphysical quadratic interactions can be avoided.

References

1. K. Palmö, L.-O. Pietilä, and S. Krimm, J. Comp. Chem, in press.
2. T.C. Cheam and S. Krimm, J. Mol. Struct. (Theochem), 188, 15 (1989).
3. T.C. Cheam and S. Krimm, J. Mol. Struct. (Theochem), 206, 173 (1990).
4. E.K. Davies and N.W. Murrall, Computers Chem. 13, 149 (1989).
5. DISCOVER (manual), Biosym Technologies Inc., 1986.
6. S. Lifson and A. Warshel, J. Chem. Phys. 49, 5116 (1968).
7. L.-O. Pietilä, J. Mol. Struct. 195, 111 (1989).

A VIBRATIONAL FORCE FIELD PACKAGE FOR MOLECULAR DYNAMICS.

P. DERREUMAUX and G. VERGOTEN.

Laboratoire de Génie Biologique et Médical, INSERM U279 Faculté de Pharmacie, 59000 Lille (France).
Groupement Scientifique Modélisation Moléculaire ' IBM-CNRS.

INTRODUCTION.

Over the last 20 years, many molecular mechanics programs have been developed for studying molecular structure and molecular conformational flexibility. The fundamental reproach which can be addressed to the currently in use empirical potential energy models is that they are unable to calculate vibrational frequencies in agreement with experiment. In order to better fit the vibrational spectra, a new molecular mechanics and dynamics package has been written.

THE POTENTIAL ENERGY FUNCTIONS.

In contrast with the more commonly used packages (CHARMM[1], AMBER[2], TRIPOS[3]), two diffferent types of vibrational force fields are used for expressing the energy surface. The modified Urey-Bradley-Shimanouchi force field[4] (UBSFF) has been slightly reformulated for molecules for which a localized electron description is applicable; that is the electrostatic potential is taken into account and the torsional energy is expressed by the classical cosine function. On the contrary, we have choosen a local symmetry force field for ring molecules with delocalized electrons; for example the side-chains of amino-acids and the nucleic acids. This local symmetry force field can be extracted from *ab initio* calculations or from classical normal mode calculations.

RESULTS

Two initial results merit to be mentioned. First of all, we are not only able to calculate the structures and energies, including conformational energies and rotational barriers, the thermodynamic functions (heat capacity, entropy and free energy), but also capable to reproduce the observed vibrational frequencies and the observed potential energy distribution of each normal mode of a series of elementary alkane molecules with a rather impressive accuracy. The test molecules are methane, ethane,propane, n-butane and cyclohexane. The standard deviation between the observed and the calculated frequencies is respectively, 8 cm^{-1} for the test molecule and 15 cm^{-1} for the deuterated sample. This clearly shows that the UBSFF force field is better suited than the rather complex MM3 force field[5], and will give reasonable results for most molecules. Then it is often admitted by people doing molecular mechanics that redundant internal coordinates must not be taken into account if the potential energy matrix is expressed in terms of 3N mass-weighted cartesian coordinates. We shall definitely show that this hypothesis is not pertinent and confirm the theoretical prediction of a few Spectroscopists.

REFERENCES.

1. Brooks, B.R. ; Bruccoleri, R.E. ; Olafson, B.D. ; States, D.J.; Swaminathan, S., Karplus, M. J. Comput. Chem. 1983, 4, 187.
2. Weiner, S. J.; Kollman, P. A.; Case, D. A.; Singh, U. C.; Ghio,C. ; Alagoma,G. ; Profeta,S.; Weiner,P. J. Amer. Chem. Soc. 1984, 106, 765.
3. Tripos 5.2 Force Field. SYBYL. Tripos associates,1699s Hanle road , suite 303 , St Louis , MO, 63144, SYBYL 5.2 1988.
4. Shimanouchi, T. Pure Appl. Chem. 1963, 7, 131.
5. Allinger, N.L. ; Yuh, Y. H.; Lii, J. J. Amer. Chem. Soc. 1989, 111, 8551.

VIBRATIONAL SPECTRA OF INDOLE AND ASSIGNMENTS FOR ITS NORMAL
MODES ON THE BASIS OF AB INITIO FORCE FIELD CALCULATIONS

M. Majoube [1*] and G. Vergoten [2]

1 DSM-DPHG-SPER, C.E.Saclay, 91191 Gif sur Yvette Cedex (France)
2 Faculté de Pharmacie, INSERM U279, 59045 Lille Cedex (France)

INTRODUCTION

The vibrational spectra of indole have been the subject of abundant
experimental data[1] . Two normal coordinate analyses have been carried
out, one[2] based on a refined empirical force field and the other[3] on a
semi-empirical force field obtained from the quantum mechanical
molecular model AM1(the Austin Method 1). Controversy still remains
concerning the assignment of several in-plane and out-of-plane
fundamentals.

  We present here new experimental and calculated data for indole.
High resolution Fourier transform infrared(FTIR) spectra for indole
vapor and its N-deuterated analogue are obtained at 45°C, using a
multi-pass cell of 20 m path-length. In addition data for solid indole
are completed with FTIR and FT Raman spectra. On the other hand,
assignments for indole normal modes are deduced from a force field
calculated by <u>ab initio</u>, using the Gaussian 88 program[4] with the 3-21G
basis set and the optimized geometry.

RESULTS

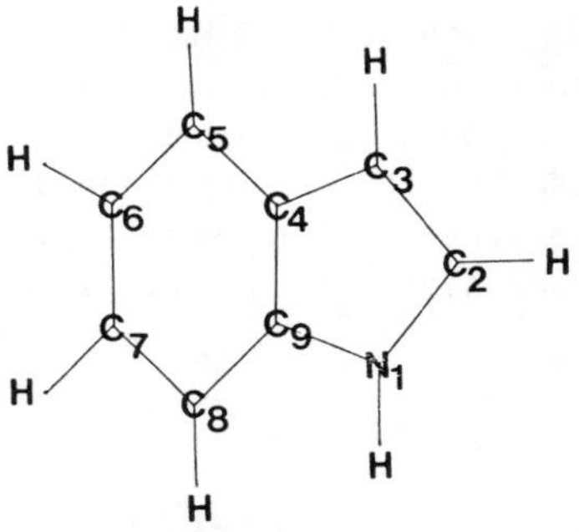

Table 1 lists observed and calculated
frequencies for selected in-plane modes
of indole vapor below 1700 cm^{-1} . These
modes correspond to strong and medium
Raman bands observed also for indole in
aqueous solutions. They may be compared
to bands observed for tryptophan at
similar frequencies.

Fig. 1 *Indole geometry*

  This table gives the intensity
(m = medium, s = strong, w = weak, and v = very) of observed bands and
their contours (A or B). The calculated frequencies are the <u>ab initio</u>

Table 1 Observed and calculated frequencies (in cm^{-1}) for indole

Mode	Obs.	$\Delta \nu$	Calc.	Potential Energy Distributions
ν_8	—		1607, I6.0	$-22C_8C_9$ $-19C_5C_6$ $+13C_4C_5$ $+13C_7C_8$
ν_9	1576.1wB	-3	1573, I3.4	$+16C_4C_9$ $+16C_6C_7$ $-13C_7C_8$ $-11C_4C_5$
ν_{10}	1519.9wB	0	1520, I8.3	$+45C_2C_3$ $+11\delta C_2H$ $-10C_8C_9$
ν_{11}	1478.6wA	+4	1493, I2.1	$-32\delta C_6H$ $+10C_4C_5$ $-10\delta NH$
ν_{12}	1457.7sA	+1	1457, I38.5	$+22\delta C_8H$ $-17\delta C_5H$ $+15C_4C_9$ $+11C_2C_3$
ν_{13}	1414.2sA	+11	1403, I35.0	$-41\delta NH$ $+14\delta C_5H$ $+14\delta C_7H$ $-11N_1C_2$
ν_{14}	1346.8sA	-11	1358, I4.1	$-28\delta C_3H$ $-25\delta C_2H$ $+12\delta C_8H$ $-14C_3C_4$
	1330.1sA			
ν_{15}	1277.5mA	+2	1276, I3.2	$+25C_4C_5$ $+15\delta C_5H$ $+13\delta C_8H$
ν_{16}	1244.0mB	-9	1253, I12.6	$-22C_8C_9$ $+20\delta C_7H$ $-12\delta C_6H$ $+13\delta NH$
ν_{17}	1227.0vwB	-1	1228, I31.4	$-31C_4C_9$ $+25N_1C_9$ $-22N_7C_8$ $-14C_5C_6$
ν_{19}	1149.5vwB	-5	1155, I6.0	$+26C_7C_8$ $-20\delta C_7H$ $+18\delta C_8H$ $-13C_6C_7$
ν_{20}	1122.3vwB	+1	1121, I6.8	$+20C_5C_6$ $+14\delta BR_1$ $-\delta C_5H$ $+10C_6H$
ν_{23}	1013.9mA	+16	998, I3.9	$-47C_6C_7$ $-10C_5C_6$ $-10C_7C_8$
ν_{24}	899.2mA	0	899, I8.0	$+40\delta PR_1$ $-29\delta PR_2$ $-21C_4C_9$
ν_{26}	744.9mA	-4	749, I3.4	$-24\delta BR_3$ $-13C_8C_9$ $-12C_4C_9$ $-11C_4C_5$

calculated ones scaled by a factor of 0.902. These are followed by IR intensities (I) in km/mole. The difference with observed frequencies is given by $\Delta \nu$ (column 3).

Normal coordinate analyses are carried out in which the ab initio F matrix in cartesian coordinates are in a first step transformed to F matrix in internal coordinates by defining a set of non-redundant internal coordinates for indole, as recommanded previously [5]. Then all the in-plane force constants are scaled by 0.813. Assignments for indole modes are obtained from potential energy distributions (column 5) amoung the internal coordinate stretchings and deformations. For instance $22C_8C_9$ and $45C_2C_3$ (see Fig.1) refer to 22 and 45% C_8C_9 and C_2C_3 bond stretchings. δ refers to in-plane deformations, BR_1 , BR_2 and BR_3 to the deformation of the benzene part of indole and PR_1 and PR_2 to that of its pyrrole part. All these motions occur in-phase for the same signs and out-of-plane for opposite signs.

REFERENCES

1. M.F. Lautié, PhD Thesis, Univ. Pierre et Marie Curie, Paris, 1978.
2. H. Takeuchi and I. Harada, Spectrochim.Acta, 1986, 42A, 1069.
3. W.B. Collier, J.Chem.Phys., 1988, 88, 7295.
4. Gaussian 88, Gaussian Inc., Pittsburg,PA.
5. P. Pulay, G. Fogarasi, F. Pang and J.E. Boggs, J.Am.Chem.Soc., 1979, 101, 2550.

A LOCALIZED MODE APPROACH TO VIBRATIONS OF LARGE MOLECULES

O. Sonnich Mortensen

Fysisk Institut
Odense Universitet
DK-5230 Odense M, Denmark

NORMAL AND LOCAL MODES

Molecular vibrations are conventionally treated in the
normal mode picture. That is, the potential is expanded
to second order in some linearized coordinates (internal or cartesian) and the normal coordinates are formed
as those linear combinations, that make the total energy (kinetic plus potential) diagonal to second order.
The vibrational wavefunction is then formed as a product of single mode harmonic wavefunctions. Higher
order terms can be dealt with by means of perturbation
theory.
The advantages of this conventional procedure are two-
fold: the treatment is exact to second order, and it is
well established. The disadvantages are that the coordinates are delocalized over the whole molecule, and
that the method does not distinguish between strong and
weak couplings between the primary coordinates. Thus a
typical normal coordinate will have components with
large differences in the coefficients of the primary
coordinates. Finally, in the linearized coordinates anharmonicities are often quite strong and typically couple
different normal modes more than they influence each
normal mode _per se_.
An alternative that has enjoyed great interest in recent
years is the local mode picture, used mainly in connection with CH stretch overtone vibrations. Here one uses
internal coordinates (length of CH bond) and expresses
the vibrational wavefunction as a product of _anharmonic_
local wavefunctions (typically Morse oscillator wavefunctions). The advantages of this procedure are that
it gives a very good and simple account of CH overtone
spectra even at high excitation, and that it is closely
connected with the way chemists like to think about molecules. The disadvantages are that the treatment is not
exact even to second order, since the various local
oscillators couple through the nondiagonal parts of the
kinetic energy operator, in particular, and that it tech-
nically is somewhat more complex because the primary
wavefunctions are anharmonic.

LOCALIZED MODES

Localized modes aim at being a compromise, using the
best features of normal and local modes. They are localized, in most cases in a single internal coordinate
(bond length or angle), but for coordinates that are
very strongly coupled through the kinetic energy we use
simple linear combinations (normal mode like). The diagonal anharmonicities are taken into account in both
cases. The couplings that remain are taken into account
in two ways: by forming linear combinations of product
wavefunctions, and by using self consistent field
methods. The former is closely related to the exciton
concept introduced by Frenkel and used in recent years
to discuss a variety of optical properties. Here, though,
we use it also for systems with non-equivalent subsystems.
We have performed model calculations on systems, where
the individual modes are rather strongly coupled, and
shown that even in this case the localized basis is
excellent, provided resonance and near-resonance couplings are dealt with through simple (pseudo-) symmetri-
zation.

LOCALIZED MODES AND PROCESSES

The localized mode description will be used to describe
optical and transport processes in large molecules.
Being localized, these modes are advantageous in particular where either an optical transition is localized,
i.e. in a chromophore, or where an interaction or the
initial stage in a process is localized. It has been
clearly demonstrated that the form of the local potential of a CH bond is strongly influenced by its surroundings, an effect that is easily quantified through
overtone spectra. Thus these, provided intensity problems are not too severe, can be used to study local
steric effects and secondary structures in large molecules.
Localized modes are equally convenient for the discussion of vibrational structure of electronic transitions.
We will show examples of this in connection with ordinary
absorption and resonance Raman spectra.

REFERENCES

1) H.G. Kjaergaard and O. Sonnich Mortensen, Chem.Phys.
   138, 237 (1989).
2) O. Sonnich Mortensen, B.R. Henry and M.A. Mohammadi,
   J.Chem.Phys. 75, 4800 (1981).
3) B.R. Henry, Acc.Chem.Res. 20, 429 (1987)

NUMERICAL SIMULATION OF MOLECULAR SPECTRA

T. Sundius

University of Helsinki, Dept. of Physics,
Siltavuorenpenger 20 D, SF-00170 Helsinki, Finland

As is well known, spectra of biological molecules tend to be very complicated, due to broad and overlapping bands. Different methods have been proposed to resolve the bands into peaks. Lately the maximum entropy method (MEM) has been suggested as an alternative to ordinary deconvolution and Fourier transformation techniques [1]. When a statistical determination of the peak parameters is desired, least squares (curve fitting) methods are usually preferred [2].

In order to compare different methods for analyzing spectra, it should be helpful to know the properties of the individual peaks of a spectrum in advance. Obviously a spectrum may be simulated by the Monte Carlo method, but it is usually much simpler to construct the spectrum directly from given peak parameters and line profiles.

We have designed a program that simulates a spectrum in this manner. The background parameters and peak parameters (positions, widths and heights) are first read from a file. These parameters are then used to calculate a spectrum from a line profile function. This function is computed by numerical convolution of an incoming spectral line shape and a spectrometer slit function, using the methods of MERLIN, a program for spectrum analysis, which has been previously described [3-4] .

To make the computed analytical spectrum more realistic, random noise has been added. The amplitude of the random fluctuations may be changed to study the effects of noise amplification. Naturally different statistical distribution functions can also be studied.

On the next page we show as an example the effects of changing the slit width in a Raman spectrum with very broad lines. As can be seen, this will not affect the resolution very much, as long as the slit dispersion is small compared with the widths of the peaks. This makes it, however, more difficult to resolve overlapping peaks.

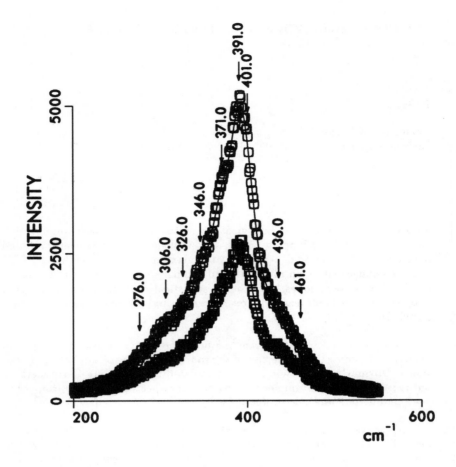

<u>Figure 1</u> Test spectrum generated from given peak parameters and a given line shape, using two different slit widths.

REFERENCES

1. F. Ni and H.A. Scheraga, <u>J. Raman Spectrosc.</u>, 1985, <u>16</u>, 337.
2. K. Palmö, B. Mannfors and L-O. Pietilä, <u>J. Mol. Struct.</u>, 1988, <u>174</u>, 101.
3. T. Sundius, <u>J. Raman Spectrosc.</u>, 1973, <u>1</u>, 471.
4. T. Sundius, <u>J. Mol. Struct.</u>, 1988, <u>175</u>, 319.

INELASTIC NEUTRON SCATTERING STUDY OF THE PROTON DYNAMICS IN POLYGLYCINE I AND II

M.H. Baron*, J.P. Fontaine*, F. Fillaux* and J. Tomkinson**

*LASIR, CNRS, 2 rue Henri Dunant, 94320 Thiais (France)
**Rutherford Appleton Laboratory, Chilton OX11 OQX (UK)

Polyglycine $(-CH_2-CO-NH-)_n$ is the simplest homopoly-peptide. It may adopt two different secondary structures refered to as I and II. Polyglycine I forms anti-parallel rippled-sheet while polyglycine II shows triple helices similar to the collagen structure. The infrared and Raman spectra of these two forms have been extensively studied and a detailedanalysis of the spectra provides a deeper understanding of the structure and interactions which are of basic importance for biochemical process.[1] In this context, realistic force-fields describing the dynamics of polyglycine I and II are extremely important. However, force-fields derived from infrared and Raman frequencies are highly underdetermined. Inelastic neutron-scattering (INS) spectra provide additional information since intensities can be calculated accurately. Therefore, force-fields giving poor agreement with the observation can be rejected. We have thus recorded the INS spectra of polyglycine I and II and their partially deuterated derivatives $(-CD_2-CO-NH-)_n$ at 20K, between 40 and 4000cm^{-1}, which have never been reported so far.

The spectra (Figure 1) were obtained on the TFXA spectrometer at ISIS, Rutherford Appleton Laboratory, Chilton, U.K. The samples were contained in aluminium foils and loaded into a cryostat at 20K.

The INS frequencies are in perfect agreement with those obtained in infrared and Raman. However, the INS intensities provide additional informations on the proton displacements involved in each mode and a preliminary analysis based on our previous results on N-methyl-acetamide[2] shows that the force-fields derived previously from infrared and Raman spectra of polyglycine should be significantly modified in order to achieve a satisfactory representation of the dynamics. The most intense INS bands in the fully-hydrogenated compounds are due to the δCH modes between 1230 and 1500cm^{-1}. Comparison with the deuterated compounds shows that the rCH_2 mode at ~1000cm^{-1} is coupled with the mode at ~550cm^{-1} (possibly Amide VI). In the deuterated compounds, the Amide II and III bands give a broad band between 1230 and 1540cm^{-1}. As for N-méthylacetamide, overtones, combinations, phonon-wings and internal vibrational coupling may contribute to the band-shape in this region. The Amide V band in the 700- 800cm^{-1} region is rather intense for both PGI(CD$_2$) and PGII(CD$_2$). As for N-methylacetamide, it is concluded that this mode corresponds to a pure γNH vibration decoupled from the (-OCNCD$_2$-). The proton displacements should be governed mainly by the external field due to the surrounding atoms. This contrasts markedly to previous forcefield analysis for which the γNH mode is supposed to contribute mainly at much lower frequency.[1] Force-field calculations including INS intensities are currently in progress. They should provide a more quantitative description of the dynamics of these macromolecules.

The enormous change of the spectra in the 200-400cm^{-1} region in polyglycine II compared to polyglycine I shows that the dynamics of the protons and skeletal bending modes is strongly affected by the secondary structure. This is also indicated by the very different low-frequency density-of-states for the external modes below 200cm^{-1}. More advanced theoretical approaches are required for a complete analysis of the dynamics.

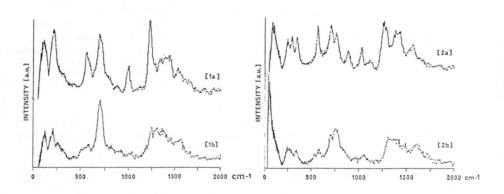

INS spectra : 1a) Polyglycine I ; 1b) Polyglycine I (CD$_2$) 2a) Polyglycine II ; 2b Polyglycine II (CD$_2$) ; All spectra at 20K

REFERENCES

1. S. Krimm and J. Bandekar, Advances in Protein Chemistry, 1986, 38, 181.
2. F. Fillaux, M.H. Baron, J.P. Fontaine, J. Tomkinson and G.J. Kearley, unpublished.

USE OF A RAMAN MICROSCOPE IN CONFORMATIONAL ANALYSIS OF A PEPTIDE WITH POLYMORPHISM

M. Tsuboi[*], T. Ueda and T. Ikeda

Iwaki-Meisei University, Iwaki, Fukushima 970, Japan
Meisei University, Hodokubo, Hino, Tokyo 191, Japan
Japan Spectroscopic Co.,Ltd.,Ishikawa, Hachioji, Tokyo
    192, Japan

By the use of a Jasco R-MPS-11 Raman microscope, polarized Raman spectra of IA, IB, and IIA crystals of aspartame (sweetener, α-L-aspartyl-L-phenylalanine methyl ester) have been observed with a 488.0 nm excitation. On removing H_2O gradually, aspartame IA ($P2_1$) is converted into IB ($P2_1$), and then into IIA ($P4_1$).[1] These three are all needle-shaped crystals. The elongation direction of the IA crystal coincides with the crystallographic b axis (b=4.89A), that of the IB crystal also with its b axis (b=4.96A), and that of the IIA crystal with the c axis (c=4.92A).[1] Among these three, IIA is a uniaxial crystal, which is built up by identical molecules, and their structure and orintation are known by an X-ray crystallographic study of Hatada et al[2]. Therefore, this was first subjected to our detailed Raman anlysis.

We attempt to determine, for each Raman band, the shape of the molecular Raman tensor and its orientation in the crystal in question. The "shape" is represented by $r_1 = \alpha xx/\alpha zz$ and $r_2 = \alpha yy/\alpha zz$, where (xyz) are the principal axes of the molecular Raman tensor. The "orientation" is represented by two angles χ and Θ. χ is the angle from ON to Oy, where ON is the intersection of the ab plane of the crystal and the xy plane of the molecule. Θ is the angle from Oc to Oz, where O is the common origin of the xyz and abc coordinate systems. The Raman intensity ratios Icc/Iaa(=Icc/Ibb) and Iac/Iaa(=Ibc/Ibb) are given as a function of r_1, r_2, χ, and Θ. The depolarization ratio ρ measured with linearly polarized light for an isotropic distribution of the molecule is given as a function of r_1 and r_2. On the basis of our present experimental results on an aspartame IIA crystal, a set of r_1, r_2, χ, and Θ values has been determined as shown in Figure 1, for each Raman band.

The Raman study on IA and IB indicated that the ester C=O and peptide C=O are both oriented nearly in parallel to their elongation directions like those of IIA. The orientation of the C(phenyl)-C_β bond, on the other hand, was found to be appreciably different; the apparent Θ= 55°, instead of 42° (in IIA, see Figure 1).

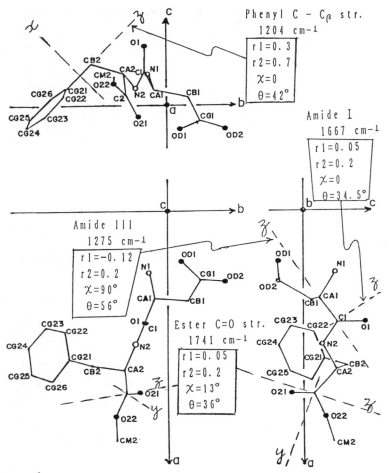

Figure 1
Structure and orientation of the aspartame molecule with respect
to the crystallographic axes in its IIA crystal, drawn on the basis
of the coordinates given by Hatada et al. [2] The shape (r_1 and r_2)
and orientation (χ and Θ) of each of the Raman tensors are
shown on the basis of our present study.

We thank Dr. N.Nagashima and Dr. E.Suzuki, Ajino-
moto Co., Inc., for the samples of aspartame and for
their valuable advices. The work in Iwaki-Meisei University was partly supported by a grant (No.02670985)
from Ministry of Education, Science and Culture of
Japan.

REFERENCES
1. N.Nagashima, C.Sano, S.Kishimoto and Y.Iitaka, Acta Cryst., A43,
   C-54, 1987; European Patent No.11983; T.Meguro, Y.Sato and N.
   Nagashima, Ann.Meeting Japan Cryst.Soc.,Nov.1990 (Sendai).
2 M.Hatada, J.Jancarik, B.Graves and S.-H.Kim, J.Am.Chem.Soc.,
   1985, 107,4279.

CONFORMATIONAL STUDY OF MYOSIN BY RAMAN SPECTROSCOPY AND MOLECULAR MECHANICS CALCULATIONS

P.J.A. Ribeiro-Claro[1], J.J.C. Teixeira-Dias[1], E.M.V. Pires[2],
L.A.E. Batista de Carvalho[1], M.Aureliano[2], and Ana Margarida Amado[1]

Department of Chemistry[1], Department of Zoology[2],
University of Coimbra
P-3049 Coimbra, Portugal.

INTRODUCTION

It has been shown that vanadate ions induce a conformational change in denatured myosin, leading to an increase of α-helix structure[1].Vanadate ions are known to compete with phosphate for the binding to proteins[2]. In particular, the myosin ATPase activity is strongly reduced by different molecular species of vanadate[3,4].

In this work, the pH and concentration dependence of vanadate oligomerization is studied by Raman spectroscopy, in order to determine the vanadate species interacting with myosin at a given concentration at physiological pH value. In addition, molecular mechanics calculations on the conformational structure of some aminoacids, serine-containing peptides and the fragment of ATP binding site of myosin light chains LC2,3 are reported. The calculated values for the ψ torsional angle (NC_α-CN) are compared with the values deduced from the analysis of the Amide III region in the Raman spectra of these peptides, on the basis of the correlation of Lord[5].

EXPERIMENTAL

Vanadate ion solutions, concentration range 100-10 mM, were prepared by dissolving sodium orthovanadate (Sigma Co.) in water. These solutions were boiled ca. 5 min., after pH adjustment with HCl solution. Aminoacids and peptide solutions were prepared from commercial products (Sigma Co.). Myosin and myosin LC2,3 were isolated from the rabbit white skeletal muscle[6].

Raman spectra were recorded on a Ramalog double spectrometer, 0.85 m, f/7.8, Spex model 1403. The light source was a Spectra-Physics Ar$^+$ laser, whose output at 514.5 nm was adjusted to provide 200 mW at the sample position. Samples solutions were kept in a cylindrical cell with inner diameter of ca. 6 mm.

Molecular mechanics calculations were performed with the program ECEPP/2[7], using a repeating cycle of sequential optimization of dihedral angles until the change in energy is less than 0.01 kJ mol^{-1}. The aminoacids were added sequentially and the last four aminoacids of the building peptide were reoptimized after each new addition.

RESULTS AND DISCUSSION

The Raman spectra of vanadate ion at different pH values ranging from 13.5 to 6.0 is dominated by the bands assigned to the symmetrical stretch of the VO bonds of different oligomers of vanadate. These band frequencies, and the corresponding pH intervals for a 50mM solution, are: 820 cm^{-1} (single band occurring at pH>12.1, disappearing at pH≈12.1), 870 cm^{-1} (from pH≈13.1 to pH≈8.3), 945 cm^{-1} (from pH≈9.6 to pH≈6.0), and a weak feature observed at the low wavenumber wing of the 945 cm^{-1} band, in the same pH interval. Despite the well known concentration dependence of vanadate oligomerization, no significant differences were found in the Raman spectra of the solutions in the concentration range 100-20 mM.

There is a good agreement between the estimated ψ values deduced from the observed Amide III frequencies and the calculated ψ values for small serine-containing peptides. For instance, in the Ser-Asn dipeptide, the Amide III band is found at ~1235 cm^{-1}, pointing to a ψ angle in the range of 80-120°, and the calculated value is ca. 85°. The bands in the Amide III region of the Raman spectra of lyophilised myosin LC2,3, are also in good qualitative agreement with the distribution of the ψ values of a 12 aminoacids peptide segment containing the serine-ATP binding site. In particular, the maximum of the Amide III band in the 1240-1280 cm^{-1} region, correlates with a large number of values of ψ in the -40 to 80° range. These results strongly support the combined use of Raman spectroscopy and molecular mechanics for conformational studies of proteins.

REFERENCES

1. J.J.C.Teixeira-Dias, E.M.V.Pires, P.J.A.Ribeiro-Claro, L.A.E.Batista de Carvalho, M.Aureliano, Ana Margarida Amado."Cellular regulation by protein phosphorylation", NATO-ASI Series, Ed. L.Heilmeyer, 1990 (in press).
2. N. D. Chasteen, Struct. Bonding (Berlin), 53 (1983) 105.
3. M. Aureliano, P. C.Silva, E.M.V. Pires, X Encontro da Sociedade Portuguesa de Química, Porto, 1987.
4. C.C. Goodno, Proc. Natl. Acad. Sci., USA, 76 (1979) 2620.
5. R. C. Lord App. Spectroscopy, 1977, 31, 187.
6. E. M. V. Pires and S. V. Perry, Biochem. J., 1977, 167, 137.
7. M. J. Browman, L. M. Carruthers, K. L. Kashuba, F. A. Momany M.S.Pottle, S. P. Rosen and S. M. Rumsey, write-up by G. F.Endres, resubmitted by H. A. Scheraga, ECEPP/2-QCPE 454, 1982.

A NEW APPROACH FOR THE ESTIMATION OF PROTEIN SECONDARY STRUCTURE FROM CIRCULAR DICHROISM SPECTROSCOPY

M. Bloemendal*, I.H.M. van Stokkum[#] and R. van Grondelle[#],

Faculty of Chemistry and [#] Faculty of Physics and Astronomy, Free University, De Boelelaan 1083, 1081 HV Amsterdam, The Netherlands.

INTRODUCTION

The secondary structure of proteins is reflected by their far-UV circular dichroism (CD) spectra. Several additional factors (*e.g.* environment and length of a particular structure, other chromophores) affect the spectra. Hence, extraction of secondary structure from the spectra is not straightforward. The most extensively used approaches for this are the "least squares analysis with regularizor" (CONTIN)[1] and "singular value decomposition" combined with "variable selection" (VS).[2] Neither of these methods yields a model based estimation of the error in the secondary structure estimates. We have developed a method, the "locally linearized model" (LLM), that produces these estimates.[3] This model is presented here.

THE LOCALLY LINEARIZED MODEL

The basic assumption of all methods to extract secondary structures from a CD-spectrum, $c(\lambda)$, is to consider it as a superposition of the contributions of the different secondary structure classes, $b_k(\lambda)$

$$c(\lambda) = \sum_{k=1}^{N_{cl}} f_k b_k(\lambda) \qquad (1)$$

f_k is the fraction of class k and N_{cl} the number of such classes, which is taken as five (α-helix [H], antiparallel [A] and parallel β-sheet [P], β-turn [T] and other [O]) following Hennessey and Johnson[4]. By definition fractions are non-negative and summate to one. These two constraints are used in the fit by CONTIN,[1] whereas in VS they are used as selection criteria.[2]

We have formulated[3] a multivariate linear model with noise, a so-called Gauss-Markoff model[5], to determine the pseudo-inverses of $b_k(\lambda)$ from a set of reference proteins, for which the secondary structure was known from X-ray diffraction. An important parameter in this approach is the number of significant singular values, N_s. The main difference between this model and others is the inclusion of a noise term, which allows estimation of errors. Van Stokkum *et al*[3] have demonstrated that only α-helix is adequately predicted by the linear model. The problems with the linear approach have been solved by adding a regulating term[1] and by removing from the reference set proteins that add conflicting information.[2] We have tried to synthesize an appropriate reference set. The hypothesis is that eq. 1 applies only to regions of the N_{cl}-dimensional f-space of secondary structure classification, which are related to regions of the N_λ-dimensional c-space of reference CD-spectra (N_λ is the number of data points of the CD-spectra considered). In this way, non-linear effects can be incorporated.

Table 1 Rms differences between f_k from CD-spectra and X-ray data (x100)[a]

Method	Class H	Class A	Class P	Class T	Class O
Linear model[b]	9	16	7	11	17
LLM (I)[c]	9	12	8	7	9
LLM (II)[d]	7	12	7	7	8
VS	9	14	7	11	14
CONTIN (I)[e]	12	21	6	11	8
CONTIN (II)[f]	9	13	6	7	8

[a] Calculated by taking one member from the reference set of proteins from ref. 2 and estimation of its structure from the remaining reference proteins. This is done for all 22 members of the reference set; [b] $N_s = 5$; [c] Selection with Σf_k nearest to 1; [d] Selection with Σf_k nearest to 1 and smallest $\sigma(\Sigma f_k)$; [e] Original criteria; [f] Selection according to ref. 2 (5 degrees of freedom).

Table 2 f_k (x100) of bovine lens-crystallins from LLM and X-ray structure[a]

Protein	Class H	Class A	Class P	Class T	Class O
α-crystallin	14 (3)[b]	26 (6)	10 (3)	13 (3)	37 (2)
β_L-crystallin	22 (3)	30 (9)	10 (4)	16 (1)	34 (8)
β_H-crystallin	15 (3)	39 (12)	19 (5)	17 (5)	22 (5)
γ-crystallin	15 (6)	24 (12)	11 (9)	14 (4)	35 (10)
γ-crystallin (X-ray)[c]	10	42[d]		11	37

[a] Measurements in collaboration with W.C. Johnson, Jr. at Oregon State University, Corvallis; [b] Values in parentheses denote estimated errors; [c] X-ray structure from ref. 6; [d] Sum of A and P.

The reference proteins are reordered according to the root mean square difference between the CD-spectrum of a specific reference protein and the protein to be analyzed, and the estimation of secondary structure, f_k, is performed as a function of the number of reference proteins and of N_s. This generates a multitude of sets of f_k. To choose from this multitude, $-0.05 \leq f_k \leq 1.05$ was used as selection criterion and the estimate with Σf_k nearest to 1 was chosen. As a refinement the estimate with the smallest standard deviation in Σf_k can be selected among the ten estimates with Σf_k closest to 1. In addition to H also T and O are adequately predicted, but results for the β-sheets remain unsatisfactory.

COMPARISON OF METHODS

A comparison of LLM with other models is given in Table 1. It is clear that with all methods overinterpretation of estimates has to be avoided. The main advantage of LLM is its ability to estimate the standard deviations in f_k.

APPLICATION

An application of LLM for bovine lens-crystallins is shown in Table 2. The result for γ-crystallin is in accordance with the X-ray structure (see table).

REFERENCES

1. S.W. Provencher and J. Glöckner, Biochemistry 1981, 20, 33.
2. P. Manavalan and W.C. Johnson, Jr., Anal. Biochem. 1987, 167, 76.
3. I.H.M. van Stokkum, H.J.W. Spoelder, M. Bloemendal, R. van Grondelle, and F.C.A. Groen, Anal. Biochem., 1990, 191, 110.
4. J.P. Hennessey and W.C. Johnson, Jr.,Biochemistry, 1981 20, 1085.
5. K.-R. Koch 'Parameter Estimation and Hypothesis Testing in Linear Models', Springer, Berlin, 1988.
6. G. Wistow, B. Turnell, L. Summers, C. Slingsby, D. Moss, L. Miller, P. Lindley, and T. Blundell, J. Mol. Biol. 1983, 170, 175.

ON THE USE OF ULTRAVIOLET RESONANCE RAMAN INTENSITIES TO REFINE MOLECULAR FORCE FIELDS: APPLICATION TO NUCLEIC ACID BASES.

P. LAGANT[1], G. VERGOTEN[1*], W.L. PETICOLAS[2]

1 Laboratoire de Génie Biologique et Médical, Faculté de Pharmacie, INSERM U279, Groupement Scientifique IBM-CNRS "Modélisation Moléculaire", CERIM, LILLE FRANCE.

2 Department of Chemistry, University of Oregon, EUGENE, OR, 97403, USA.

Introduction

It has been shown that Ultraviolet Resonance Raman intensities (UVRRI) of nucleic acid bases can be used to refine their in-plane molecular force field [1-3] and to deduce the correct corresponding force constants.

The UVRRI for the Jth normal mode is approximately proportional to the sqare of its displacement in the first two excited electronic states. This displacement is however proportional to the Jth bond length change during the ground to the excited state transition to the corresponding change of the bond order.

The bond order changes are obtained from semi-empirical calculations (CNDO and AM1 for 1-methylcytosine and 9-ethylguanine and 1-methyluracile).

Results and discussion

The harmonic general force constants for the model compounds 1-Me-cytosine, 9-Me-Guanine and 1-Me-uracil were obtained by fitting the experimental UVRR spectra recorded at two wavelengths, i.e. 266 and 213 nm, for 5'CMP and 5'GMP and at 266 nm for 5'UMP. The vertical solid lines represent the calculated intensities.

Using the linear relationship between the displaced bond lengths and angles and the displaced normal coordinates, a set of general valence force constants for the in-plane normal modes of each compoumd is obtained.

Application to the crystal structure.

The calculated in-plane force field is transferred to the crystal structure to improve the effect of the crystal force field on normal modes and check the validity of the force field. From the X-ray data of O'BRIEN [4] for 1-Me-cytosine 9-Et-guanine a normal mode treatment was performed [5]; rather good assignments of the observed Raman and Infrared spectra are obtained.

The effects of substituents and intra- and intermolecular hydrogen bonds are analyzed over the whole spectral range. Low frequencies were examined and a first set of out-of-plane force constants was deduced.

For 1-Me-uracil extension to the crystal structure [6] gives a very good agreement with previous works [7,8]. New observed very weak bands were revealed on the basis of the normal mode analysis.

References

1. W.L. PETICOLAS, D.P. STROMMEN, V. LAKSHMINARAYANAN, J. Chem. Phys., 73, 4185, (1980).
2. W.L. PETICOLAS, P. LAGANT, G. VERGOTEN, Proceedings of the SPIE meeting, 620, 89, (1986).
3. P. LAGANT, P. DERREUMAUX, G.VERGOTEN, W.L. PETICOLAS, J. Comput. Chem. In press.
4. E.J. O'BRIEN, Acta Cryst., 23, 92, (1967).
5. P. LAGANT, G. VERGOTEN, P. DERREUMAUX, R. DHENNIN, J. Raman Spectr., 21, 215, (1990).
6. P. LAGANT, G. VERGOTEN, W.L. PETICOLAS, to be published.
7. L. COLOMBO, D. KIRIN, Spectrochimica Acta, 42A, 557, (1986).
8. M.J. WOJCIK, J. Mol. Struct., 189, 239, (1988).

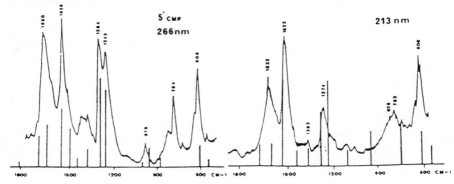

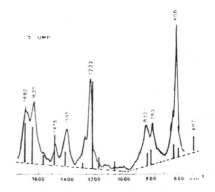

UV RESONANCE RAMAN SPECTROSCOPY OF SOLID SAMPLES USING A QUASI-CONTINUOUS WAVE LASER.

KPJ WILLIAMS and D KLENERMAN

BP Research Centre, Chertsey Road,
Sunbury on Thames, Middx TW16 7LN

INTRODUCTION

We are developing the resonance Raman method to study a wide range of materials possessing UV chromophores. Previous work, has used low repetition rate, high peak power lasers. Many operational problems ensue from these systems in that sample degradation and saturation effects limit the average power. The ideal CW laser does not exist to provide wavelength tunability between 220 and 400 nm. However, in recent years quasi CW sources (3.8 MHz rep. rate), which offer the potential for this range of tunability, have become commercially available. This paper describes our initial experiments using such a source with a UV optimised spectrometer. We assess the potential of the method as a general analytical technique, in particular for the study of solids which previously have not been amenable to analysis at these wavelengths. Whilst our aims have been driven by samples of industrial significance the method has implications for a wide range of solids (eg. biological samples) which possess UV chromophores.

EXPERIMENTAL

We have used a Coherent Antares Model 76S Nd:YAG mode locked laser providing 26 W, 1.064 um at 76 MHz with an 80 ps pulse width. This is doubled in KDP to provide 2.5W at 532nm and then further doubled in BBO giving 12 mW at 266 nm. For tunable UV radiation the 532 nm and 1.064 μm are mixed to produce 1 W at 355 nm. This pumps a stilbene3 dye laser giving radiation from 430 to 454 nm which is doubled in BBO to give tunable light between 215 and 227 nm. The Raman system used is a Dilor Model XY spectrometer fitted with 2400 groove/mm gratings and an intensified diode array detector. The optics have been optimised for UV performance. Sampling from solids has been either in a 180° backscattering geometry or in an off- axis configuration. Both systems have used a spherical mirror for collecting and focusing the Raman scatter.

RESULTS AND DISCUSSION

Figure 1 illustrates the Raman spectrum obtained from polyethylene using 266 nm excitation. Comparison between this spectrum and previously published data shows significant changes. In particular additional intense features are observed at 1646 and 1132 cm^{-1}. The former is attributed to the presence of unsaturation within the polymer. In this particular polyethylene sample vinyl end groups are present at 2 per 1000 carbon atoms.

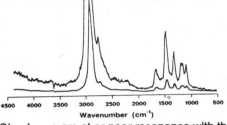

Figure 1
Polyethylene, λ_{exe}, 266 nm

Clearly, we are at or near resonance with the chromophore of this moiety, hence producing a significant signal enhancement. The origin of the 1132 cm^{-1} band is less obvious but may arise from a skeletal C-C vibration associated with some delocalisation of in-chain unsaturation, similar to the situation that exists in polyene containing systems.

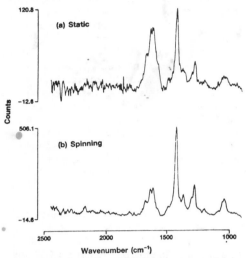

Figure 2
9-Methyl anthracene in KCl,
λ_{exe}, 266 nm

An important consideration in all of our studies has been monitoring any laser induced sample decomposition. For the example cited above no sign of laser damage was detected either visually, by inspection of the sample after the experiment, or spectroscopically, in that the sequential spectra acquired were identical. Figure 2 shows resonance Raman spectra obtained from 9-methyl anthracene (0.1%) in potassium chloride. The data from the sample when held static in laser beam is shown in Figure 2(a). It was clear that the band observed at 1600 cm^{-1} became more intense with time whereas other bands diminished in intensity. This feature we attribute to the thermal decomposition of the sample to form carbon. Figure 2(b) illustrates the data obtained when the sample was rotated to minimise the laser damage. This illustrates that sample decomposition can be largely overcome by spinning.

ACKNOWLEDGEMENT
Permission to publish has been given by BP Research

SOLID PHASE PEPTIDE SYNTHESIS INVESTIGATED BY NIR-FT-RAMAN SPECTROSCOPY

O. Faurskov Nielsen,[a*] B. Due Larsen,[a] A. Mortensen,[a] E. Pedersen[b] and D. H. Christensen[a]

[a]Chem. Inst., University of Copenhagen, 5, Universitetsparken, DK-2100 Copenhagen, Denmark. [b]Haldor Topsøe A/S, Nymøllevej 55, DK-2800 Lyngby, Denmark.

In solid phase peptide synthesis a polypeptide is synthesized on a polymer support-resin. The method is used to synthesize polypeptides with a well-defined chain length and secondary structure. However, synthetic problems are often caused by an interaction between the growing peptide chains attached to the resin. It has been proposed that a beta sheet formation occurs.[1] Recently some of us have found that incomplete removal of the protection group[2] in some cases is an additional difficulty. In order to elucidate these problems we have used Raman spectroscopy in studies of the peptides directly bound to the resin.

Raman spectroscopy by ordinary Ar-ion excitation on these samples always results in a broad fluorescence band. However, NIR-FT-Raman spectroscopy by Nd/YAG-excitation at 1064 nm is a powerful tool in a direct analysis of the resin-bound peptide. This method yields valuable information about the secondary structure of the peptide. Also the degree of deprotection can be estimated.

Raman spectra were obtained in a 180°-scattering configuration on a BRUKER FRA-106 Raman module coupled to an IFS-66 interferometer. A cooled Ge detector was used. A Nd/YAG laser (1064 nm) with an output up to 300 mW was the exciting source. Figure 1 shows a Raman spectrum of the resin-bound oligopeptide Fmoc-$(Ala)_3$Pro$(Ala)_3$Lys(Boc)-OR, where formulas for the Fmoc and the Boc-groups are given in Figure 2. The polymer support R is polydimethylacrylamid. Strong bands in the amid-III region are observed. Based on the fact that the polydimethylacrylamid does not give rise to any bands in this region, the observed bands could give information about the peptide secondary structure. Recently we have also studied homo-oligo-peptides with the general formula H-$(Ala)_n$Lys-OH. We have shown that the intensity of the band at 1025 cm^{-1} (Fig. 2) is proportional to the degree of deprotection of the Fmoc-group.

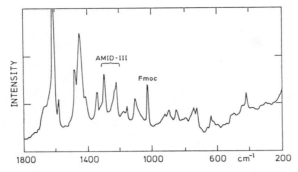

Figure 1. NIR-FT-Raman spectrum of a resin-bound peptide.

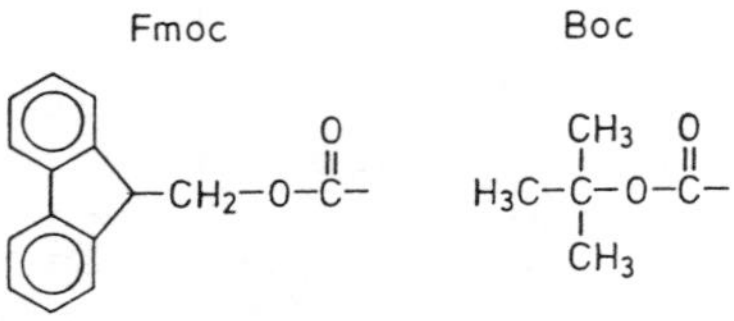

Figure 2. Chemical formulas for Fmoc and Boc.

The spectrum in Figure 1 was obtained from a sample in the dry state, but also NIR-FT-Raman spectra in the swelled state (in e.g. dimethylformamide) can easily be obtained.

Analysis of peptide synthesis on other commonly used resins are in progress. Unfortunately silicon based resins give rise to some broad background features. These might be caused by heating of the sample at high laser power. Another possibility is that the background is caused by ionic impurities or by easily polarizable hydrogen bonds. Recently we have found similar problems in investigations of powdered Al_2O_3 and Al_2O_3/MoO_3 samples.[3] Surprisingly excitation by Ar-ion excitation of the same samples did not give rise to a broad background band.[3] We believe that the broad background observed by excitation in the NIR-region is of a general nature for certain types of hydrogen-bonded systems and a better understanding of this phenomenon is necessary in order to obtain the full advantage of NIR-FT-Raman spectroscopy as an analytical tool.

REFERENCES

1. S.H.B. Kent, Ann. Rev. Biochem., 1988, 57, 957.
2. B. Due Larsen, C. Larsen and A. Holm. To be published.
3. A. Mortensen, D.H. Christensen, E. Pedersen and O. Faurskov Nielsen, J. Raman Spectrosc., 1991, 22, 47.

II

PHOTOSYNTHETIC

REACTION

SYSTEMS

FTIR DIFFERENCE SPECTROSCOPY OF MENAQUINONE PHOTOREDUCTION IN BACTERIAL REACTION CENTERS

J. BRETON[1], M. BAUSCHER[2], C. BERTHOMIEU[1], D. THIBODEAU[1], S. ANDRIANAMBININTSOA[1], D. DEJONGHE[1], W. MÄNTELE[2] and E. NABEDRYK[1]

[1] SBE/DBCM, CEN-Saclay 91191 Gif-sur-Yvette Cedex, France
[2] Universität Freiburg, Albertstraße 23, W7800 Freiburg, FRG

The elucidation by X-ray crystallography of the structure of the reaction center (RC) of the purple photosynthetic bacteria *Rhodopseudomonas viridis* [1] and *Rhodobacter sphaeroides* [2] has provided essential information on the bonding interactions between the cofactors and the protein. Nevertheless, the understanding of the remarkable efficiency of the photosynthetic process requires the knowledge at the molecular and submolecular levels of the subtle modifications of this structure brought about by the primary electron transfer reaction. This charge separation occurs in about 200 picoseconds between the primary electron donor P, a dimer of bacteriochlorophyll and the primary acceptor Q_A, a quinone molecule. In *Rb. sphaeroides*, Q_A is a ubiquinone (UQ) while in *Rps. viridis* and in other species like *Chloroflexus aurantiacus* and *Chromatium vinosum*, Q_A is a menaquinone (MQ) [3,4].

In previous reports, we have established that light-induced FTIR difference spectroscopy is a sharp tool to investigate at a vibrational level the molecular changes that occur in P and Q_A, as well as in their respective protein environment when the charge separated state $P^+Q_A^-$ is photogenerated [5-7]. In these conditions the contributions of P and P^+ dominate the $P^+Q_A^-/PQ_A$ spectrum and impede the analysis of vibrational modes that are specifically affected by the reduction of Q_A. However, it has been recently demonstrated that when the electron transfer step from Q_A to the secondary acceptor is blocked by a suitable inhibitor, the addition to the RC sample of reductant and mediator, which rapidly reduce P^+, allows a pure Q_A^-/Q_A spectrum to be photogenerated [8]. Furthermore, isolated quinone model compounds *in vitro* have been investigated by FTIR spectroelectrochemistry [9], so that a comparison of the vibrational modes of the Q^-/Q couple *in situ* and *in vitro* is possible. In the present report, we compare the FTIR difference spectra associated to the reduction of MQ *in vitro* and in the chromatophore membranes of three different species of photosynthetic bacteria.

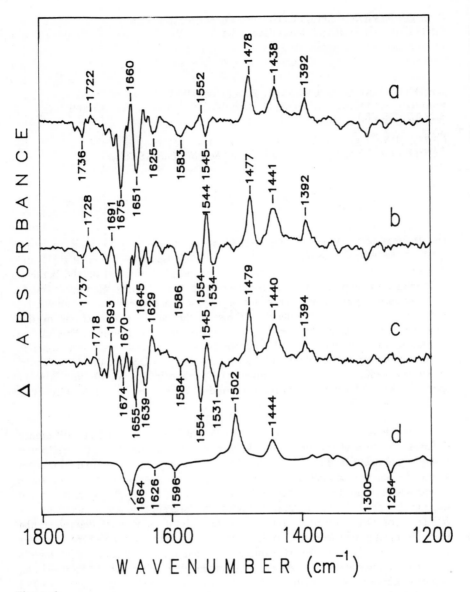

Figure 1

a–c: Light–induced FTIR difference QA⁻/QA spectra obtained at 10°C with chromatophores purified from *Rhodopseudomonas viridis* (1a), *Chromatium vinosum* (1b) and *Chloroflexus aurantiacus* (1c). All these spectra were recorded in the presence of diaminodurene (10 mM), Na–ascorbate (10 mM) and orthophenanthroline (10 mM), as previously described [8].

d: Electrochemically–induced FTIR difference spectrum of 2–methyl–menaquinone reduction in acetonitrile performed as described in [9]. Applied potential was –1.2 V versus Ag/Ag⁺.

Fig.1 shows the light–induced FTIR difference Q_A^-/Q_A spectra obtained with chromatophores purified from *Rps. viridis* (1a), *Chr. vinosum* (1b) and *C. aurantiacus* (1c). All these spectra were recorded in the presence of diaminodurene, Na–ascorbate and orthophenanthroline, as previously described [8]. The negative bands correspond to the disappearance of the vibrations of the neutral quinone and the positive ones to those of the appearing anionic species, keeping in mind that perturbations of amino acid residues concomitant with Q_A reduction will also lead to either positive or negative bands. The presence of the triplet of positive bands at about 1478 cm^{-1}, 1440 cm^{-1} and 1392 cm^{-1} demonstrates a striking analogy between these spectra. On the other hand, large differences can be noticed in the 1600–1700 cm^{-1} region and in the amide II region around 1550 cm^{-1}.

Compared to these light–induced FTIR difference spectra of MQ photoreduction *in vivo*, the FTIR difference spectrum of 2–methyl–menaquinone reduction in acetonitrile performed electrochemically (1d) is considerably simpler. The broad negative band at 1664 cm^{-1} with a shoulder at 1672 cm^{-1} is due to the C=O vibrations of the neutral MQ [10, 11] while the negative band at 1596 cm^{-1} is tentatively assigned to its C=C vibrations. The positive band at 1502 cm^{-1} arises from the C$\cdots$O vibrations of the anion and agrees well with that found by Clark *et al.* in DMSO [11]. The other positive band at 1444 cm^{-1}, which could not be detected in [11] due to the high background absorption of DMSO, is tentatively assigned to the C$\cdots$C vibrations of the anion [12]. Although these assignments appear reasonable, they should be considered as tentative until detailed normal mode calculations and experiments with isotopically–labelled MQ become available. The C$\cdots$O anion band observed at 1502 cm^{-1} for MQ in acetonitrile (1d) thus appears downshifted by 24 cm^{-1} in the Q_A^-/Q_A spectra (1a–c). This shift is most probably explained by the difference of environment *in vivo* and *in vitro*, notably the existence of strong hydrogen bonds in the state Q_A^-, although the presence of the isoprenoid chain *in vivo* (MQ9 in *Rps. viridis*) could also play a role. An even larger downshift is observed for the band assigned to the anion C$\cdots$O vibrations of UQ0 (1500 cm^{-1} in acetonitrile [9]) and of UQ10 in the RC of *Rb. sphaeroides* (1468 cm^{-1} [8,13]).

The observation that the band assigned to the C$\cdots$O of Q_A^- is located at 1478 ± 1 cm^{-1} in all three species of bacteria (1a–c) implies that the environment and the bonding of the C=O groups of the neutral Q_A should also be very comparable in all these bacteria. The C=O band at 1664 cm^{-1} *in vitro* has almost the same amplitude as that of the anion at 1502 cm^{-1} (1d). Even considering a possible broadening effect linked to a distribution of H–bonds, the spectra 1a–c show no common features which could unambiguously be assigned to the C=O vibrations of Q_A in the 1600–1700 cm^{-1} region (see also 7–9, 13–16).

In the 1570–1520 cm^{-1} region, where the MQ difference spectrum *in vitro* (1d) is essentially flat, the difference spectra recorded *in vivo* (1a–c) show significant features. Comparable signals have been previously reported for Q_A reduction in photosynthetic membranes of Photosystem II [16], of *Rb. sphaeroides* and *Rhodospirillum rubrum* [8] as well as in isolated RCs of *Rb. sphaeroides* [8, 13–15]. Note that in all these cases the chemical nature of the quinone is different from

MQ (plastoquinone in Photosystem II and UQ in bacteria). The observation that the presence of the 1570–1520 cm^{-1} bands is independent of the chemical nature of QA and the position of these signals under the amide II band, provide compelling evidence for an assignment to contributions from the protein moiety (peptide N–H and/or aromatic amino acid side chains). Consequently, contributions from the protein are also expected to appear in the amide I region (1600–1700 cm^{-1}). Due to differences in the primary sequences of the protein forming the QA binding niche, one can expect some variations in the shape of the differential signals of the protein.

Our present interpretation of the large variations observed in the 1600–1700 cm^{-1} region of the QA^{-}/QA spectra *in vivo* is thus related more to the multiplicity of overlapping bands than to a real difference in the bonding interactions between QA and the protein. Due to the overlap of positive and negative bands in this region, the peak frequency of the observed bands probably contains little information on the precise frequency and thus the nature of the molecular vibrations involved. A more complete interpretation of the QA^{-}/QA spectra calls for additional experiments with chemically–modified and/or isotopically–labelled quinones.

1. H. Michel, O. Epp, J. Deisenhofer, 1986, EMBO J. 5, 2445.
2. J.P. Allen, G. Feher, T.O. Yeates, H. Komiya, D.C. Rees, 1988, Proc. Natl. Acad. Sci. USA 85, 8487.
3. R.E. Blankenship, J.T. Trost, L.J. Mancino, 1988, in: "The Photosynthetic Bacterial Reaction Center", NATO ASI Series, A149, (J. Breton, A. Verméglio, eds) Plenum, 119.
4. G. Feher, M.Y. Okamura, 1978, in "The Photosynthetic Bacteria", (R.K. Clayton, W.R. Sistrom, eds) Plenum, 349.
5. W. Mäntele, E. Nabedryk, B.A. Tavitian, W. Kreutz, J. Breton, 1985, FEBS Lett. 187, 227.
6. W. Mäntele, A.M. Wollenweber, E. Nabedryk, J. Breton, 1988, Proc. Natl. Acad. Sci. USA 85, 8468.
7. E. Nabedryk, K.A. Bagley, D.L. Thibodeau, M. Bauscher, W. Mäntele, J. Breton, 1990, FEBS Lett. 266, 59.
8. J. Breton, D.L. Thibodeau, C. Berthomieu, W. Mäntele, A. Verméglio, E. Nabedryk, 1991, FEBS Lett. 278, 257.
9. M. Bauscher, E. Nabedryk, K.A. Bagley, J. Breton, W. Mäntele, 1990, FEBS Lett. 261, 191.
10. M.L. Meyerson, 1985, Spectrochim. Acta 41A, 1263.
11. B.R. Clark, D. H. Evans, 1976, J. Electroanal. Chem. 69, 181.
12. D.M. Chipman, M.F. Prebenda, 1986, J. Phys. Chem. 90, 5557.
13. W. Mäntele, M. Leonhard, M. Bauscher, E. Nabedryk, J. Breton, D.A. Moss, 1990, in: "Reaction Centers of Photosynthetic Bacteria", (M.E. Michel–Beyerle, ed.), Springer, 31.
14. D.L. Thibodeau, J. Breton, C. Berthomieu, K.A. Bagley, W. Mäntele, E. Nabedryk, 1990, in: "Reaction Centers of Photosynthetic Bacteria", (M.E. Michel–Beyerle, ed.), Springer, 87.
15. D.L. Thibodeau, E. Nabedryk, R. Hienerwadel, F. Lenz, W. Mäntele, J. Breton, 1990, Biochim. Biophys. Acta 1020, 253.
16. C. Berthomieu, E. Nabedryk, W. Mäntele, J. Breton, 1990, FEBS Lett. 269, 363.

NEAR-IR FT RESONANCE RAMAN SPECTROSCOPY OF THE PRIMARY DONOR IN BACTERIAL REACTION CENTER PROTEINS

T.A. MATTIOLI[*1], A. HOFFMANN[2], B. ROBERT[1], B. SCHRADER[2], and M. LUTZ[1]

[1] Département de Biologie Cellulaire et Moléculaire, CE Saclay, 91191 Gif-sur-Yvette cedex, FRANCE.

[2] Institut für Physikalische und Theoretische Chemie, Universität Essen, 4300 Essen, GERMANY.

INTRODUCTION

The primary light reactions in bacterial photosynthesis occur in membrane proteins known as reaction centers (RC). Because of its obvious relevance to the understanding of photoinduced charge separation, the structure of the primary electron donor, P, has received much attention. Resonance Raman spectroscopy has proven to be a powerful tool in this regard because many chlorophyll-protein interactions are readily observable[1].

The absorption spectrum of the bacterial RC from *Rhodobacter* (*Rb.*) *sphaeroides*, strain R26 exhibits a broad near-infrared (NIR) band at 865 nm which corresponds to the first excited singlet state of P which is a dimer of bacteriochlorophyll *a* (BChl *a*) molecules. The characterization of this state has been the subject of recent studies (reviewed in Ref. 2). When P undergoes one-electron chemical or photochemical oxidation, this NIR absorption band bleaches and a new, weaker band appears at ca. 1250 nm which presumably belongs to the oxidized primary donor cation radical, P^+.

With the recently introduced technique of NIR Fourier transform (FT) Raman spectroscopy the Raman effect is excited using NIR laser radiation - generally, the 1064 nm emission of a Nd:YAG laser - and the scattered light is analysed using an interferometer which, in part, compensates for the low scattering cross-sections at this wavelength. We have exploited this NIR excitation to obtain, at room temperature, preresonance Raman spectra of P and resonance Raman spectra of P^+, according to the redox potentials at which the RCs are poised.

Reduced Reaction Centers.

Figure 1 contains the room temperature NIR FT Raman spectrum of RCs from *Rb. sphaeroides* R26 in the presence of ascorbate. It remarkably resembles the NIR FT preresonance Raman spectrum[3] of BChl *a* as signalled by the characteristic 728, 894, 1014, 1163, and 1528 cm^{-1} bands and thus indicates that the bacteriochlorin modes dominate the FT Raman spectrum of the RCs. The five distinct carbonyl bands (each ca. 14 cm^{-1} broad, FWHM) in the 1620-1740 cm^{-1} region clearly indicates that more than one bacteriochlorin species is observed. In this spectral region, the protein modes are expected to contribute mainly at ca. 1655 cm^{-1}, the region of the Amide I band. Because this band is expected to be much broader than 14 cm^{-1} and is not clearly visible in the spectrum, we conclude that the contributions of the RC protein, if sizable at all in the present FT Raman spectrum, should be extremely weak as compared to those of BChl a.

The C_aC_m stretching mode, observed at 1607 cm^{-1} (Fig. 2), unambiguously indicates the presence of a BChl a species with one axial ligand.[1] The 1620 cm^{-1} frequency matches that of a H-bonded acetyl carbonyl only, while the 1653 cm^{-1} band can arise from either a H-bonded keto carbonyl group or an acetyl carbonyl free from interaction.[1] The 1691 and 1679 cm^{-1} bands correspond to free keto carbonyl groups in different environments. There is a weak shoulder observed at 1663 cm^{-1} (about one-third the intensity of the other carbonyl bands as estimated by a simple curve-fitting routine). Because of its weakness, and also because it seems to be present in the oxidized RC spectrum, we conclude that this band arises from a species other than P (its frequency matches well those reported for the acetyl carbonyl groups of either of the accessory BChl molecules).[4] The C_{10} carbomethoxy carbonyl groups likely contribute a weak band at 1741 cm^{-1}.

Oxidized Reaction Centers.

The FT Raman spectrum of oxidized RCs poised in the P$^+$ state by treatment with ferricyanide markedly differs from that of reduced RCs (Fig. 1). Normalizing these two spectra using the 645 cm^{-1} sapphire band from the sample holder reveals a significant loss of intensity (75-80% as estimated from the 728 cm^{-1} band of BChl a) for almost all Raman bands in the reduced RC spectrum. Bands which have been bleached should primarily correspond to those preresonantly enhanced modes associated with the 865 nm absorption band (which bleaches upon oxidation of P, whilst the new bands are associated with P$^+$ formation. Indeed, the addition of either ferricyanide or ascorbate is not expected to induce drastic change of the Raman cross-section of chromophores other than P. The extent of bleaching is such that it is clear from the computed oxidized-minus-reduced RC spectrum (Fig. 1) that the bands arising from P in the reduced RC spectrum dominate and give *direct* vibrational information about P in its neutral, ground state.

In the carbonyl region, five bands are observed to bleach upon P oxidation, at 1620, 1653, 1679, 1691, and 1741, indicating that at least two BChl molecules are involved. These bands are assigned to neutral, ground state P. Upon oxidation, a new band appears at 1717 cm^{-1} which can be considered as a marker for a monomeric BChl a^+ species.[5] A weak 1641 cm^{-1} band is also present in the oxidized RC spectrum which was not present in that of the reduced RC spectrum. The weak 1657, 1681, and 1697 cm^{-1} bands present in the oxidized spectrum cannot be exclusively assigned to P$^+$ because their frequencies correspond to those of the accessory BChl and BPhe molecules in the RC[4] and whose contributions may have been masked in the reduced spectrum by the P bands. Moreover, the contributions of neutral BChl is signalled in other regions of the oxidized RC spectrum by the presence of the 728, 894, 1014, and 1136 cm^{-1} bands.

The Primary Donor

Four distinct carbonyl bands have been attributed to the primary donor: 1620, 1653, 1679, and 1691 cm^{-1}. The similar intensities of these bands reveal that both components of P, namely P_L and P_M, are preresonantly enhanced to the same degree, indicating that the P excited state corresponding to the 865 nm absorption band is associated with these two molecules (strongly excitonically coupled). Because the 1679 (weakly interacting with a H-bond) and 1691 cm^{-1} (free from H-bond interaction) bands are consistent only with keto carbonyls, we attribute the 1620 and 1653 cm^{-1} bands to acetyl carbonyls of P, one of them being strongly H-bonded and the other free. This interaction pattern is in total agreement with X-ray crystallographic studies.[6] For *Rb. sphaeroides*, no suitable amino acid which could act as a H-bond donor has been found in the local environment of any of the keto groups.

Structure of P$^+$

The spectrum of P$^+$ is dominated by an intense 1600 cm^{-1} band which likely corresponds to the 1607 cm^{-1} band observed for the neutral species. This 7 cm^{-1} downshift is smaller than the 12 cm^{-1} downshift observed in the one-electron oxidation of monomeric BChl *a, in vitro.*[7] In the carbonyl stretching region, the 1641 and 1717 cm^{-1} bands can be unambiguously attributed to P$^+$. The 1717 cm^{-1} band can be assigned to a free keto carbonyl group that has upshifted with respect to its neutral counterpart. The 1641 cm^{-1} band can be assigned to the stretching mode of a H-bonded acetyl carbonyl of an oxidized BChl *a* species having upshifted from 1620 cm^{-1}. Such a 1636 cm^{-1} band has been observed in the FTIR spectrum of BChl a^+ in a H-bonding solvent.[5] We may thus conclude that the acetyl carbonyl of P$_L$ is still H-bonded in the P$^+$ state, most likely with the same amino acid residue.

No other carbonyl modes which can be attributable to P$_M$ are observed in the P$^+$ spectrum. As it is unlikely that the P$_L$ and P$_M$ bands are degenerate, we conclude that P$_M$ only weakly contributes in the P$^+$ spectrum. The magnitude of the upshift of the resulting 1717 cm^{-1} band of P$^+$ (+26 or +38 cm^{-1}, as compared to +32 cm^{-1} for BChl a → BChl a^+ *in vitro* [5]) strongly suggests that one BChl in P carries nearly the full +1 charge. The probable +21 cm^{-1} upshift of the acetyl carbonyl supports this conclusion. Indeed, if the unpaired electron were delocalized over both BChl molecules in P, a shift of about one-half of that observed for monomeric BChl would be expected. It thus appears that the unpaired electron is preferentially localized over one BChl only, on the time scale of the resonance Raman effect i.e. the unpaired electron does not share a common redox orbital of P$_L$ and P$_M$. The absence or weakness of P$_M{}^+$ contributions may be explained by sizable differences in residence times of the net +1 charge over P$_L$ and P$_M$ molecules. An alternative or complementary origin for the weakness of the P$_M{}^+$ contribution may be a difference in the RR scattering cross-sections of the P$_L{}^+$ and P$_M{}^+$ species at 1064 nm. Such a difference could result from the vibronic structure of the 1250 nm transition, or from a larger participation of a molecular orbital (MO) of P$_L$ in one of the dimer MOs responsible for the 1250 nm transition.

REFERENCES

1. M. Lutz and B. Robert, "Biological Applications of Raman Spectroscopy", Vol. 3 (T.G. Spiro, Ed.), Wiley-Interscience, New York, 1988, p.347.
2. R.A. Friesner and Y. Won, Biochim. Biophys. Acta, 1989, 977, 99.
3. T.A. Mattioli, A. Hoffmann, M. Lutz, and B. Schrader, C.R. Acad. Sci. Paris, 1990, t310 Série III, 441.
4. B. Robert, Biochem. Biophys. Acta, 1990, 1017, 99.
5. W.G. Mäntele, A.M. Wollenwebber, E. Nabedryk, and J. Breton, Proc. Natl. Acad. Sci. USA, 1988, 85, 8468.
6. D.M. Tiede, D.E. Budil, J. Tang, O. El-Kabbani, J.R. Norris, C.H. Chang, and M. Schiffer, "The Photosynthetic Bacterial Reaction

Center, Structure and Dynamics" (J. Breton and A. Verméglio, Eds.), Plenum, New York, 1988, p. 13.

7. T.M. Cotton, K.D. Parks, and R.P. Van Duyne, <u>J. Am. Chem. Soc.</u>, 1980, <u>103</u>, 6020.

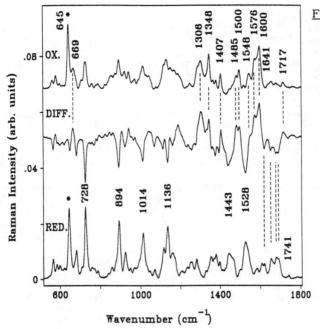

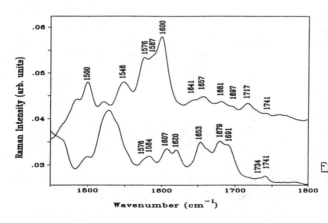

<u>Figure 1</u> Below: FT preresonance Raman spectrum of RC from *Rb. sphaeroides* R26 in the presence of ascorbate (reduced RC) using 280 mW laser power. Above: FT Raman spectrum of RC in the presence of ferricyanide (oxidized RC). Middle: The difference spectrum P^+-minus-P is calculated by subtracting the Below spectrum from the Above spectrum normalizing on the 645 cm^{-1} band (denoted with *) of sapphire cell sample holder. In the difference spectrum, negative peaks represent the preresonance Raman bands of P while the positive bands represent the resonance Raman bands of P^+. Spectral resolution is 4 cm^{-1}; room temperature; co-addition of 1000 scans.

<u>Figure 2</u> Same as Fig. 1: Blow-up of carbonyl stretching region for oxidized (above) and reduced (below) RCs.

ELECTRON TRANSFER IN PHOTOSYSTEM 2 REACTION CENTRES

David R.Klug[*], James R.Durrant[#*], Gary Hastings[*], Qiang Hong[*], James Barber[#] and George Porter[*].

[*]Photochemistry Research Group, Dept. of Biology and [#]AFRC Photosynthesis Research Group, Dept. of Biochemistry, Imperial College, London SW7.

Photosynthesis is the principal source of energy for biological activity. The conversion from sunlight to a transmembrane electrical potential is the primary role of a photosynthetic reaction centre, the first step in this process being the formation of a primary radical pair by an electron transfer mechanism. In order to maintain high efficiency, it is necessary for this reaction to proceed very rapidly (~1ps), and for the electron to be transferred over a relatively large distance (~17Å).

Most studies of primary electron transfer have been restricted to the reaction centres of purple bacteria (1). This is largely because well characterised chromophore-protein complexes have been available for over 20 years. Studies of electron transfer in purple bacterial reaction centres (PBR's) have also benefitted from the availability of a high resolution structure for these complexes, yet despite the wealth of available information, the mechanism of primary electron transfer has not been conclusively established.

Higher plants dominate photoautotrophic activity in most ecosystems, yet despite the local homologies between Photosystem 2 reaction centres and those of purple bacteria (2), it is far form certain that they function in the same manner.

We have been studying a relatively new form of Photosystem 2 reaction centre from higher plants, known as the D1/D2 reaction centre (3-7). Photosynthetic activity in this complex is almost completely restricted to primary electron transfer and related photochemistry, but this quality makes the D1/D2 reaction centre an ideal system in which to study primary electron transfer.

In this paper we summarise our findings regarding the primary photochemistry of the D1/D2 reaction centre and present our most recent measurements on primary electron transfer using femtosecond transient absorption spectroscopy.

The D1/D2 reaction centre is far more labile than its PBR counterparts, and this lability is essentially a by-product of the stringent photochemical requirements for a reaction centre which drives the water splitting process. The lability can be overcome to some extent by appropriate biochemistry (4,6), but photochemical effects must also be taken into account. Back reactions from the radical pair state cause the formation of triplet states which readily react with oxygen to produce singlet oxygen, which in turn reacts with the reaction centre and rapidly destroys its ability to function (5). PBR's suffer less from this problem as the triplet states are too low in energy to react with oxygen (8). Primary electron transfer in PS2 reaction centres produces a highly oxidising form of chlorophyll known as P680$^+$. If left in the oxidised state, the ~1eV potential of this species is sufficient to drag electrons from surrounding chromophores, and this eventually leads to the destruction of the reaction centre (Telfer et al. these proceedings). The D1/D2 reaction centre also exhibits a form of proteolytic activity usually associated with thylakoid membranes (9). This 'self-destruct' mechanism allows the plant to remove and subsequently repair damaged reaction centres, presumably in response to the high rate of photochemical attrition suffered by the reaction centre protein and/or chromophores.

Despite the problems described above, it has proved possible to perform very accurate spectroscopic measurements on D1/D2 RC's. Over the past two years we have made a number observations regarding the primary photochemistry of this particle. We have been able to show that the triplet states, produced by back reaction from the primary radical pair, reside on the chlorophyll special pair (P680) (5) and determine the quantum yield of this process. We have also been able to observe charge recombination fluorescence (3) and have used this to determine the temperature dependence of the free energy gap between the radical pair state and the special pair singlet state (4). These measurements of the radical pair thermodynamics have demonstrated a remarkable similarity to those made on PBR's (10), and suggested that in both cases the free energy gap is dominated by an entropic contribution. D1/D2 reaction centres extracted by different biochemical methods have been shown to be essentially identical (7), and we have recently been able to observe multiple forms of the radical pair state (6). A number of other time resolved studies have been made (11-13) including an attempt to deconvolute the absorption spectrum of the reaction centre (14).

One major and unanswered question is whether primary electron transfer proceeds via an accessory chlorophyll, or whether the electron hops directly to a pheophytin molecule assisted by a superexchange coupling. This question when applied to PBR's has been the subject of extensive theoretical and experimental debate, and obtaining an answer is central to developing an understanding of the form and function of reaction centres.

Although attempts have been made to measure electron transfer rates in the D1/D2 reaction centre (11), the task is far harder than in PBR's due to the extensive spectral overlap of the reaction centre chromophores and the relative lability of the complex. In order to overcome this problem we have developed a transient absorption apparatus using an Oxford Lasers copper vapour laser and a femtosecond dye laser. This apparatus provides both the time resolution and the sensitivity required to monitor the electron transfer kinetics under non-saturating conditions.

An example of our preliminary results on the D1/D2 reaction centre can be seen in figure 1. The data was produced by exciting the reaction centres at 625nm and the apparatus had a time resolution of 150fs. This data clearly exhibits a kinetic component of 1.2ps±0.2ps. The spectrum of the 1.2ps component is shown in figure 2 along with the spectrum of all longer lived components.

The 1.2ps component has a spectrum which is very similar to the spectrum of fluorescence emitted by the singlet state of the primary electron donor of PS2 (P680*). We therefore believe that this component is caused by the loss of a stimulated emission band associated with the P680 singlet state. We conclude that the lifetime of this state is 1.2ps, with this lifetime presumably being limited by forward electron transfer from P680. It is important to point out the relatively large contribution from the long lived states in these data. This suggests that a lot of energy is 'trapped' in molecules other than the chlorophyll special pair, and consequently great caution should be exercised in interpreting this data. Although the 1.2ps lifetime almost certainly corresponds to an electron leaving the excited state of the chlorophyll special pair, the kinetics may be 'distorted' due to the use of an inappropriate excitation wavelength which tends to excite all chromophores in the reaction centre rather than just the lowest exciton state of P680.

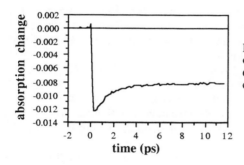

Figure 1: The absorption change observed at 690nm following excitation of the PS II reaction centres at 625nm.

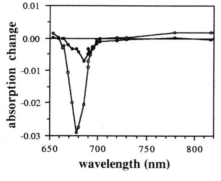

Figure 2: The spectra of the 1.2ps component (closed symbols), and the longer lived absorption changes (open symbols) observed in figure 1.

In order to counter this problem, our most recent measurements involve exciting the chlorophyll special pair (P680) directly, under conditions which result in far less absorption by the other chromophores. The same 1.2ps component is observed, but with an increased amplitude relative to other, longer lived absorption changes. Under these conditions few long lived stated are produced, other than the radical pair itself, and it is possible to discern other kinetic components as well.

In conclusion we have been able to observe the spectrum of a kinetic component which probably corresponds to an electron leaving the P680 singlet state under conditions where only some of the chlorophyll special pairs are directly excited.

This research is supported by the SERC, AFRC and Royal Society. We wish to thank Oxford Lasers for the loan of a copper vapour laser during this work.

References.

(1) Kirmaier, C. and Holten, D. (1987) Photosynth. Res. 13, 225-260.

(2) Barber, J. (1987) Trends in Biochem. Sci. 12, 321-326.

(3) Crystall, B., Booth, P.J., Klug, D.R., Barber, J. and Porter, G. (1989) FEBS Lett. 249, 75-78.

(4) Booth, P.J., Crystall, B., Giorgi, L., Barber, J., Klug, D.R. and Porter, G. (1990) Biochim. Biophys. Acta 1016, 141-152.

(5) Durrant, J.R., Giorgi, L.B., Barber, J., Klug, D.R. and Porter, G. (1990), Biochim. Biophys. Acta 1017, 167-175.

(6) Booth, P.J., Crystall, B., Ahmad, I., Barber, J., Porter, G. and Klug, D.R. (1991) submitted to Biochemistry. Also Booth et al. these proceedings.

(7) Gounaris, K., Chapman, D.J., Booth, P., Crystall, B., Giorgi, L.B., Klug, D, Porter, G. and Barber, J. (1990) FEBS Lett. 265, 88-92.

(8) Takiff, L. and Boxer, S.G. (1988) Biochem. Biophys. Acta 932, 325-334.

(9) Shipton, C.A.. & Barber, J. (1991) submitted to Proc. Natl. Acad. Sci.

(10) Woodbury, N.W.T. & Parson, W.W. (1984) Biochem. Biophys. Acta 767, 345-361.

(11) Wasielewski, M.R., Johnson, D.G., Seibert, M. and Govindjee (1989) Proc. Natl. Acad. Sci. USA 86, 524-528.

(12) Takahashi, Y., Hansson, O., Mathis, P. and Satoh, K. (1987) Biochim. Biophys. Acta 893, 49-59.

(13) Danielius, R.V., Satoh, K., Van Kan, P.J.M., Plijter, J.J., Nuijs, A.M. and Van Gorkom, H.J. (1987) FEBS Lett. 213, 241-244.

(14) van Kan, P.J.M., Otte, S.C.M., Kleinherenbrink, F.A.M., Nieven, M.C., Aartsma, T.J. and van Gorkom, H.J. (1991) Biochim. Biophys. Acta 1020, 146-152.

REDOX ACTIVE AMINO ACIDS IN PHOTOSYSTEM II AND IN OTHER ENZYME SYSTEMS

G. T. Babcock, I. D. Rodriguez, C. Hoganson, P. O. Sandusky, and M. El-Deeb

Department of Chemistry
Michigan State University
East Lansing, Michigan 48824, USA

INTRODUCTION

The occurrence of redox active amino acids has been recognized recently in several protein systems. These species are used as both electron-transfer cofactors and as catalytic sites to effect substrate metabolism. In this regard, they share similarities with organic cofactors, such as quinones and flavins, whose participation in enzymatic processes has been recognized for some time. In general, however, substantially higher redox potentials are required to generate amino acid radicals and, *a priori*, they are expected to be chemically more reactive. Our major focus in this work has been on tyrosine free radicals,[1] which occur in Photosystem II in photosynthetic systems,[2] in ribonucleotide reductase,[3] in prostaglandin synthase,[4] and, in a modified form, in galactose oxidase.[5]

In this report, we summarize some of the factors that control the magnetic resonance spectroscopy of these radicals, show how the spectroscopy can be used to gain information on the structure of the radical in the biological milieu, and present initial results on isotopic substitution at the phenol oxygen.

THE EPR LINESHAPE OF TYROSINE RADICALS

Figure 1 shows EPR spectra of three different tyrosine radicals: Y_D^+, the stable radical in Photosystem II; the immobilized tyrosine radical in frozen aqueous solution; and the tyrosine radical in ribonucleotide reductase (RDPR). A detailed study of these species by Electron Nuclear Double Resonance (ENDOR) and by selective deuterium substitution has provided a rationale for the differences in lineshape exhibited by these three radicals.[1,3] Our results have shown that the couplings to the protons *ortho* to the phenol oxygen are essentially constant in this class of radicals but that the hyperfine coupling to the methylene protons at the *para* position varies in this family of radicals. From these data, we have concluded that the spin-density distribution in biologically occurring and model tyrosine radicals is very similar and that the conformation of the β-CH_2-protons, with respect to the phenol ring plane, determines the spectral shape. Because the β-CH_2 coupling varies as $B\rho \cos^2 \theta$, where B is a constant, ρ is the unpaired electron-spin density at the C_4 ring carbon, and θ is the dihedral angle between the β-C-H bond and the p_z orbital on the C_4 ring carbon (Fig. 2), the coupling to the two β-CH_2 protons can vary. Figure 3 shows the variation in $\cos^2 \theta$ with θ_R for these two protons and for the R group (i. e., the α-carbon in tyrosine) and summarizes

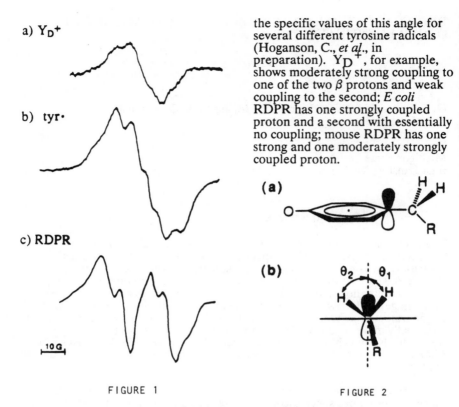

a) Y_D^+

b) tyr·

c) RDPR

|— 10 G

the specific values of this angle for several different tyrosine radicals (Hoganson, C., *et al*., in preparation). Y_D^+, for example, shows moderately strong coupling to one of the two β protons and weak coupling to the second; *E coli* RDPR has one strongly coupled proton and a second with essentially no coupling; mouse RDPR has one strong and one moderately strongly coupled proton.

(a)

(b)

FIGURE 1

FIGURE 2

The β-CH_2 twist angles vary strongly from one protein system to another. For a given protein, however, the angle is highly conserved. In Photosystem II, for example, the Y_D^+ and Y_Z^+ twist angles are the same in cyanobacteria, in green algae, and in higher plants. We are now in the process of assessing the protein factors that determine these angles and the effect that the twist angle may have in fine-tuning the reactivity and redox potential of the tyrosine free radical.

THE MEMBRANE ORIENTATION OF THE TYROSINE RADICALS IN PHOTOSYSTEM II

Chloroplast membranes and oxygen-evolving Photosystem II particles can be oriented on mylar sheets to form two-dimensionally ordered systems.[6] Figure 4 shows spectra with the membrane planes in these samples oriented perpendicular (left) and parallel (right) to the applied magnetic field. The spectra vary markedly with orientation in the laboratory field and have allowed us to do orientation selected ENDOR on these samples.[7] From the ENDOR spectra, we have been able to extract the orientation of the hyperfine tensors, with respect to the membrane plane, and from these to estimate the orientation of the molecular axes of the radical in the membrane. Our best fit orientation is shown in Figure 5; using this orientation and the hyperfine data we have obtained on these samples, we have simulated the oriented spectra as shown in Fig. 4.

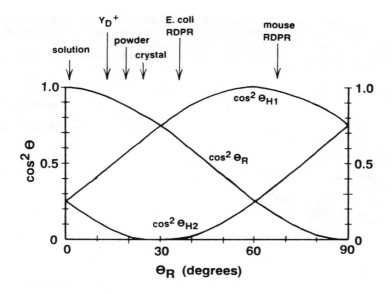

FIGURE 3

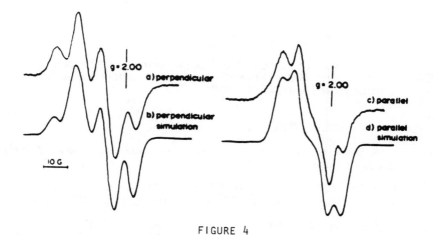

FIGURE 4

O^{17}ENRICHMENT IN MODEL TYROSINE RADICALS AND EFFECTS ON THE EPR SPECTRUM

Our ENDOR data and analysis for ribonucleotide reductase[3] indicated that there should be substantial unpaired electron-spin density on the tyrosyl oxygen. This is of interest, as we have shown that the hydrogen-bonding status of this oxygen varies from one protein system to another: the RDPR tyrosine oxygen is not H bonded,[3] whereas those in PSII[7] and in galactose oxidase (Sandusky, P., *et al.*, in preparation) are H bonded. In order to monitor the spin density on the oxygen and to assess the effects of hydrogen bonding on ρ directly, we have recently initiated a series of studies with ^{17}O labeled tyrosine.

The initial results from this work are shown in Figure 6, where we compare the lineshape of natural abundance and ^{17}O labeled immobilized model tyrosine radicals. The broadening induced by ^{17}O substitution is clear, and we are in the process of measuring the oxygen-spin density in this and in the protein systems directly by ENDOR.

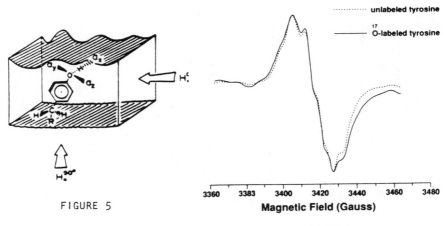

FIGURE 5

.... unlabeled tyrosine

———— ^{17}O-labeled tyrosine

Magnetic Field (Gauss)

FIGURE 6

ACKNOWLEDGEMENTS

This work was supported by the U. S. Department of Agriculture, Competitive Research Grant Office, Photosynthesis Program and by the National Institutes of Health (GM37300).

REFERENCES

1. B. A. Barry, M. K. El-Deeb, P. O. Sandusky, and G. T. Babcock, J. Biol. Chem., in press.

2. G. T. Babcock, B. A. Barry, R. J. Debus, C. W. Hoganson, M. Atamian, L. McIntosh, I. Sithole, and C. F. Yocum, Biochemistry, 1989, 28, 9557.

3. C. J. Bender, M. Sahlin, G. T. Babcock, B. A. Barry, T. K. Chandrashekar, S. P. Salowe, J. Stubbe, B. Lindstrom, L. Petersson, A. Ehrenberg, and B.-M. Sjoberg, J. Am. Chem. Soc., 1989, 111, 8076.

4. a) R. Karthein, R. Dietz, W. Nastainczyk, and H. H. Ruf, Eur. J. Biochem., 1988, 171, 313; b) R. J. Kulmacz, A.-L. Tsai, and G. Palmer, J. Biol. Chem., 1987, 262, 10524.

5. M. M. Wittaker, J. W. Whittaker, J. Biol. Chem., 1990, 265, 9610.

6. A. W. Rutherford, Biochim. Biophys. Acta, 1985, 807, 189.

7. I. D. Rodriguez, Ph.D. Thesis, Michigan State University, E. Lansing MI (1990).

PICOSECOND ABSORPTION MEASUREMENTS IN PHOTOSYTEM 1 OF *CHLOROGLOEA FRITSCHII*

R.W.Sparrow[a], E.H.Evans[a], R.G.Brown[a]*, M.C.W.Evans[b], R.Chittock[c] and W.T.Toner[c].

[a]Departments of Applied Biology and Chemistry, Lancashire Polytechnic, Preston. Lancs. UK.

[b]Department of Biology, University College, London. UK.

[c]Central Laser Facility, Rutherford-Appleton Laboratory, Chilton, Didcot. Oxon. UK.

Measurements have been made of fast absorption transients in Photosystem 1 (PS1) preparations from the cyanobacterium *Chlorogloea fritschii* in the picosecond timescale. Absorption changes were measured over the wavelength range 670-730 nm following excitation at 600 nm, from 0.6 ps up to 300 ps after excitation and their saturation behaviour (effect of excitation pulse intensity) and redox dependence were determined.

We observe four transient species in the wavelength range and timescales studied. Three of the transients saturate with increasing excitation intensity. The fourth transient, which we ascribe to light-harvesting chlorophyll on the basis of its spectrum and non-saturating behaviour, rises in <0.6 ps and decays with a lifetime of 4.0 ps. This lifetime, which is three orders of magnitude smaller than that of chlorophyll *in vitro*, pays testimony to the efficiency of energy transfer from the light-harvesting system to the PS1 reaction centre.

The three saturating species are

a) P_{700}. This is observed to have maximum absorption at 706 nm in *Chlorogloea fritschii* and is only observed under reduced conditions (redox potential <200 mV) as would be expected of the reaction centre chlorophyll. The transient has either a prompt (<0.6 ps) rise or a risetime of 30-40 ps and a slow 2-3 ns decay.

b) F_{706} This is observed in both oxidised and reduced samples, rises promptly (<0.6 ps) and decays with a lifetime of 30-40 ps.

c) F_{685} This is observed in both oxidised and reduced samples, rises promptly (<0.6 ps) and decays with a lifetime of 25 ps.

We interpret F_{706} and F_{685} as 'funnel' chlorophylls on a common energy transfer pathway to P_{700}. The saturation behaviour indicates that all three components are present in similar concentration. In parallel fluorescence measurements we have found that this PS1 preparation fluoresces predominantly at 690 and 730 nm, the decay being dominated by a fast component with a lifetime of some 30 ps[1]. These fluorescent species are likely to be F_{685} and F_{706}. The 'funnel' signals both appear at the earliest times measured and F_{706} obscures early measurements of the P_{700}-P_{700}^* transition. P_{700}^* is therefore either excited on a similar timescale as predicted by Wasielewski *et al*[2] or is a daughter product of F_{685}/F_{706} recovery, in agreement with the predictions of Holzwarth *et al*[3].

We suggest the following scheme for energy transfer within PS1

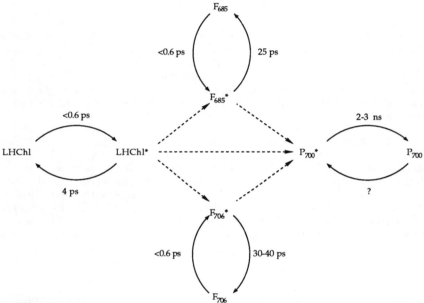

References

1.   R.W.Sparrow, E.H.Evans, R.G.Brown, and D.Shaw, J.Photochem. Photobiol. B, 1990, 5, 445.

2.   M.R.Wasielewski, J.M.Fenton and Govindjee, Photosynth. Res., 1987, 12, 181.

3.   A.R.Holzwarth, H.G.Schatz, H.Brock and E.Bittersmann, Physiol. Plant, 1989, 76, 1A pA85.

MULTIPLE RADICAL PAIR STATES IN PHOTOSYSTEM 2 REACTION CENTRES

Paula J. Booth, Ben Crystall, Iqbal Ahmad, *James Barber, George Porter and David R Klug.
Photochemistry Research Group, Department of Biology and *AFRC Photosynthesis
Research Group, Deparment of Biochemistry, Imperial College, London SW7 2BB.

Photosystem 2 (PS2) is responsible for the splitting of water to produce oxygen. A reaction centre complex from PS2 known as the D1/D2 reaction centre has been successfully isolated (1,2). In the absence of modifications the photosynthetic activity of the D1/D2 reaction centre is limited to primary electron transfer which results in the formation of the primary radical pair $P680^+Ph^-$ .Previous measurements have indicated that the radical pair decays with a lifetime of the order of 36ns (3,4), and we have used measurements of the charge recombination fluorescence to calculate the free energy gap between the excited singlet state of P680 and the radical pair state $P680^+Ph^-$ (5).

In this paper we present more detailed time resolved absorption and fluorescence measurements of the primary radical pair in D1/D2 reaction centres. These data indicate that the radical pair dynamics are not fully represented in earlier studies, and that there is more than one form of the radical pair.

D1/D2 reaction centres were isolated as described by Chapman et al. (6). Time resolved fluorescence measurements were made using time correlated single photon counting, while the transient absorption measurements were made using a modified form of the flash photolysis apparatus probing the sample at 820nm. Further experimental details can be found in ref. (7). Data were analysed both globally and individually and simulated data studies were performed to test for overparameterisation.

The fluorescence data includes a contribution from inactive reaction centres. A maximum of 6% of the reaction centres are inactive in electron transfer yet this population accounts for almost half of the total emission from the samples as 'free chlorophyll' has a much higher fluorescence quantum yield than chlorphyll which is active in a reaction centre. The 'free chlorophyll' component can be easily distinguished from charge recombination fluorescence by its spectrum as shown in figure 1. Both time resolved emission spectra and globally analysed kinetic spectra show that the 6ns component has a bluer spectrum than the charge recombination fluorescence. The free chlorophyll component is not seen in the transient absorption data as would be expected, as it only contributes a maximum of 6% of the optical density at 820nm.

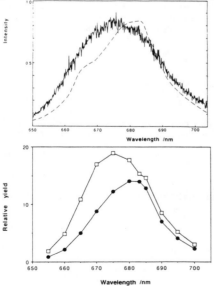

FIG. 1 a) Time resolved emission spectra, showing the spectrum of the radical pair emission (- - -) and the 'Free chlorophyll' (-----).

b) Global analysis of the time resolved emission showing the spectrum of the 52ns radical pair component (filled circles) and the 5.3ns 'free chlorophyll' component (open squares).

Table 1 summarises the results of this study. The short lifetimes are obtained by data collection over a much narrower time window than that used for assigning the long lifetimes.

Assignment	Fluorescence kinetics	Absorption kinetics
Radical Pair	52ns	56ns
Radical Pair	20ns	20ns
'Free Chlorophyll'	5.3ns	-
Short component 1	1.5ns	2ns
Short component 2	0.1ns	-

Table 1. Kinetic components observed in this study. Short components 1 and 2 can only be accurately determined by analysing emission data collected over a range of timescales.

It is clear that at least two lifetimes are needed to adequately model radical pair recombination kinetics in PS2, and that consistent kinetics are obtained for transient absorption and fluorescence measurements. The emission data can be used to calculate the radical pair free energy gap (5) and this is displayed as a function of temperature in figure 2. Figure 2 suggests that although the free energy gap is dominated by entropy, the difference between the two radical pair states is enthalpic.

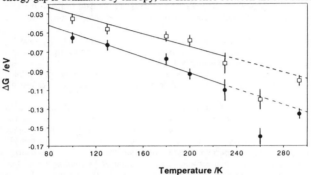

FIG.2 The temperature dependence of the free energy gap associated with the two radical pair states. Filled circles represent the 52ns component while open squares represent the 20ns component.

Although there are many possible explanations for the results presented here, most of them come down to one of two models. Either there are two or more conformational forms of the reaction centre which presumably interconvert and demonstrate different kinetics as in myoglobin and haemoglobin (8), or the two forms of the radical pair are formed sequentially by a relaxation process in which case our kinetic model represents two quasistatic equilibria.

This research was supported by the SERC, AFRC and Royal Society.

References
1) Nanba, O., & Satoh, K. (1987) Proc. Natl. Acad. Sci. USA, 109-112.
2) Barber, J.,Chapman, D.J., & Telfer, A. (1987) FEBS Lett., 220, 67-73.
3) Danelius, R.V., Satoh, K., Van Kan, P.J.M., Plijter, J.J., Nuijs, A.M., & van Gorkom, H.J. (1987) FEBS Lett. 213, 241-244.
4) Crystall, B., Booth, P.J., Klug, D.R., Barber, J., & Porter, G. (1989) FEBS Lett. 249, 75-78.
5) Booth, P.J., Crystall, B., Giorgi, L.B., Barber, J., Klug, D.R., & Porter, G. (1990) Biochim. Biophys. Acta 1016, 141-152.
6) Chapman, D.J.,Gounaris, K., & Barber J. (1991) in 'Plant Biochemistry Amino Acids, Proteins and Nucleic Acids (Rogers , L.J. ed.) Vol. 5, pp 171-193, Academic Press, London.
7) Booth, P.J., Crystall, B., Ahmad I.,Barber, J., Porter G., & Klug D.R.(1991) submitted to Biochemistry.
8) Frauenfelder, H., Parak, F., & Young, R.D. (1988) Ann. Rev. Biophys. Biophys. Chem. 17, 451-479.

UNDERSTANDING THE ROLE OF β-CAROTENE IN PHOTOSYSTEM TWO: LIGHT DRIVEN OXIDATION AND BLEACHING OF THIS PIGMENT IN THE ISOLATED PHOTOSYSTEM TWO REACTION CENTRE

A. Telfer*, J. De Las Rivas and J. Barber

AFRC Photosynthesis Research Group
Department of Biochemistry
The Wolfson Laboratories
Imperial College of Science, Technology and Medicine
London SW7 2AY, UK

INTRODUCTION

The primary electron donor, P680, in isolated Photosystem Two (PSII) reaction centres is rapidly destroyed by light under aerobic conditions. Addition of the artificial acceptor, silicomolybdate (SiMo) delays photodamage to P680, at the expense of a chlorophyll species absorbing maximally at 670 nm[1]. We proposed that this chlorophyll, designated Chl670, was oxidised by P680+ and was then bleached irreversibly. Here we have extended our investigations on the ability of intrinsic electron donors to protect P680 against photodamage.

RESULTS AND DISCUSSION

Fig. 1 shows the effect of the presence of SiMo and 2,5-dibromo-3-methyl-6-isopropyl-p-benzoquinone (DBMIB) on photodamage difference spectra of PSII reaction centres. These acceptors stimulate a rapid bleaching in the carotenoid absorbing region, ~485 nm. Bleaching at 485 nm is considerably more rapid with SiMo than with DBMIB. Unlike the situation with SiMo (Fig.1a), with DBMIB there is no initial, preferential bleaching at 670 nm as compared to 680 nm (Fig. 1b). Bleaching at 485 nm also occurs in the presence of ferricyanide under anaerobic conditions (Fig. 1c). The chromophores photobleached in the presence of electron

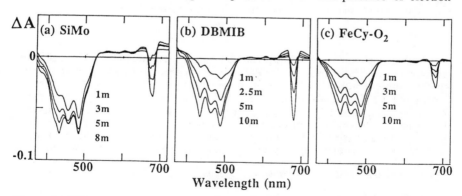

Figure 1 Difference spectra of irreversible changes (light treated - dark) in absorbance of PSII reaction centres (4 μg ml^{-1} chl) induced by 600 μE m^{-2} s^{-1} red light for various times. (a) 0.1 mM SiMo (b) 0.1 mM DBMIB (c) 0.5 mM ferricyanide plus an oxygen trap. Temp. 10°C. Reaction centres isolated and spectroscopy as in [1].

acceptors were identified by HPLC of the extracted pigments (Fig. 2). Using DBMIB, we found that approx. 50% of the total pool of β-carotene is lost rapidly followed by a slower bleaching of the remainder. In contrast there was essentially no loss of pheophytin *a* and only a slow loss of some chlorophyll *a*. The "*in vivo*" spectrum of β-carotene is shown in Fig. 3 with the spectrum of the extracted pigment in organic solvents. The former spectrum was obtained by subtraction of the kinetically distinct absorption changes due to chlorophyll bleaching. The more polar environment of a protein brings about a shift to the red and a sharpening of the absorption maxima. In order to determine whether carotenoid is oxidised prior to its bleaching we measured flash-induced transient absorption changes at 950 nm (Car$^+$, ref. 3) and 820 nm (P680$^+$). The slow rise component, we observe at 950 nm, was seen only in the presence of added electron acceptor and we attribute it to Car$^+$. This is confirmed by its spectrum which peaks at 980 nm (Kleinherenbrink, personal communication). The decay of the 820 nm signal is multiphasic but includes an intermediate component that correlates with the slow rise at 950 nm ($\tau \sim 1$ ms).

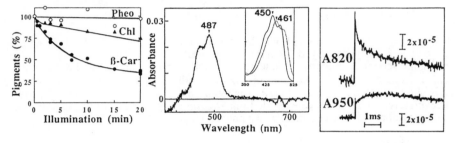

Figure 2 Time course of loss of β-carotene (●), chlorophyll *a* (▲) and pheophytin *a* (o) during photo-bleaching of PSII reaction centres (50 μg ml^{-1} chl) in the presence of 0.1 mM DBMIB as assayed by HPLC.

Figure 3 "*In vivo*" spectrum of ß-carotene in PSII reaction centres compared to the *in vitro* spectrum in n-hexane (solid line) and chloroform (dotted line). "*In vivo*" spectrum obtained by normalisation of the DBMIB 1 and 4 min photodamage spectra (as in Fig. 1) at 680 nm and subtraction (4 min - 1 min).

Figure 4 Flash-induced transient absorption changes, measured as in R2 at 950 and 820 nm in PSII reaction centres (5 μg ml^{-1} chl) plus 0.1 mM DBMIB. Temp. 4°C.

CONCLUSION

We conclude that when isolated PSII reaction centres are illuminated plus an electron acceptor, β-carotene is oxidised by P680$^+$. Car$^+$ appears to be an unstable species which is subsequently bleached in an oxygen-independent reaction. An accessory chlorophyll, Chl670, is also oxidised and then bleached irreversibly [1]. Once these donors are inactivated, P680$^+$ itself is destroyed. This *in vitro* evidence suggests that β-carotene and chlorophyll molecules in the reaction centre are involved in protection against photodamage to the PSII primary donor. We suggest that electron donation from these chromophores may play an important role in prevention of photoinhibition *in vivo*.

We thank the AFRC for funding and JDLR holds a FEBS Research Fellowship.

REFERENCES
1. A. Telfer, W.-Z He and J. Barber Biochim. Biophys. Acta 1990, 1017, 143
2. J.R. Durrant, L.B. Giorgi, J. Barber, D.R. Klug and G. Porter Biochim. Biophys. Acta 1990, 1017, 167
3. C.C. Schenck, B. Diner, P. Mathis and K. Satoh Biochim. Biophys. Acta 1982, 680, 216

TEMPERATURE DEPENDENCE OF THE INFRARED DIFFERENCE SPECTRUM OF PHOTOSYNTHETIC BACTERIAL REACTION CENTERS

E. H. Morita,[*] H. Hayashi[**] and M. Tasumi[*]

[*] Department of Chemistry, Faculty of Science,
The University of Tokyo
Bunkyo-ku, Tokyo 113, Japan
[**] National Institute for Basic Biology
Okazaki-shi, Aichi 444, Japan

INTRODUCTION

Response of photosynthetic reaction centers (RCs) to the absorption and transfer of light energy has been studied for many years by spectroscopic methods. The three-dimensional arrangements of the pigments and protein subunits in the RCs from a certain species of photosyn-thetic bacteria have been recently determined by X-ray analysis. However, many features, *e.g.*, interactions between pigments and protein main and/or side chains, still remain to be elucidated.

Infrared spectroscopy is a relatively new technique for studying structural and dynamic aspects of RCs. A joint research group at Freiburg (W. Mäntele *et al.*) and at Saclay (J. Breton *et al.*) has published a series of papers, in which they have clarified many important points regarding assignments of bands observed in the light-induced IR difference spectra. We focus our attention on the structures of bacteriochlorophyll-a (BChl-a) special pair and its photogenerated radical cation in the RC.

MATERIALS AND METHODS

RCs and chromatophores were isolated and purified from *Rb. sphaeroides* wild type. Light-induced IR difference spectra were observed for RCs in D_2O, chromatophores in D_2O, and dry films of chromatophores at 300, 150, 100, and 80K.

RESULTS AND DISCUSSION

In the light-induced IR difference spectrum of RCs in D_2O observed at room temperature, two bands at 1714 and 1702 cm^{-1} are assignable to the C_9-keto CO stretch of the BChl-a special pair radical cation, $(BChl-a)_2^+$. On the other hand, the electrochemically generated radical cation, $BChl-a^+$ (monomer), shows a band at 1716 cm^{-1} in

THF solution, which is shifted to 1708 cm^{-1} in CD$_3$OD solution. This shift is considered to be due to hydrogen-bond formation between the keto CO and the methanol OD groups. It is notable that this downshift (8 cm^{-1}) due to hydrogen bonding is much smaller than that (35 cm^{-1}) observed for neutral BChl-a (from 1687 to 1652 cm^{-1}).

The 1714 cm^{-1} band of (BChl-a)$_2^+$ is very close in wavenumber to the 1716 cm^{-1} of BChl-a$^+$. This means that the positive charge in (BChl-a)$_2^+$ is localized on either one BChl-a of the special pair (which is not hydrogen-bonded) at least in the timescale of infrared absorption ($\sim 10^{-13}$ s). According to an electron-spin-echo (ESE) experiment (M.K. Bowman and J.R. Norris, J.Am.Chem.Soc., 1982, 104, 1512), the timescale of this charge transfer is estimated to be of the order of 10^{-12} s. The IR study seems to be consistent with this ESE result.

The other band at 1702 cm^{-1} of (BChl-a)$_2^+$ is more difficult to assign. The following two assignments are possible; one is to assign this band to a hydrogen-bonded C$_9$-keto CO, and the other is to consider that this band corresponds to a charge-delocalized state which may be represented as (BChl-a$^{0.5+}$)$_2$. The former assignment is based on the proximity of the 1702 cm^{-1} band to the 1708 cm^{-1} band of BChl-a$^+$ in CD$_3$OD. However, existence of a hydrogen bond between the C$_9$-keto CO group of either one of BChl-a's and an OH and NH group in the protein main or side chain is not indicated in the three-dimensional structure of the RC determined by X-ray analysis. The latter assignment comes from an expectation that the non-hydrogen-bonded C$_9$-keto CO stretching band of (BChl-a$^{0.5+}$)$_2$ should be located halfway between that of BChl-a$^+$ (1716 cm^{-1}) and that of neutral BChl-a (1687 cm^{-1}). About this assignment, a question is raised as to whether both the charge-delocalized state and the charge-localized state really coexist in the timescale of $\sim 10^{-13}$ s.

In order to obtain more information on the origin of the 1702 cm^{-1} band of (BChl-a)$_2^+$, we have been trying to observe temperature dependences of the light-induced IR difference spectra for RCs in D$_2$O, chromatophores in D$_2$O, and its dry film. Temperature dependences are clearly observable for all these samples. In addition, differences are found between the RCs and the chromatophores. The 1702 cm^{-1} band seems to consist of two components at 1707 and 1702 cm^{-1}, whose relative intensities change with temperature. At lower temperatures, the intensity of the 1707 cm^{-1} band increases. This intensity increase occurs gradually for the chromatophores. In the case of the RCs, it occurs between 300 and 150K, and the relative intensity of the 1707 and 1702 cm^{-1} bands at 150, 100, and 80K stay constant. These results suggest that interactions between the special pair and its environment play a key role in determining the intensities of the bands at 1707 and 1702 cm^{-1}.

FTIR INVESTIGATION OF PROTEIN SECONDARY STRUCTURES IN PHOTOSYNTHETIC BACTERIAL REACTION CENTERS

E. NABEDRYK, J. BRETON and D.L. THIBODEAU

SBE / DBCM, CEN Saclay 91191 Gif-sur-Yvette cedex, France

The broad amide I band (1700–1610 cm^{-1}), which arises primarily from stretching vibrations of peptide backbone C=O groups, is a sensitive probe of specific secondary structures types (e.g. α–helix, β–sheet, turns). Until Fourier self–deconvolution and derivative methods were developed (1,2), the inability to separate the individual component peaks of the composite amide I band, however, severely limited the quantitative interpretation of IR spectra of complex biological systems. In the present study, we report the investigation of the membrane protein structure of bacterial photosynthetic reaction centers (RC) by FTIR spectroscopy and compare these IR data i) to the ones we have previously obtained by UV circular dichroism and ii) to the X–ray crystallographic RC model.

FTIR absorbance spectra of air–dried films of *Rb. sphaeroides* and *Rps. viridis* RC (Fig. 1) show the amide I band at 1657–1658 cm^{-1} characteristic of α–helical structure (3). Second–derivative as well as deconvolved spectra reveal additional peaks at 1680 cm^{-1}, 1670 cm^{-1}, 1635 cm^{-1} and 1626 cm^{-1} for *Rb. sphaeroides* (4) and at 1684 cm^{-1}, 1672 cm^{-1}, 1633 cm^{-1} and 1626 cm^{-1} for *Rps. viridis*. These components revealed in both second–derivative and deconvolved spectra were considered, at first, in the curve–fitting analysis of the RC amide I band which was performed according to the method described in ref.5. For both RC, the curve–fitting analysis using 4 Lorentzian is displayed on Fig. 1; it results in a main component at 1657 cm^{-1} which corresponds to 52 ± 3% of the integrated intensity of the amide I band for *Rb. sphaeroides* RC. The corresponding content for *Rps. viridis* RC is 50 ± 5% at 1658 cm^{-1}. These values agree well with the α–helical content deduced from i) UV circular dichroism measurements –51% and 47%– for *Rb. sphaeroides* and *Rps. viridis* RC, respectively (3,6) and ii) X–ray data –51% and 49%–, respectively (7,8). The curve–fitting analysis supports the presence of other components at 1677 cm^{-1}, 1638 cm^{-1} and 1618 cm^{-1} for *Rb. sphaeroides* and at 1680 cm^{-1}, 1637 cm^{-1} and 1615 cm^{-1} for *Rps. viridis* RC. The 1625–1615 cm^{-1} region is characteristic of the absorption from side chains and β–structure. The $\approx$1680 cm^{-1} peak ($\approx$10–15% of the amide I area) reflects contributions from β–turns and/or antiparallel β–sheets, while the 1638 cm^{-1} peak (20–25% of the amide I band) mainly arises from β–structure. From the X–ray crystallographic models (7,8), it can be calculated that the percentage of β–sheets in *Rb. sphaeroides* and *Rps. viridis* is 15% and 9%, respectively. The β–turns content is not reported in ref. 7 & 8. From UV circular dichroism measurements, a $\approx$15% β–structure value in both RC was estimated (6). It thus appears that the β–sheet content determined by IR (20–25%) might be overestimated. We do not exclude, however, that unordered structure may also contribute to the 1655–1640 cm^{-1} IR

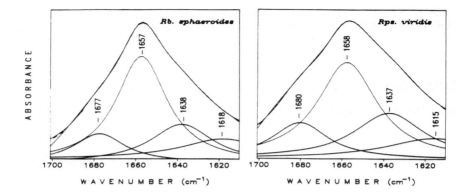

Figure 1: FTIR spectra of air–dried films of *Rb. sphaeroides* and *Rps. viridis* RC. Spectra (512 interferograms co–added) were recorded on a Nicolet 60SX spectrometer at $4\,cm^{-1}$ resolution. Curve–fitting analysis showing 4 Lorentzian components. The sum of the 4 peaks is represented by the dotted spectra.

domain (1,2), and therefore, ^1H–^2H exchange experiments were performed to distinguish the amide components arising from several peptide conformations. The rate of deuteration of unordered segments is expected to be faster than that of ordered segments (α–helices, β–sheets). The curve–fitting analysis of the amide I band for RC incubated in ^2H_2O for 24 hours at 10°C (data not shown) displays an additional component, at $1646\,cm^{-1}$ for *Rps. viridis* and at $1642\,cm^{-1}$ for *Rb. sphaeroides*, which can be assigned to deuterated unordered structures ($\approx15\%$ of the amide I). For both RC, the component at 1657–$1658\,cm^{-1}$ still accounts for 45–50% of the integrated amide I intensity, indicating that peptide protons in α–helices have not been significantly exchanged in these conditions. It should be kept in mind that, under our experimental conditions (the IR beam is perpendicular to the film), the orientation of the RC α–helical segments preferentially perpendicular to the air–dried film may lead to an overestimation of the β–sheet and unordered structures *versus* α–helices. The analysis of polarized IR spectra is in progress.

The agreement between IR and X–ray determined structures is very good for globular proteins (1,2). Here, we demonstrate that for two membrane proteins for which the 3-D structure is known, the α–helical content determined by IR on RC films is in remarkable agreement with that obtained from X–ray on RC crystals (7,8).

1. Byler & Susi (1986) Biopolymers 25, 469–487
2. Surewicz & Mantsch (1988) Biochim.Biophys.Acta 952, 115–130
3. Nabedryk et al. (1982) Biochim. Biophys. Acta 682, 273–280
   Nabedryk et al. (1985) Biochim. Biophys. Acta 809, 271–276
4. Nabedryk et al. (1989) in "Techniques and New Developments in Photosynthesis Research (Barber, J. & Malkin, R. eds.), Vol. 168, pp. 17–34, Plenum Publishing Corporation
5. Griffiths & Pariente (1986) Trends Anal. Chem. 5,209–215
6. Breton & Nabedryk (1987) in "The Light Reactions" (Barber,J. ed.) Vol.8, pp.159–195, Elsevier Science Publishers
7. Allen et al. (1987) Proc.Natl.Acad.Sci. USA 84,6162–6166
8. Deisenhofer et al. (1985) Nature 318, 618–624

LIGHT–INDUCED POLARIZED FTIR SPECTROSCOPY OF ORIENTED PHOTOSYNTHETIC BACTERIAL REACTION CENTERS

D.L. THIBODEAU, E. NABEDRYK and J. BRETON

SBE / DBCM, CEN Saclay, 91191 Gif–sur–Yvette cedex, France

Polarized FTIR spectroscopy has been used to investigate the orientation, relative to the membrane plane, of the vibrational modes of cofactors which are affected by the light–induced charge separation in *Rb. sphaeroides* and *Rps. viridis* reaction centers (RC). This method combines the sensitivity of FTIR difference spectroscopy with the ability of polarized spectroscopy to probe the orientation of chemical bonds. Pertinent structural information can be gained concerning the orientation of specific molecular groups which are affected by the photooxidation of the primary electron donor, P (a dimer of bacteriochlorophyll), and the photoreduction of the primary acceptor quinone, Q_A. This polarization spectroscopic method finds itself particularly useful in the light of previous FTIR difference measurements which have helped to identify specific groups involved during the $PQ_A \rightarrow P^+Q_A^-$ charge separation. Information about the orientation of the individual bonds in the resting state of the chromophores/cofactors can be deduced from the atomic coordinates obtained by X–ray crystallography[1,2], which in turn makes it possible to assess earlier FTIR assignments.

Figure 1 shows the light–induced difference spectra (ΔA) and the linear dichroism of the absorbance changes (ΔLD) for the $PQ_A \rightarrow P^+Q_A^-$ transition measured at 100K for *Rps. viridis* (Fig. 1a), and the corresponding spectra are shown for *Rb. sphaeroides* in Fig. 1b. In these figures, the positive bands in the ΔA spectra represent contributions from $P^+Q_A^-$, the negative bands arise from PQ_A; also for a given transition moment when ΔA and ΔLD have the same sign, it indicates that this transition is parallel (less than 35°, *i.e.* $\phi > 55°$) to the membrane. Table 1 lists the qualitative orientation relative to the membrane plane, ϕ, for bonds represented by selected IR difference bands. For their assignments in the difference spectra, the readers may wish to consult Ref. 3.

The dichroism of the band at 1757/1747 cm^{-1} in *Rps. viridis* previously assigned to the 10a ester C=O of P+/P is characteristic of a transition moment rather perpendicular to the membrane plane. This result is in agreement with the angle, ϕ_x, calculated from atomic coordinates: 29° and 45° for P_M and P_L, respectively. Moreover, this is   compelling evidence that no change of orientation of this group occurs upon the formation of P+. The X–ray coordinates indicate that the 9 keto C=O of P in *Rps. viridis* are both oriented at $\phi_x \approx 80°$. The positive dichroism of the bands at 1711 cm^{-1} and 1700 cm^{-1} supports the previous assignment[3] to 9 keto C=O of P+. Consequently, the band at 1674 cm^{-1} is no longer consistent with its previous assignment of a 9 keto C=O of P, but the dichroism observed suggests that the band at 1684 cm^{-1} is a more plausible candidate. Comparison of the X–ray

Frequency (cm^{-1})	ΔA	ΔLD	ϕ	Frequency (cm^{-1})	ΔA	ΔLD	ϕ	Assignments	
Rps. viridis				*Rb. sphaeroides*					
1757	>0	<0	⊥	1753	>0	>0	//	10a ester C=O	P+
1747	<0	>0	⊥	1739	<0	<0	//		P
1711	>0	>0	//	1714	>0	>0	//		P+
1700	>0	>0	//	1703	>0	>0	//	9 keto C=O	P+
1684	<0	<0	//	1683	<0	<0	//		P

Table 1. Orientations relative to the membrane of selected IR transitions

descriptions of *Rps. viridis*[1] and *Rb. sphaeroides*[2] shows that, in both RC, the 9 keto C=O of P have the same orientation. In the FTIR difference spectrum of *Rb. sphaeroides*, the contributions of the 9 keto C=O have been assigned to the 1683 cm^{-1} band for P and to the 1703 cm^{-1} and 1714 cm^{-1} bands for P+. Their respective dichroism clearly agrees with these earlier assignments. The ΔA of the 10a ester C=O region is much more complex for *Rb. sphaeroides* than for *Rps. viridis*, as it has been noted before[3] and as it is confirmed by the complex shape of the corresponding ΔLD. This result leads us to propose that the 10a ester C=O in *Rb. sphaeroides* is rather parallel to the plane of the membrane, as opposed to its orientation in *Rps. viridis*. For both RC, the dichroism of the bands in the 1680–1600 cm^{-1} region shows that the bonds are oriented rather perpendicular to the plane of the membrane which might indicate contributions from protein conformation changes. Although the 1500–1380 cm^{-1} region comprises bands where P+ contribute significantly, the positive dichroism observed in this region is consistent with the calculated ϕ_X of $\approx 88°$ for the C=O of Q_A.

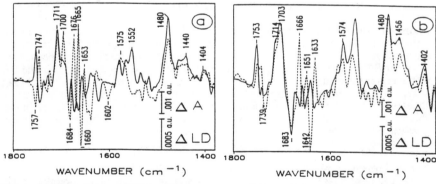

WAVENUMBER (cm^{-1}) WAVENUMBER (cm^{-1})

Figure 1 Light-induced FTIR difference spectra, $P^+Q_A^-/PQ_A$, at 100K for (a) *Rps. viridis* chromatophores and (b) *Rb. sphaeroides* RC reconstituted in liposomes. The oriented films were tilted at 60° from the incident infrared light. The spectra were recorded at 4 cm^{-1} resolution with alternate periods of illumination and darkness under parallel and perpendicular polarization[4]. Light-induced FTIR absorbance (– ΔA) and linear dichroism (··· ΔLD) spectra, a.u. absorbance units.

1. H. Michel, O. Epp, and J. Deisenhofer, EMBO J., 1986, 5, 2445.
2. J.P. Allen, G. Feher, T.O. Yeates, H. Komiya and D.C. Rees, Proc. Natl. Acad. Sci. USA, 1988, 85, 8487.
3. W. Mäntele, A. Wollenweber, E. Nabedryk and J. Breton, Proc. Natl. Acad. Sci. USA, 1988, 85, 8468 and refs. therein.
4. E. Nabedryk and J. Breton, FEBS, 1986, 202, 356.

RESONANCE RAMAN CHARACTERIZATION OF MUTANT *Rb.* *sphaeroides* REACTION CENTERS

T.A. MATTIOLI[*1], K.A. GRAY[2], J. WACHTVEITL[2], J.W. FARCHAUS[3], M. LUTZ[1], D. OESTERHELT[2], and B. ROBERT[1]

[1] Département de Biologie Cellulaire et Moléculaire, CE Saclay, 91191 Gif-sur-Yvette cedex, FRANCE.
[2] Department of Membrane Biochemistry, Max-Planck-Institut für Biochemie, D-8033 Martinsried bei München, GERMANY.
[3] Central Research and Development Station, E.I. DuPont de Nemours Co. Wilmington, DE 19880, USA

INTRODUCTION

Resonance Raman (RR) and low temperature absorption spectroscopies have been used to investigate structural changes in the reaction center (RC) from *Rhodobacter* (*Rb.*) *sphaeroides* resulting from site-directed mutagenesis of amino acids in the vicinity of the primary electron donor, P.

Tyrosine M210

Tyrosine M210 and phenylalanine L181 in *Rb. sphaeroides* form a pair of symmetry-breaking residues related by the pseudo-C_2 symmetry axis of the RC[1]. Both residues are in van der Waals contact with P. Recent room temperature femtosecond kinetic studies[2] reveal that the electron transfer rate from P to the bacteriopheophytin acceptor decreases about five-fold when tyrosine (Y) M210 is genetically replaced with phenylalanine (F) or leucine (L).

The general similarity of the RR spectra of the wild type RCs and of the two mutants (Y→F and Y→L) indicates that upon mutation no global structural changes occur at the level of the six bacteriochlorin pigments in the RC[3]. Comparisons of the difference RR spectra of the primary electron donor, P, show that tyrosine M210 does not bind any of the conjugated carbonyls of P and that it does not likely bind any of those of the accessory $BChl_L$. Furthermore, neither the Y→F nor the Y→L mutation resulted in any large change in the protein pockets around P, $BChl_L$, and $BPhe_L$. This strongly suggests that the marked effect of Tyr M210 on the electron transfer kinetics occur through direct, nonbonding interactions with P and/or $BChl_L$ (e.g. in lowering the $P^+BChl_L^-$ state energy, cf ref. 4, or in affecting charge repartition within the P^* state, cf ref. 5) rather than indirectly through structural rearrangement of the protein pockets of these sites.

Phenylalanine M195

In *Rhodopseudomonas* (*Rps.*) *viridis*, this residue is a tyrosine molecule and forms a H-bond with the acetyl carbonyl of P_M which has been assigned to the 1628 cm^{-1} Raman band in the RR difference spectrum of P in this RC[6]. In *Rb. sphaeroides* the equivalent position is occupied by a phenylalanine residue unable to form a H-bond. As with the M210 mutants (above), there is a conspicuous absence of the 1704 cm^{-1} band in the RR spectrum of the M195 F→Y mutant. This mode arises from the $BPhe_M$ molecule which is far removed from the mutation site(s). This spectral change most probably results from a local

71

rearrangement of the protein around BPhe$_M$ and can also be induced by other causes (e.g. mild treatment with chaotropic chemicals). This rearrangement is also manifested by changes in the electronic energy levels of BPhe$_M$, such as an upshift in the Q_x absorption band.

The difference RR spectrum of P for the M195 Y→F mutant of *Rb. sphaeroides* as compared with that of the wild type spectrum (Fig. 1) reveals that structural changes have occurred in the binding pocket of P. This shows that the mutation of the M195 site affects the structure of the P binding pocket more than does the mutation of the M210 site. As compared to the wild type, the M195 F→Y mutant difference RR spectrum exhibits a new 1625 cm^{-1} shoulder, similar to that which was observed for the case of *Rps. viridis*[6]. This could indicate the formation of a H-bond with an acetyl carbonyl (H-bonded keto carbonyls are not likely to exhibit frequencies lower than 1650 cm^{-1}), conceivably with tyrosine M195. In this mutant, there is also an additional effect with the weakening of the 1684 cm^{-1} band as compared to the difference spectrum of the wild type, indicating that a keto carbonyl of P has been affected. Work is in progress using near-IR Fourier transform Raman spectroscopy as well as other RC preparations to determine if the 1684 cm^{-1} band has shifted or decreased in relative intensity.

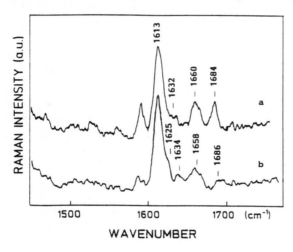

Figure 1 Difference RR spectra (363.8 nm excitation) obtained from ascorbate-treated minus ferricyanide-treated RR spectra *of Rb. sphaeroides* reaction centers at 14 K for a) wild type and b) F M195 → Y mutant.

REFERENCES

1. D.M. Tiede, D.E. Budil, J. Tang, O. El-Kabbani, J.R. Norris, C.H. Chang, and M. Schiffer, "The Photosynthetic Bacterial Reaction Center, Structure and Dynamics" (J. Breton and A. Verméglio, Eds.), Plenum, New York, 1988, p. 13.
2. U. Finkele, C Lauterwasser, W. Zinth, K.A. Gray, and D. Oesterhelt, Biochemistry, 1990, 29, 8517.
3. T.A. Mattioli, K.A. Gray, M. Lutz, D. Oesterhelt, and B. Robert, Biochemistry, 1990, 30, 1715.
4. W.W. Parson, Z.T. Chu, and A. Warshel, Biochim. Biophys. Acta, 1990, 1017, 251.
5. M. Plato, F. Lendzian, W. Lubitz, E. Tränkle, and K. Möbius. In Ref. 1, p. 378.
6. Q. Zhou, B. Robert, and M. Lutz, Biochim. Biophys. Acta, 1989, 977, 10.

STUDYING RECOMBINATION OF FLASH LIGHT- INDUCED CHARGE SEPARATION IN PHOTOSYSTEM II MEMBRANES BY TIME- RESOLVED INFRARED SPECTROSCOPY

R. Hienerwadel, C. Berthomieu[*], E.. Nabedryk[*], J. Breton[*], W. Kreutz and W. Mäntele
Institut für Biophysik der Univ. Freiburg, Albertstr.23, 78 Freiburg (FRG)
[*] DBCM/SBE, CEN Saclay, 91191 Gif-sur-Yvette Cedex (France)

The photosynthetic complex of higher plants or bacteria is a very efficient light-driven motor, which in its first steps transforms absorbed photon energy into a transmembrane charge separation. In order to understand these mechanisms, the interaction between cofactors embedded in their specific protein environment and the role of functional side chains of amino acids need to be understood. Using Infrared(IR) spectroscopy as a tool to investigate these mechanisms might help to understand the high efficiency of this vectorial charge separation.

In order to get access to the specific vibrations of cofactors or protein side chains, we have developed a highly sensitive kinetic IR spectrophotometer using laser diodes as a powerful IR measuring light source. These laser diodes can be wavelength-tuned by variation of the temperature from 80 to 120 Kelvin. With two laser diodes, it is possible to get an almost continuously tuned wavelength region from 1450 to 1800 cm^{-1}. Actually the time resolution of the spectrophotometer is around 500 nsec. This setup is characterized by very high sensitivity, i. e. the signal to noise ratio is below 10^{-3} absorbance units per turnover with appropriate electrical filters. This is due to the high measuring intensity of the laser diode. A detailed description of the setup can be found in [1,2]. For measurements of Photosystem II membranes, charge separation was induced by a 15 nsec light pulse originated by a dye laser at the wavelength of 670 nm, the red absorbance maximum of the antenna chlorophylls.

Photosystem II-membranes from spinach were prepared as described in [3]. The oxgen evolving complex was inhibited by 0.8 M Tris washing at pH 8.5. The final buffer for measurements was kept at pH 6.5 with MES in the presence of 1 mM EDTA, 0.1 mM DCMU and 2 mM PMS. With these PS II-enriched membranes IR samples were prepared as in [4]. Under these conditions charge recombination should follow the light induced charge separation, mainly between the reduced primary acceptor Q_a^- (Plastoquinone-9) and the oxidized donor Z^+ (Tyrosin), the physiological donor to the primary donor P680, should be observed [5]. Charge recombination between Z^+ and Q_a^- should occur in the time range from 100 to 200 msec range, while donation from Z to $P680^+$ occurs in 2 to 20 μsec, measured at the absorption band of $P680^+$ in the near IR at 820 nm [6]. A typical time -resolved absorbance change at 820 nm with samples obtained as described above is shown in fig. 1, with a decay component of 5 μsec. for 70% of the particles, the other 30% are decaying mainly in the msec time range. In fig. 2 IR kinetic traces at 1507 cm^{-1} and at 1522 cm^{-1} show the same fast phase of 5 μsec. We thus attribute the IR-signals to the states $ZP680^+Q_a^-$ or $Z^+P680Q_a^-$. The kinetic signal at 1507 cm^{-1} is most likely due to $P680^+$, because the signal is arising instantaneously within the instrument time resolution. We cannot exclude a contribution of Q_a^- for the slow decaying signal, since static Q_a^-/Q_a FTIR spectra from Berthomieu et al. [4] show also a small positive band in this region. Measurements on a longer time scale show a component in the 100 to 200 msec range (not shown). The situation at 1522 cm^{-1} might be more complex, since according to the Q_a^-/Q_a difference spectra [4], Q_a might have a negativ contribution as well, so that we cannot decide at this stage, whether we observe a single negative signal, arising in the μsec. range and decaying in the msec. range, or a contribution of two independent signals. Further investigations will have to be done to clarify this point.

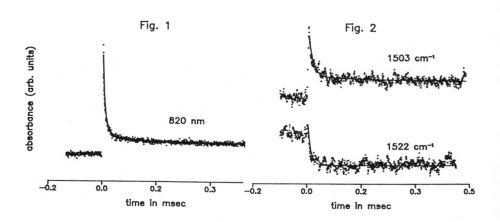

Fig. 1 Absorbance change of the oxidized primary Donor P680$^+$
Fig. 2 IR absorbance changes at specified wavenumbers

REFERENCES

1. Hienerwadel et al. (1989), Spectroscopy of Biological Molecules
   A. Bertoluzza, C. Fagnano, P. Monti eds. (Soc. ED. Esculapio, Bologna), 315
2. Mäntele et al. (1990), spectroscopy, Vol. 6, No. 1, 25-30
3. Ford et. al. (1983), FEBS Lett. 160, 160,159-164
4. Berthomieu et al. (1990), FEBS Lett. 269, 269, 363-367
5. Gerken et al. (1989), Biochim. Biophys. Acta, 977, 52-61
6. Conjeaud et al. (1979), Biochim. Biophys. Acta, 546, 280-291

CHARACTERIZATION OF THE PRIMARY ELECTRON DONOR "P" OXIDATION IN BACTERIAL PHOTOSYNTHETIC REACTION CENTERS BY REDOX-TRIGGERED INFRARED DIFFERENCE SPECTROSCOPY

M. Leonhard, D. Moss, M. Bauscher, E. Nabedryk[*], J. Breton[*], W. Mäntele

Institut für Biophysik und Strahlenbiologie der Universität Freiburg, Albertstr. 23, 7800 Freiburg FRG
[*]Service de Biophysique, CEN Saclay, 91191 Gif-s-Yvette, France

Introduction. In bacterial photosynthetic rection centers (RC), about 40 % of the energy of the absorbed photon or trapped exciton are stored in a state of charge separation between the primary electron donor P (a bacteriochlorophyll dimer) and the quinone electron acceptor (Q_A or Q_B). X-ray crystallography of Rp. viridis[1,2] and of Rb. sphaeroides RC crystals[3] have provided detailed protein structures, reflecting a static picture of the quiescent state of the RC. In order to obtain dynamic or functional information on the level of individual bonds of the protein, we have used infrared (IR) differential techniques. In previous studies, intrinsic photochemical reactions in RC were used for the creation of intermediates, implying charge-separated states such as P^+Q^-. Consequently, the resulting spectra were composed of bands from several species, a fact which complicates an interpretation due to overlap or cancellation of difference bands. Our previous work[4,5] has shown that the combination of Fourier transform IR (FTIR) difference spectroscopy and electrochemistry as a trigger for redox-reactions in a spectroelectrochemical cell is a successful tool for the monitoring of molecular changes at the redox-site. We now report that this technique is also applicable to a large, detergent-solubilized protein such as bacterial RC in order to determine the molecular processes upon formation of the P cation radical, comprising the bonds of the cofactor itself and those of its binding site.

Materials and Methods. Redox-induced FTIR difference spectra of RC were generated as described[6]. RC were concentrated to approx. 0.5 mM in phosphate buffer (12.5 mM, pH 7.4) with 0.1 % octylglycoside as detergent, 62.5 mM KCl as supporting electrolyte and 0.25 mM ferrocyanide as mediator for Rb. sphaeroides and in phosphate buffer (50 mM, pH 7.4) with 1 % octylglycoside, 250 mM KCl and 0.25 mM ferrocyanide for Rp. viridis.

Results and Discussion. Electrochemical redox poising of the primary electron donor P, monitored by chronoamperometry and by spectroscopy in the visible/near infrared region[7] has shown that the redox-induced difference spectra can be precisely triggered by the selection of the potential range and are completely reversible. Figure 1a shows the IR difference spectra of the primary electron donor oxidation (P^+(+0.4 V)/P(0 V); positive bands corresponding to P^+) of Rb. sphaeroides obtained between 0 V and +0.4 V (all potentials are quoted versus Ag/AgCl in 3 M KCl). Figure 1b shows the IR difference spectra of P (P^+(+0.45 V)/P(+0.24 V)) of Rp. viridis obtained between +0.24 V and +0.45 V. Under these conditions control spectra in the UV/VIS spectral region (data not shown) do not exhibit any contribution of the quinone or the cytochromes to the difference spectra. The highly structured spectra represent the direct changes of bond energies of the BChl dimers and their protein environment upon P_{866}^+

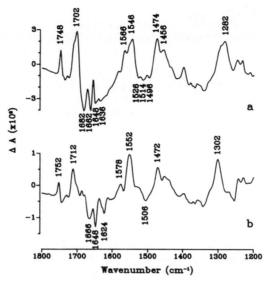

Figure 1 a: P^+/P of *Rb. sphaeroides* b: P^+/P of *Rp. viridis*

or P_{960}^+ formation, without any involvement of other cofactors. For *Rb. sphaeroides*, the strong difference bands in the C=O frequency region, at 1748 cm^{-1}, at 1702 cm^{-1} and at 1682 cm^{-1}, can be interpreted as changes of position and intensity of the 13^3-ester and the 13^1-keto C=O groups, respectively, of the BChl a dimer. They can be closely approximated by electrochemically generated BChl a model difference spectra[8]. The P^+/P difference spectra of *Rp. viridis* (BChl b instead of BChl a, but a strongly conserved amino acid sequence), however, are not modelled as perfect by BChl b$^+$/BChl b difference spectra[9] as are the *Rb. sphaeroides* spectra. The strong difference bands at 1752 cm^{-1} and at 1712 cm^{-1} can be assigned to changes of the 13^3-ester and the 13^1-keto C=O groups, respectively, of the BChl b dimer. In this spectral region, band features very close to those previously obtained by photochemical charge separation (P^+Q^- /PQ)[10] appear.

1. H. Michel, O. Epp and J. Deisenhofer, The EMBO Journal, 1986, 5, 2445.
2. J. Deisenhofer and H. Michel, EMBO J., 1989, 8, 2149.
3. J. P. Allen, G. Feher, T. O. Yeates, H. Komiya and D. C. Rees, Proc. Natl. Acad. Sci. USA, 1988, 85, 8487.
4. D.A. Moss, F. Fritz, W. Haehnel, J. Breton and W. Mäntele, Spectroscopy of Biological Molecules - 'State of the Art', A. Bertoluzza, C. Fagnano and P. Monti eds., Societa Editrice Esculapio, Bologna, 1989, p. 355.
5. D. Moss, E. Nabedryk, J. Breton and W. Mäntele, E. J. B., 1990, 187, 565.
6. W. Mäntele, M. Leonhard, M. Bauscher, E. Nabedryk, J. Breton and D.A. Moss, 'Reaction Centers of Photosynthetic Bacteria', M.-E. Michel-Beyerle ed., Springer-Verlag, Berlin Heidelberg, 1990, p. 31.
7. D. A. Moss, M. Leonhard, M. Bauscher and W. Mäntele, FEBS, in the press.
8. W. G. Mäntele, A. M. Wollenweber, E. Nabedryk and J. Breton, Proc. Natl. Acad. Sci. USA, 1988, 85, 8468.
9. W. Mäntele, A. Wollenweber, E. Nabedryk, J. Breton, F. Rashwan, J. Heinze and W. Kreutz, 'Progress in Photosynthesis Research', J. Biggins ed., Martinus Nijhoff Publishers, Dordrecht, 1987, p. 177.
10. W. Mäntele, E. Nabedryk, B. A. Tavitian, W. Kreutz and J. Breton, FEBS Lett., 1985, 187, 227.

INFRARED DIFFERENCE SPECTROSCOPY OF ELECTROCHEMICALLY AND PHOTOCHEMICALLY GENERATED QUINONE ANIONS IN BACTERIAL REACTION CENTERS

M. Bauscher[*], D.A. Moss, M. Leonhard, E. Nabedryk[+], J. Breton[+], W. Mäntele

Institut für Biophysik und Strahlenbiologie der Universität Freiburg, Albertstrasse 23, D-7800 Freiburg, FRG
[+]Service de Biophysique, CEN Saclay, 91191 Gif-sur-Yvette, France

The bacterial photosynthetic reaction center (RC) provides a simple system to study the primary reactions occuring upon absorption of a photon producing a charge-separated state. The electron is transferred from the primary electron donor P (a bacteriochlorophyll dimer, $BChl_2$) in a series of very fast electron transfer steps to quinone acceptors (Q_A or Q_B, if present) to produce $P^+Q^-_{A/B}$. The efficiency and directionality of this charge separation and stabilization is governed by the arrangement of the cofactors in the RC and by their respective cofactor-protein interactions. Although the crystallization of RC has provided a detailed picture at an atomic level[2], there is still a lack of information on the binding of the pigments in the respective state and the dynamic behaviour of the RC following charge-separation. Fourier transform infrared (FTIR) difference spectroscopy is able to give those information because of its ability to study the RC during its function and to monitor changes of individual bonds.

We have previously obtained FTIR difference spectra between the charge-separated and the relaxed state (P^+Q^-/PQ) of RC. To decide whether absorbance changes arise from pigments or quinones, 'model' spectra of electrochemically generated BChl cations and quinone anions in different organic solvents to account for different polarity and proticity, were recorded[3,4]. Although this approach has led to a clear assignment of some bands arising from P, absorbance changes due to quinone reduction remained uncertain.

To clarify this, we have used three independent methods to record FTIR difference spectra between the reduced and the neutral quinone acceptor Q_A bound to the RC. The first approach was the direct electrochemical reduction of Q_A in an electrochemical cell (fig. 1 a). Benzylviologen (BV) served as a redox mediator to shuttle electrons from the electrode to Q_A. Since BV has a much more negative midpoint potential (-0.36 V vs. SHE) than Q_A (-0.05 V vs. SHE) reduction of BV leads to an almost complete reduction of Q_A. To eliminate the absorbance changes arising from BV, the potential was increased to a value where BV is reoxidized and Q_A remains reduced (-0.15 V vs. SHE). The second approach was the light-induced reduction of Q_A in presence of electrochemically reduced cytochrome (cyt) c_2 (the native, fast electron donor to P^+), so that the Q_A^- state is trapped (fig. 1 b). The resulting spectrum also contained contributions from cyt c_2 oxidation which were subtracted. The third approach was the subtraction of an electrochemically generated P^+/P spectrum from a light-induced $P^+Q_A^-$/PQ_A spectrum (fig. 1 c). The electrochromic shift of the bacteriopheophytin (BPheo) absorption (fig. 1 d), as well as time-resolved (TR) measurements at 866 nm of flash-induced charge separation and recombination (fig. 1 e), served as two independent controls for Q_A reduction.

The bands present between 1668 cm^{-1} and 1600 cm^{-1} can be regarded as C=O or C=C vibrations arising from Q_A itself, or from amino acid vibrations in its vicinity. The bands at 1668 cm^{-1} and 1652 cm^{-1} are clearly associated with the $Q_A \rightarrow Q_A^-$ transition (for detailed discussion see ref. 5). At present, we rather assign both bands to amide I C=O vibrations in the environment of Q_A. The bands at 1630 cm^{-1} and 1600 cm^{-1} have been assigned to Q_A C=O and C=C vibrations[4,5], respectively and the band at 1462 cm^{-1} to Q_A^- C=O vibration[4,5]. The bands between 1560 cm^{-1} and 1520 cm^{-1} are possibly the amide II signals corresponding to the 1668 cm^{-1} and 1652 cm^{-1} bands. Alternatively they might arise from aromatic amino acids in the vicinity of Q_A. The differential feature at 1734/1726 cm^{-1} could arise from a carboxyl group side chain residue although the nearest one is 10 Å away from Q_A[2].

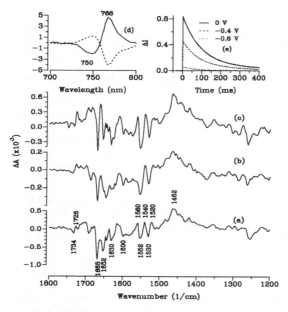

Figure 1 Q_A^-/Q_A IR difference spectra obtained by different methods: (a) Direct electrochemistry, (b) photochemically with cyt c_2 reduced, (c) $(P^+Q_A^-/PQ_A)/(P^+/P)$ difference, (d) BPheo shift upon reduction of Q_A (full line) and reoxidation (dotted line), (e) TR measurements at 866 nm of flash-induced charge separation and recombination

REFERENCES

1. W.W. Parson, Annu. Rev. Biophys. Bioeng., 1982, 11, 57.
2. J.P. Allen, G. Feher, T.O. Yeates, H. Komiya and D.C. Rees, Proc. Natl. Acad. Sci. USA, 1988, 85, 8487.
3. W. Mäntele, A. Wollenweber, E. Nabedryk and J. Breton, Proc. Natl. Acad. Sci. USA, 1988, 85, 8482.
4. M. Bauscher, E. Nabedryk, K. Bagley, J. Breton and W. Mäntele, FEBS Lett., 1990, 261, 191.
5. W. Mäntele, M. Leonhard, M. Bauscher, E. Nabedryk, J. Breton and D.A. Moss, 'Reaction Centers of Photosynthetic Bacteria', Springer Verlag, Heidelberg, 1990, 31.

REDOX-INDUCED INFRARED DIFFERENCE SPECTROSCOPY OF C_2 CYTOCHROMES: COMPARISON OF HEME ENVIRONMENT AND PROTEIN CONFORMATIONAL CHANGES

F. Baymann, D.A. Moss, and W. Mäntele

Institut für Biophysik und Strahlenbiologie
der Univerität Freiburg
Albertstraße 23
7800 Freiburg FRG;

Cytochromes c_2 act as water soluble electron carriers between membrane bound proteins in bacterial photosynthesis. They have one covalently bound heme. Their molecular weight is about 12 kD and the midpoint potentials are around 100 mV (vs. Ag/AgCl). In spite of the successful crystallisation of some c-type cytochromes, the molecular mechanisms that regulate the redox-dependent affinities to the reaction center and the bc_1-complex are not fully understood.

Infrared-difference spectroscopy, able to detect small reaction-induced molecular changes in protein complexes, should help to elucidate these mechanisms.

In the present work reduced-minus-oxidized infrared difference spectra were obtained from the water soluble c_2 -type cytochromes of *Rb. sphaeroides* (Y strain), *Rps. viridis* and *Rs. rubrum* by electrochemically triggered redox reactions in a spectroelectrochemical cell suitable for infrared spectroscopy (figure 1).

In this context we developed an assay for the purification procedure. In order to make sure that the cytochromes c_2 were free from redox-active contaminents, there was a need to combine spectroscopic tests with electrochemical tests. As a quick and extremely sensitive method, this combination should reveal contaminants through their VIS absorption as well as through the electrode current upon the electrochemically-induced redox transition. In addition, we thus obtained information on the reactivity of the cytochrome at the electrode, and the cytochrome midpoint potentials.[3]

The highly detailed IR difference spectra reflect subtle protein conformational changes, changes of heme bonding and interaction, but also perturbations of heme vibrations upon reduction.

We compared the IR-difference spectra, obtained from the c_2-cytochromes with those from cytochrom c_{551} of *Ps. aeruginosa* and horse cytochrome c (data not shown). The latter was measured at different pH values.

The amplitudes of the difference spectra in the amide I and amide II regions (at most 1% of the total amide I absorbance) exclude large and "global" conformational changes upon reduction of the *Rp. viridis* and *Rb.*

sphaeroides c_2 cytochromes. In the *Ps. aeruginosa* cytochrome c_{551} difference spectra, however, a strong (approx. 5% of the total amide I absorption) difference band centered at 1650 cm^{-1} indicates a larger extent of structural rearrangement, in contrast to predictions from X-ray structural analysis.[2] Amazingly enough, the IR-difference spectra of the cytochrome from this procaryotic organism looks very similar to that of horse cytochrome c at pH 11.5.

On the basis of the conserved and non-conserved amino acid residues for the cytochromes of the three different bacterial species and by comparison to IR difference spectra of horse and tuna cytochrome c,[1] assignments of bands in the 1750 cm^{-1} to 1200 cm^{-1} spectral region and interpretations in terms of local modifications of the protein structure and heme interactions can be obtained.

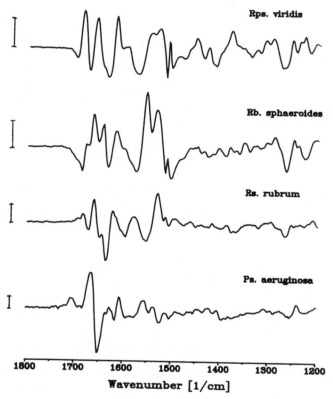

Figure 1 IR-difference spectra of cytochromes c_2 and Ps. aeruginosa cytochrome c_{551}. bar length indicates 0.002 AU.

REFERENCES
1. Moss,D.A., Nabedryk,E., Breton,J., Mäntele,W. Eur.J.Biochem. 187:565-572,1990
2. Matsuura,Y., Takano,T., Dickerson,R.E., J.Mol.Biol. 156,389-401,1982
3. Baymann,F., Moss,D.A., Mäntele,W., Manuscript in preparation.

DETERMINATION OF THE MIDPOINT POTENTIALS OF THE TETRAHEME CYTOCHROME C SUBUNIT OF RHODOPSEUDOMONAS VIRIDIS BY REDOX-TRIGGERED OPTICAL DIFFERENCE SPECTROSCOPY

F. Fritz, D.A. Moss, W. Mäntele

Institut für Biophysik und Strahlenbiologie der Universität Freiburg, Albertstr. 23, 7800 Freiburg FRG

Introduction. The photosynthetic reaction center (RC) of *Rp. viridis* contains a tetraheme cytochrome c subunit which is responsible for the re-reduction of the photooxidized primary electron donor. The four hemes are characterized by their different redox midpoint potentials (E_m) and their different optical absorption (α - band). Previous chemical redox titrations have resulted in variations of the determined E_m values[1,2]. Chemical titrations suffer from several disadvantages, such as dilution and incomplete reversal on the same sample. Electrochemical titrations, in which the potential is applied electrically, offer a number of advantages. These include fast equilibration with the applied potential, access to a wider potential range, repeated reduction and oxidation of the sample without dilution problems, direct measurements of the oxidative / reductive current and the amount of charge transferred. The different α - band absorptions have been used in a potential titration to determine the respective E_m-values. Furthermore, it is possible to record infrared (IR) difference spectra to obtain information of the molecular processes caused by electron transfer.

Materials and Methods. *Rp. viridis* RC were prepared as described[3]. RC were concentrated to approximatly 0.5 mM in phosphate buffer (pH 7.4, 50 mM phosphate, 250 mM KCl, 1 % octylglycoside) and a mixture of mediators (E_m values in brackets): ferrycyanide (+0.218 V); 1,1'-dimethyl-ferrocene (+0.133 V); tetrachloro-benzoquinone (+0.74 V); rutheniumhexamine-chloride (-0.8 V); 1,2-naphtoquinone (-0.63 V); trimethyl-hydroquinone (-0.108 V); 2-methyl-menaquinone (-0.220 V); 2-OH-1,4-naphtoquinone (-0.333 V). All potentials are quoted vs. Ag/AgCl. The thin-layer-electrochemical cell, the filling procedure and the procedure to record redox-poised VIS and IR spectra have been described[4]. PATS-4 has been used to modify the gold electrode[5].
A redox titration in the range of -0.390 V to +0.26 V at 279 K has been performed, monitoring the absorbance between 500 and 600 nm.

Results and Discussion. Figure a and b show the difference spectra of the oxidative titration of the two low-potential and the two high-potential hemes, respectively. Plots of the absorbance changes vs. applied potentials are shown in figure c and d. The E_m values of the respective hemes have been obtained by approximation with two Nernst equations. For better peak separation different reference spectra have been used (-0.39 V and 0 V, figure a and b, respectively). Figure e shows the difference spectra of the reductive titration of all hemes (spectrum at +0.23 V used as reference).

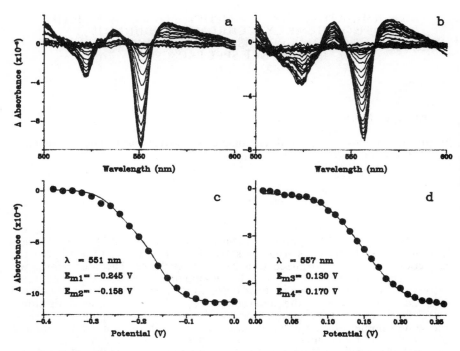

Figure: a, b: Difference spectra of the oxidative titration of a) the low-potential hemes, b) the high-potential hemes.   c, d: Redox-titration and Nernst curve of c) the low-potential hemes, d) the high-potential hemes.

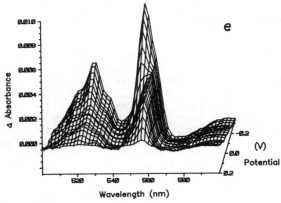

Figure e: Difference spectra of the reductive titration of all hemes

REFERENCES

1. S. M. Dracheva et al., FEBS Lett., 1986, 205, 41.
2. G. Fritzsch et al., BBA, 1989, 977, 157.
3. W. Welte et al., J.Chrom., 1983, 259, 341.
4. D. A. Moss et al., Eur. J. Biochem., 1990, 187, 565.
5. H. A. O. Hill et al., J. Electroanal. Chem., 1985a, 187, 315.

ELECTROCHEMICALLY-TRIGGERED OXIDOREDUCTION OF THE BIS-(METHYLIMIDAZOLE) COMPLEX OF IRON PROTOPORPHYRIN STUDIED BY FTIR DIFFERENCE SPECTROSCOPY: A MODEL FOR CYTOCHROME b-559 PHOTOOXIDATION IN PHOTOSYSTEM II.

C. BERTHOMIEU[1], D. MOSS[2], W. MÄNTELE[2], A. BOUSSAC[1], J. BRETON[1] and E. NABEDRYK[1].

[1]SBE/DBCM, CEN–Saclay, 91 191 Gif–sur–Yvette Cedex, France.
[2]Institut für Biophysik, Albertstraße 23, W–7800 Freiburg, F.R.G.

In photosystem II, absorption of a photon results in the oxidation of the primary donor P_{680} and reduction of a primary electron acceptor Q_A. Below 77 K, P_{680}^+ is reduced either by a chlorophyll or by the cytochrome b559 tightly bound to the photosynthetic reaction center. The light–induced charge separated state Cyt b559+ Q_A^- has been studied by FTIR difference spectroscopy (1). Q_A^-/Q_A FTIR difference spectra, corresponding to Q_A photoreduction have recently been obtained (2). In the Cyt b559+Q_A^-/Cyt b559 Q_A spectrum, we intend now to analyse cytochrome contributions. The structural organisation of cyt b559 is not known. Nevertheless, a model of the b–type heme, arranged in a cross linked structure with two different polypeptides through electrostatic interactions between the heme iron and two histidine residues has been proposed (3). In order to determine the relative contributions of the protein and heme moieties to charge stabilisation of the cytochrome subunit *in vivo*, we have used electrochemistry to investigate a model of the six–coordinated cytochrome heme *in vitro*. We report here infrared absorption changes characteristic of the oxidation state of the bis(methylimidazole) complex of iron protoporphyrin (FePP(MImH)2).

The FePP(MImH)2 complex was obtained by dissolution of hemine chloride in a phosphate buffer (pH 8) in the presence of the cationic detergent cetyltrimethyl-ethylammonium bromide (CTABr) and of 4,5–methylimidazole (MImH) in excess. Linear sweep voltammetry performed in the thin–layer electrochemical cell described by Moss et al. (4), showed that the oxidoreduction was a one–electron process. The midpoint potential was observed at –360 mV (*versus* Ag/AgCl).

The sample IR absorption is dominated by water and MImH modes. To reveal the vibration frequencies associated with the reduced and oxidised states of the complex, difference spectroscopy between must be performed. Fig. 1 shows the electrochemically–triggered oxidised–minus–reduced (and reverse) FTIR difference spectra of the FePP(MImH)2 complex in CTABr. This figure shows a number of well resolved reversible signals. In the reduced–minus–oxidised spectrum (···), negative bands are characteristic of $Fe^{III}PP(MImH)_2$, positive ones of $Fe^{II}PP(MImH)_2$. No absorption changes are observed above 1660 cm^{-1} from the C=O mode of a protonated propionic group, as expected, since at pH 8 the propionic groups are deprotonated. The most prominent features are observed at 1552 cm^{-1} and 1535 cm^{-1} for the reduced form. In this region, C–O stretching modes from COO$^-$ groups are expected to contribute. A more probable assignment, however, would be to heme C=C stretching modes. The $\nu(C_bC_b)$ $\nu 11$ and $\nu 38$ (E_u) modes have been reported in Resonance Raman (RR) spectra of FePP(ImH)2 and FePP(MImH)2 at 1563–1547 cm^{-1} and 1556–1535 cm^{-1} for oxidised and reduced forms respectively (5–7). We thus tentatively assign the changes observed at 1574/1552 cm^{-1} and 1535 cm^{-1} in Fig. 1 to heme skeletal C=C stretching modes. Fig. 2 shows the IR absorption spectrum of MImH in water. Absorption is observed at 1572 cm^{-1}. This mode could also be affected by the change of electrostatic interaction with the heme iron and thus account for the change observed at 1574/1552 cm^{-1} in Fig. 1.

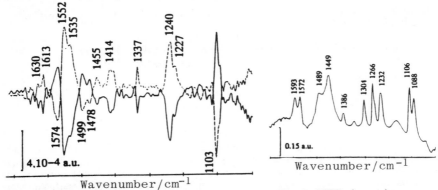

Fig. 1 FTIR difference spectra of
... Fe^{II}PP(MImH)$_2$-minus-Fe^{III}PP(MImH)$_2$,
—— reverse, resolution 4 cm^{-1}

Fig. 2 FTIR absorption
spectrum of 4,5-methyl
imidazole in water

The redox sensitive RR mode at 1375 cm^{-1} in the oxidised and 1360 cm^{-1} in the reduced form (5–7), assigned to $\nu(C_aN)=\nu 4$ is not observed in the FTIR difference spectrum. The change at 1337 cm^{-1}, however, could account for this mode. Another possible assigment is to a $\nu s(=CH_2$ vinyl), as observed at 1347 cm^{-1} and 1334 cm^{-1} in RR spectra of oxidised and reduced forms of the complex, respectively. The positive signals observed at 1630 cm^{-1} and 1613 cm^{-1} upon reduction are in the C=C stretch frequency range. These signals have possible assignments to $\nu(C_aC_m)$ skeletal mode (observed at 1639/1635 cm^{-1} and 1621/1616 cm^{-1} in RR) or to $\nu(C\alpha C\beta)$ vinyl modes (8,9). Finally, the strong band at 1103 cm^{-1}, associated to the oxidised form could arise from contributions of MImh, since an IR absorption mode is observed at 1106 cm^{-1} in Fig. 2. Modifications of the difference spectrum were observed in this region when the electrochemistry was performed in Tx100 and OGP as other detergents and we expect MImh modes to be more sensitive to the environment than heme skeletal modes.

Comparison of these studies with *in vivo* photooxidation of cyt b559 leads us to tentatively assign the negative band observed at 1543 cm^{-1} (1) to a heme contribution. Electrochemistry of the isolated cyt b559 unit will be performed for further study of the protein rearrangement following oxidation.

1. Berthomieu, C. et al. (1990) Biophys. J. 57, 556a.
2. Berthomieu, C. et al. (1990) FEBS Lett. 269, 363–367.
3. Widger W.R. et al. (1985) FEBS Lett. 191, 186–190.
4. Moss, D. et al. (1990) Eur. J. Biochem. 187, 565–572.
5. Babcock, G.T. et al. (1985) Biochemistry 24, 3638–3645.
6. Choi, S. et al. (1982) J. Am. Chem. Soc. 104, 4345–4352.
7. Desbois, A. et al. (1989) Biochemistry 28, 8011–8022.
8. Choi, S. et al. (1983) J. Am. Chem. Soc. 105, 3692–3707.
9. Applications of Infrared, Raman and Resonance Raman Spectroscopy in Biochemistry (1983) Parker, F.S. Plenum Press N.Y. Chap.6 p. 273–274 and 278–279.

FOURIER TRANSFORM INFRARED (FTIR) DIFFERENCE SPECTROSCOPY OF ELECTROCHEMICALLY INDUCED BACTERIOCHLOROPHYLL CATION RADICALS IN ISOLATED LIGHT HARVESTING COMPLEX OF PHOTOSYNTHETIC BACTERIA

M.P. Skatchkov[*], M. Leonhard, W. Mäntele

Institut für Biophysik der Universität Freiburg, Albertstr. 23, D-7800 Freiburg

Light-harvesting (LH) complex is a bacteriochlorophyll (BChl)-protein complex which does not exibit any photochemical activity *in vivo*. In order to determine the factors responsible for the spectral shifts of the lowest singlet transition of BChl molecules in different BChl-protein complexes, FTIR and IR resonance Raman spectroscopy have been employed[1-3].

We tried to force LH 2 complex, obtained from photosynthetic bacteria, to participate in reversible processes of oxidation and reduction *in vitro*. Our main goal was to monitor changes in the absorbance of the LH complex while varying the potential applied to the electrochemical cell containing the sample of LH 2. Spectra were recorded simultaneously in the 650-950 nm region (fig. e and f) and in the 1620-1730 cm^{-1} region (fig. a, b and c).

LH 2 complex was obtained from *Rb. sulfidophilus* (strain W 4) as described[3]. Protein concentration was 10 mg/ml. All potentials are quoted versus the Ag/AgCl reference electrode. The electrochemical cell and IR equipment have been described[2,4]. All difference spectra were obtained by subtracting the initial absorbance spectrum of the sample equilibrated at 0 V.

The absorbance spectrum of LH 2 complex is presented in fig. d. It is assumed that the peak at 800 nm is due to the absorbance of monomer BChl molecules and that the peak at 850 nm is due to the presence of BChl dimers in the complex[3].

Mild oxidation of LH 2 complex was performed by applying a potential of +0.2 V for 120 sec. The reduction was then started by applying a potential of -0.6 V for 300 sec. Mild oxidation (oxidation of about 10 % of BChl 800) was almost reversible. Strong and irreversible oxidation (oxidation of ca. 100 % of BChl 850) was performed applying a potential of +0.55 V for 120 sec. IR absorbance peaks in the 1800-1400 cm^{-1} region corresponding to the carbonyl stretching vibrations in the BChl molecules of LH 2 complex from *Rb. sulfidophilus* have been studied[3], leading to the following assignments:

peak at 1630 cm^{-1} 2a-acetyl C=O group of BChl 850
peak at 1652 cm^{-1} 2a-acetyl C=O group of BChl 800
peak at 1673 cm^{-1} 9-keto C=O group   of BChl 850
peak at 1690 cm^{-1} is present in both BChl 800 and BChl 850.

The dashed line in fig. e shows that mild oxidation affects both BChl 800 and BChl 850. The ratio of the negative IR bands at 1652 and 1673 cm^{-1} in

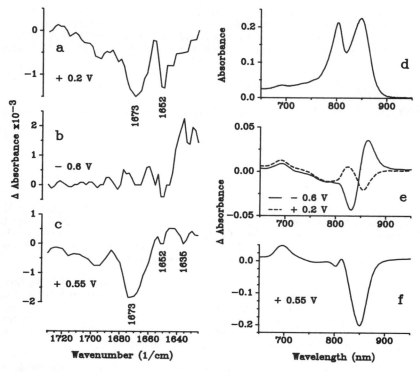

Fig. a, b and c: IR difference spectra of LH 2 complex at different potentials, fig a and b are 3 fold enlarged.
Fig. d: Absorbance spectra of LH 2 complex.
Fig. e and f: Absorbance changes in LH 2 complex in the 650-950 nm region obtained at different potentials. (e (-0.6 V) was obtained by subtracting the spectrum at +0.2 V)

fig. a is in good agreement with this assumption. Strong oxidation does not change the absorbance of LH 2 at 800 nm. One possible explanation for this feature is that BChl 800 and only one molecule of BChl in the BChl 850 dimer are oxidized, the other half of the dimer remaining in the neutral state. The latter probably has an absorbance maximum at about 810 nm which contributes to the absorbance of LH 2 at 800 nm in the case of strong oxidation of the complex. We assume that the IR absorbance of this unoxidized part of the BChl dimer in the 1730-1620 cm^{-1} region is the same as that of BChl 800.

The authors are grateful to Mr. N. Gad'on for providing LH 2 complex and to Prof. Dr. G. Drews for stimulating discussions.

REFERENCES
1. M. Lutz, W. Mäntele, 'Chlorophylls', H. Scheer ed., CRC Press, 1991, Ch. 5.
2. W. Mäntele et al., 'Reaction Centers of Photosynthetic bacteria', M.-E. Michel-Beyerle ed., Springer-Verlag, Heidelberg, 1990, Vol. 6, 31.
3. W. Mäntele, J. Savatzky et al., BBA, 1991, in the press.
4. D. Moss, E. Nabedryk et al., Eur. J. Biochem., 1990, 187, 565.

CAROTENOID DIMERS IN THE LIGHT-HARVESTING SYSTEM OF PURPLE PHOTOTROPHIC BACTERIA

J. Zurdo, C. Fernández-Cabrera and J. M. Ramírez*

Centro de Investigaciones Biológicas, CSIC
Velázquez 144
28006 Madrid, Spain

The antenna or light-harvesting system of purple phototrophic bacteria is formed by a few types of intrinsic
membrane proteins that have noncovalently bound pigments as
their prosthetic groups. The minimal units of the light-
harvesting proteins contain a small number of low molecular
weight polypeptides, 2 or 3 molecules of either bacteriochlorophyll *a* or *b*, and 1 or 2 molecules of a carotenoid.
The fixed composition of each type of light-harvesting protein is indicative of the existence of well defined molecular interactions among their pigment and polypeptide constituents.

A dual function has been demonstrated for the carotenoid in these proteins. While as a light-harvesting pigment
it absorbs light that will be used to drive the primary
photosynthetic processes, as a protective pigment it provides harmless deexcitation pathways for triplet state
chlorophyll and singlet oxygen, which are toxic byproducts
of the primary photoprocesses. An additional function in
the dissipation of excess energy has also been proposed for
some photosynthetic carotenoids. Besides these functional
roles, which involve steps of intermolecular energy transfer, the specific molecular interactions between the carotenoid and other constituents of the pigmented protein
influence to some definite extent the properties of the interacting constituents and contribute therefore to the global properties of the complex.

In previous work[1] it was shown that the visible circular dichroism (CD) of bacteriochlorophyll in the core
light-harvesting protein (LH1) of purple phototrophic bacteria was induced by the carotenoid. Therefore, it seemed
to arise either from nondegenerate interactions between
both pigments or from carotenoid-elicited protein bacteriochlorophyll interactions. In contrast, the shape and location of the visible carotenoid CD band suggested that it
was due to exciton interaction between two carotenoid molecules. In order to check those interpretations, it seemed
convenient to carry out a quantitative analysis of the influence of the carotenoid on the visible CD bands of the

LH1 protein. To this aim we have developed a method to partly extract the carotenoid from isolated membrane vesicles in which LH1 is the only or major holochrome. In such preparations the carotenoid dichroism changes with the square of the residual carotenoid level, as expected from its being caused by the random dimerisation of the pigment. In contrast, the visible optical activity of bacteriochlorophyll depends linearly on the carotenoid content, indicating that the aggregation of this pigment is not significant for the type of interaction that induces bacteriochlorophyll dichroism.

The vesicle preparations partially depleted of carotenoids have been also used to show that the carotenoid-elicited red shift of the near infrared transition of bacteriochlorophyll does not depend on carotenoid dimerisation. On the contrary, the quantum yield of singlet-singlet carotenoid to bacteriochlorophyll energy transfer, a process directly involved in the carotenoid light-harvesting function, is significantly increased by the dimerisation of the pigment. Thus, carotenoid aggregation is a new factor among those already recognized to influence energy transfer between the LH1 pigments[2,3].

REFERENCES

1. R.M. Lozano, C. Fernández-Cabrera and J.M. Ramírez, Biochem. J., 1990, 270, 469.
2. T. Noguchi, H. Hayashi and M. Tasumi, Biochim. Biophys. Acta, 1990, 1017, 280.
3. J.M. Ramírez in 'Trends in Photosynthesis Research' (J. Barber, ed.), Intercept, Newcastle-upon-Tyne, in press.

SPECTROSCOPIC CHARACTERIZATION AND ENERGY TRANSFER IN CYANOBACTERIA

I. B. Jha and S. Mahajan

School of Life Sciences
Jawaharlal Nehru University
New Delhi - 110 067
INDIA

INTRODUCTION

Cyanobacteria are prokaryotes and contain brilliantly coloured phycobiliproteins that are involved in light harvesting and Excitation Energy Transfer to reaction centres (1). These biliproteins together with linker polypeptides constitute a supra-molecular structure called phycobilisome (PBS), which are attached to the thylokoid membranes (2). The association of proteins in PBS is weak and high concentration of phosphate is required to keep them in an aggregate form (4).

Experimental Methods

Phycobilisomes were isolated according to (3) but with modification where we could finish ultracentrifugation in an hour using vertical rotors at 55,000 r.p.m. The isolated intact PBS were dissociated by dializing against 5mM Potassium Phosphate buffer overnight. Spectra were recorded in UV-3000 spectrophotometer, RF-540 Shimadzu spectrophotofluorometer.

Results & Discussion

In Fig. 1, the peculiar absorption peak at 565nm and a previous shoulder at 545nm in the absorption spectrum indicates the presence of phycoerythrin (PE) and the broad absorption band in the red region is due to phycocyanin (PC) and Allophycocyanin (APC).

In intact Phycobilisome, the excitation spectrum nearly resembles with the absorption spectrum. This indicates that various components (PE, PC and APC) of the intact Phycobilisomes are energetically coupled. Thus, the Excitation Energy Transfer from PE to PC to APC is not affected when Phycobilisomes are excited at 545nm. Hence we observe, a final fluoresence emission with its maxima at 672nm for Intact PBS (Fig. 2). Here the 545nm excitation energy is first captured by PE and then efficiently transferred to APC through PC.

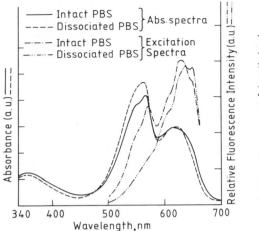

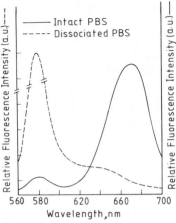

Fig. 1 Absorption &
Excitation spectra of
Intact and dissociated
PBS.

Fig. 2 Fluorescence Emission
spectra of Intact and
dissociated PBS excited
at 545nm.

APC being the final emitter, cannot transfer excitation
energy to reaction centres in isolated conditions
therefore fluoresces at 672nm. Upon dissociation, the
components of PBS are no more energetically coupled as
we see the excitation and absorption spectra do not
resemble each other. Thus here we observe fluorescence
emission with its maxima at 580nm when excited at 545nm.
This emission is actually PE emission. Because of the
uncoupling of PE, PC and APC, the energy transfer from
PE to the subsequent acceptors is not favoured. Hence
the 580nm emission (Fig. 2). We finally conclude that
the dissociation process destablizes the aggregation of
biliproteins within PBS thereby affecting the energy
transfer process.

REFERENCES

1.   H. Scheer, 'Excitation Transfer in Phycobili-
     proteins'. In: Encyclopedia of Plant Physiology;
     edited by L.A. Staehelin and C.J. Arntzen, Springer,
     Berlin, 1986, Vol.19, Photosynthesis-III, p.327.
2.   E. Gantt, 'Phycobilisome'. In: Encyclopedia of
     Plant Physiology; edited by L.A. Straehelin and
     C.J. Arntzen, Springer, Berlin, 1986, Vol.19,
     Photosynthesis-III, p.260.
3.   E. Gantt, C.A. Lipschultz, J. Grabowski and B.K.
     Zimmermann, Plant Physiol., 1979, 63, 615.
4.   B.A. Zilinskas and R.E. Glick, Plant Physiol.,
     1981, 68, 447.

REVERSIBLE PHOTOCHEMISTRY, AN ANALYTICAL TOOL TO DETECT THE PRESENCE OF THE BILIPROTEIN PHYCOERYTHROCYANIN IN CYANOBACTERIA.

P.S. Maruthi Sai,[1] S. Siebzehnrübl,[2] S. Mahajan[1] and H. Scheer.[2]

[1]School of Life Sciences, Jawaharlal Nehru University, New Delhi-110067, INDIA and [2]Botanisches Institute der Universitat, 67 Menzinger Strasse, 8000 München 19, GERMANY.

INTRODUCTION

Cyanobacteria are photosynthetic prokaryotes, which perform oxygen evolving photosynthesis similar to that of higher plants. These organisms possess brilliantly colored fluorescent macromolecular protein pigment complexes called phycobiliproteins as antenna pigments which are involved in the excitation energy transfer to the photosynthetic reaction centres. The major phycobiliproteins present in cyanobacteria are allophycocyanin (APC), phycocyanin (PC), phycoerythrin (PE), and phycoerythrocyanin (PEC). All the biliproteins carry linear tetrapyrroles as chromophores on their constituent subunits.[1] Among these biliproteins, PEC shows a reversible photochemistry in isolated native form, reminiscent of that of the higher plant photopreceptor phytochrome, but with an orange/green rather than a red-far/red action spectrum. The α-subunit carries an unusual phycoviolobilin, instead of the phycocyanobilins which are present on the β-subunit only. PEC from various species of cyanobacteria has been studied for its reversible photochemistry.[2-4]

The present paper deals with the reversible photochemistry of PEC and its application as a technique to detect its presence in the cyanobacteria possessing it in minute amounts.

EXPERIMENTAL PROCEDURES:

PEC or fractions of biliproteins enriched in PEC were obtained from different species of cyanobacteria described as previously.[2] In a similar fashion PE was also obtained. Photochemistry was induced by saturating pre-irradiation with green light for ten minutes followed by eight minute orange illumination. Reversibility was checked by ten minutes post-irradiation with green light. The details of the experiment are published elsewhere.[2]

DISCUSSION AND CONCLUSIONS

The absorption of the biliproteins in the visible range is due to the chromophore-protein interactions. Although some of the features are similar to that of the higher plant photoreceptor phytochrome, they are all involved in the harvesting of light energy and its final transfer to the chlorophyll a. The resemblances of phytochrome to that of cyanobacterial biliproteins prompted various workers to look for the reversible photochemistry, if any, from these pigments. Attempts were made to induce reversible photochemistry in the purified forms of PC and APC in the native state but without any success, but it has been observed only after partial denaturation of the biliproteins. After people have failed to induce photochemistry in the native forms of PC and APC,[5,6] experiments were performed on another biliprotein, PEC. It was reported to show reversible photochemistry in the native form which is due to the presence of a violobilin chromophore on the α-subunit from a few species of cyanobacteria. The organisms used for the purpose were, Mastigocladus laminosus[2,3] and Tolypothrix distorta[3], Westiellopsis prolifica[4] and Nostoc rivulare[4]. In all cases, different preparations of PEC exhibited a reversible photchemistry in the isolated native forms. In Nostoc rivulare, the PEC was present in a very minute amount, but still the fractions enriched in PEC exhibited total reversible photochemistry. These studies prompted us to perform similar experiments on PE which absorbs closely to that of the biliprotein PEC. We could not find any photosignal for PE isolated from Anabaena variabilis ARM 310 in the orange/green action spectrum.

After testing for this property for PEC from different organisms, we now suggest that the reversible photochemistry will enable us to study its presence in cyanobacteria even when found in minute amounts. Further this would also enable us to differentiate PEC from the other closely absorbing biliprotein PE.

REFERENCES

1. R. MacColl and D. Guard-Friar, 'Phycobili-proteins',CRC Press, Boca Raton, Florida, 1987.
2. S. Siebzehnrübl, R. Fischer, W. Kufer and H. Scheer, Photochem. Photobiol. 1989, 49, 753.
3. W. Kufer and Björn, Physiol. Plant, 1989, 75, 389.
4. P.S. Maruthi Sai, S. Siebzehnrübl, S. Mahajan and H. Scheer, 1990, communicated to Photochem. Photobiol.
5. J. de Kok, S.E. Braslavsky, C.J. Spruit, 1981, Photochem.Photobiol. 34, 705.
6. K. Okhi and Y. Fujitha, 1979, Plant Cell Physiol. 20, 483.

FT-RAMAN CONTRIBUTION FOR THE VIBRATIONAL INVESTIGATION OF MODEL ANTHOCYANIDINS.

C. DEPECKER[a] , **M.F. LAUTIE**[b] , **P. LEGRAND**[a] , **J.C. MERLIN**[a] and
B. SOMBRET[a] : Laboratoire de Spectrochimie Infrarouge et Raman
[a] (CNRS UPR A 2631 L), UST Lille Flandres Artois, Bât. C5,
59655 Villeneuve d'Ascq cedex, France.
[b] (CNRS UPR A 2631 T), 2 rue Henri Dunant, 94320 Thiais, France.
A. STATOUA : Laboratoire de Spectroscopie Infrarouge, Faculté des Sciences,
Université Mohamed V, Rabat, Maroc.
R. BROUILLARD : Laboratoire de chimie des Pigments des plantes, (CNRS URA 31),
Université L. Pasteur, Institut de Chimie, 1 rue Blaise Pascal, 67008 Strasbourg, France.

Infrared and Raman spectroscopies are suitable for the structural analysis of anthocyanins, the most important group of plant pigment, but very few vibrational data are available for such structures. Because of the color of these compounds, the Raman investigations by using conventional visible laser radiation can present some difficulties: (a) By exciting near an electronic absorption band, a fluorescence emission is often obtained and in some cases hinders the observation of the Raman spectrum. (b) The resonance Raman (RR) effect, difficult to overcome for colored samples, gives a vibrational picture of only a small part of the molecule. The great selectivity of the RR effect, presented as an advantage for some applications, can be a severe limitation for the complete knowledge of the vibrational properties of the investigated molecule. In the RR spectra of natural anthocyanins and anthocyanidins, the line positions and relative intensities are very similar from one spectrum to another[1], and an interpretation is difficult to propose. The FT-Raman technique offers many advantages over conventional Raman methods for colored samples[2]. It is well known that the long wavelength of the IR excitation does not promote energy levels that give rise to fluorescence and allows the vibrational modes of all parts of the molecules to be observed.

Galanginidin (3,5,7-trihydroxyphenyl,2-benzopyrylium cation) is a synthetic anthocyanidin which can be used as a model for the understanding of the vibrational properties of natural anthocyanidins. Figure 1 presents the vibrational spectra of solid samples. One can easily see that significant changes appear in the relative intensities of lines if we compare the RR and the FT-Raman spectra in the solid state. In order to assign the characteristic phenyl ring modes, Galanginidin,D5 where the phenyl ring is fully deuterated, was synthesized and the spectra recorded in similar conditions. It is apparent that many strong lines in the IR and FT-Raman spectra are shifted while no drastic changes appear for the main lines in the RR spectrum. This effect clearly indicates that the chromophore is mainly located on the benzopyrylium moiety and no significant coupling is assumed between the two parts of the aglycone.

By using literature data[3], the medium RR line near 1350 cm^{-1}, backshifted upon deuteration, is assumed to contain a large contribution of the C-C inter-ring stretching mode. Previous investigations of anthocyanins and anthocyanidins both in solid and acidic solution, have shown that the wavenumber of this line appears to be dependant on the hydroxylation and methoxylation patterns of the phenyl ring, which is one of the parameters affecting the color. This line could perhaps be taken as a key band for the establishment of a relation between structure and the color of these pigments. However, as some intense Raman lines appear shifted in the FT-Raman spectrum and unshifted in the RR spectrum, the superposition of phenyl and benzopyrylium modes can be considered and can be taken into account for a more precise assignment.

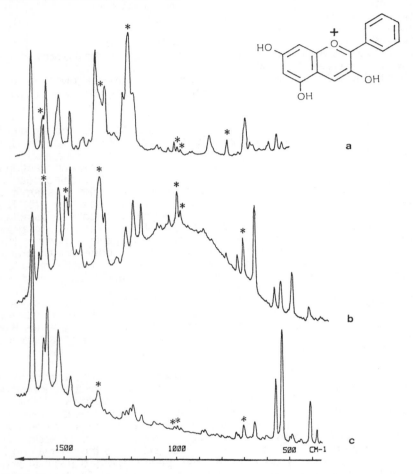

Figure 1 : IR diffuse reflectance (a), FT-Raman (b) and RR, 488 nm excitation (c) spectra of solid Galanginidin in KBr (2% w/w). * indicates lines shifted by more than 10 cm^{-1} upon phenyl ring deuteration.

REFERENCES

1. J.C. Merlin, A. Statoua and R. Brouillard, Phytochemistry, 1985, 24(7), 1575.
2. B. Schrader, A. Hoffman, R. Podschadlowski and A. Simon, 'Spectroscopy of Biological Molecules, State of Art', A. Bertoluzza, C. Fagnano and P. Monti Eds, Societa editrice Esculapio, Bologna, 1989, 327.
3. M. Forster, R.B. Girling and R.E. Hester, J. Raman Spectrosc. 1982, 12(1), 36.

ENERGY TRANSFER TIMES IN PHYCOBILISOMES OF *Mastigocladus Laminosus* AS DETERMINED BY SPECTRAL HOLE–BURNING

A. Feis[1], J. Friedrich[*1], L. Gottschalk[2], and H. Scheer[2]

[1] Institut für physikalische Chemie, J.–Gutenberg–Universität,
D–6500 Mainz FRG

[2] Botanisches Institut, Universität München,
D–8000 München 19 FRG

Phycobilisomes are protein aggregates acting as antenna systems in the photosynthesis of cyanobacteria and red algae. In particular, phycobilisomes of *Mastigocladus laminosus* contain three different kinds of hexameric biliproteins, namely phycoerythrocyanin (PEC), C–phycocyanin (PC) and allophycocyanin (APC), with electronic transitions spreading over the visible light range (*see Fig. 1*). Light absorption at shorter wavelengths is followed by stepwise energy transfer to the chromophores at longer wavelengths. Energy transfer times, that dominate the energy relaxation rates from excited electronic states in the present case, can be measured by means of persistent spectral hole–burning.[1] In persistent hole–burning experiments, samples are irradiated with narrow–line laser light at sufficiently low temperatures. The subsequently measured absorption spectrum shows a hole at the laser wavelength. Under certain conditions, the width of the hole is twice the homogeneous transition linewidth. The relation between homogeneous linewidth (Γ_{hom}) and energy relaxation (T_1) and pure dephasing (T_2^*) times is given by the following equation:

$$2\pi\Gamma_{hom} = \frac{1}{2T_1} + \frac{1}{T_2^*} \qquad (1)$$

We performed hole–burning in phycobilisomes suspended in an aqueous saccharose/phosphate buffer solution in a temperature range from 1.45 to 4.2 K. The samples were irradiated with a narrow band (<10 MHz) CW ring dye laser at 637.7 nm in the PC band, which shows an inhomogeneous broadening of about 150 cm^{-1}. Narrow holes, with temperature–dependent widths ranging from 3×10^{-2} to 10^{-1} cm^{-1} could be detected by scanning the laser over the burn wavelength (*see insert in Fig. 1*). The hole width appears to be independent from temperature between 1.45 and 2 K. Thus, in this range, the pure dephasing contribution to the homogeneous linewidth ($1/T_2^*$ in Eq. (1)) can be neglected and the measured hole width is directly related to T_1, i.e. to the PC→APC energy transfer time. Moreover, spectral diffusion processes, that may give rise to additional broadening of the holes on the time scale of the measurement, seem to be negligible at these temperatures.[2] Saturation broadening due to chromophore bleaching was taken into account by extrapolating the measured hole width to zero burning intensity. A first estimate of the energy transfer time PC→APC is in the order of 100 ps.

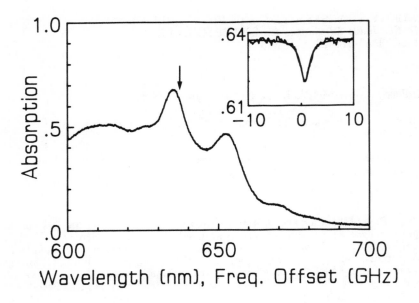

Figure 1 – Absorption spectrum of phycobilisomes of *Mastigocladus laminosus* at 4.2 K. The bands at ~610, 635, and 652 nm are due to PEC, PC, and APC, respectively. The insert shows a spectral hole after irradiating the sample at 1.45 K with the wavelength indicated by the arrow (637.7 nm). The hole was detected by scanning the laser over the burn wavelength. A Lorentzian fit is overlayed.

REFERENCES

1. W. Köhler, J. Friedrich, R. Fischer and H. Scheer, in "Photosynthetic Light–Harvesting Systems", H. Scheer and S. Schneider, Eds., W. de Gruyter & Co., Berlin–New York, 1988, p. 293.
2. W. Köhler and J. Friedrich, J. Chem. Phys., 1989, 90, 1270.

LIGHT-INDUCED CHANGES IN THE ATR/IR SPECTRA OF CHROMATO-PHORE FROM RB.SPHAEROIDES

K.Okada, E.Nishizawa, Y.Koyama, T.Kakuno[*], and Y.Ozaki

Department of Chemistry, Kwansei Gakuin University, Nishinomiya 662, JAPAN, and [*]Institute for Protein Research, Osaka University, Suita 565, JAPAN

In order to provide new insight into the mechanism of bacterial photosynthesis, we have measured attenuated total reflection (ATR)/FT-IR spectra of the chromatophores from Rb.shaeroides G1C with ('dark' spectrum) and without ('light' spectrum) the irradiation of a fluorescent light and calculated 'light-minus-dark' and 'dark-minus-light' difference spectra from them. The obtained difference spectra, which are significantly different from those obtained earlier by other research groups probably due to the use of different light source,[1] suggest that the state of ubiquinone-10 alters upon the light illumination.

I INTRODUCTION

Light-induced structural changes in constituents of chromatophore of photosynthetic bacteria have been investigated by a couple of groups using FT-IR spectroscopy[1,2]. The purpose of the present study is to try a FT-IR study on the light-induced changes by using different infrared technique (ATR) and different irradiation source (a fluorescent light).

II RESULTS AND DISCUSSION

Figure 1 shows ATR/IR spectra of chromatophore from Rb.sphaeroides G1C before (a), under (B), and after (C) irradiation of fluorescent light. The spectra in Figure 1 largely reflect those of light-harvesting(LH) complex. Bands at 1738, 1651, 1543, and 1240 cm^{-1} may be assigned to an ester C=O stretching mode of bacteriochlorophyll a (BChl a), and amide I,II,III modes of proteins, respectively, and a broad feature near 1065 cm^{-1} is probably due to C-O stretching modes of suger groups. The three specrta seem to be very similar but the calculation of difference spectrum between Figure 1(A) and 1(B) and that between Figure 1(B) and 1(C) demonstrates that they are significantly different from each other.

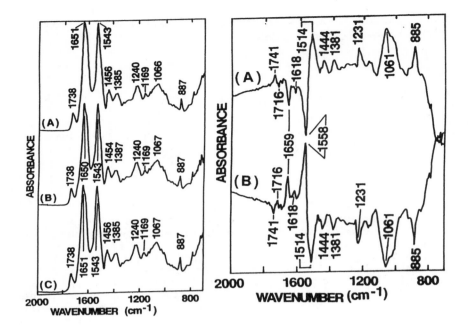

Figure 1 ATR/IR spectra
of chromatophore from
Rb.sphaeroides G1C
(A) before the irradiation
(B) under the irradiation
(C) after the irradiation.

Figuer 2 (A) 'light-dark'
difference spectrum (Fig.1(B)
-Fig.1(A)).(B) 'dark-light'
difference spectrum (Fig.1(C)
- Fig.1(B)).

Fugure 2 (A) and (B) present the calculated 'light-minus-dark' (Figuer 1(B)-1(A)) and 'dark-minus-light' (Figure 1(C)-1(B)) difference spectra. The two spectra are nearly symmetrical, indicating that the light-induced changes in the chromatophore are reversible. A C=O groups of some BChl $\underline{a}$, which engage in hydrogen bondings, become free upon the irradiation. The origin of the changes in the 1700-1500 cm^{-1} region is unclear at the moment, but comparison of spectra in Figure 2 with those of ubiquinone-10[3] leads us to conclude tentatively that they reflect alteration of the state of ubiquinone-10.

References
1 W.Mantele, E.Nabedryk, B.A.Tavitian, W.Kreutz, and
  J.Breton, FEBS Lett., 1985, 187, 227.
2 H.Hayashi, M.Go, and M.Tasumi Chem.Lett., 1986,1511.
3 M.Bauscher,E.Nabedryk, J.Breton, and W.G.Mantele,
  'Spectroscopy of Biological Molecules-State of the
  Art',(A.Bertoluzza, C.Fagnano, and P.Monti Eds.),
  Societa Editorice Esculapio, Bologna, 1989, p.397.

III

CRYSTALLOGRAPHY

TIME-RESOLVED SPECTROSCOPY

CHIRALITY

THE ANALYSIS OF STRUCTURAL MOVEMENTS IN PROTEIN BY X-RAY ANALYSIS

G. G. Dodson

Department of Chemistry
University of York
York
YO1 5DD

INTRODUCTION

Generally speaking globular proteins, although held together mostly by non-covalent forces, have definite 3 dimensional structures. In the last 25 years or so the crystal structures of 300 or more proteins (and other macromolecules) have been determined by X-ray analysis. The knowledge of these structures has been a major factor in the phenomenal activity in biology and biochemistry. Proteins often crystallise and thus can be sufficiently well defined to be arranged in exact and regular lattices; they possess however intrinsic flexible characteristics to a greater or lesser degree, depending on their function. Many proteins (enzymes particularly) under go considerable conformational changes during activity. These occur on a time scale of typically 10^{-2} - 10^{-4} seconds, and are thus not usually detectable by X-ray analysis. A reasonably accurate description of these structural changes and their associated interactions with ligands is essential for a fundamental understanding of the relationships between sequence, structure and function.

Spectroscopic methods such as fluorescence and nuclear magnetic resonance can signal the presence of conformational changes sometimes identifying residues and structures involved, but of course yield little 3 dimensional detail directly. Nuclear magnetic resonance has however become a successful technique in structure determination though still limited to molecules of molecular weight < 30-40,000. This technique thus sidesteps the uncertain stage of crystallisation and is consequently of real practical importance. It has also been argued that, because in spectroscopic experiments the proteins are in solution, their intrinsic motions are not frozen by the crystal lattice and a more relevant structural information can be obtained.

Although X-ray analysis of the crystalline state may seem unpromising for the study of structural changes it has proved an extremely valuable approach. All that is necessary is that the states under study can be crystallised and in many cases they can. The research of Max Perutz into haemoglobin structure revealed very early on the differences in the molecule's two allosteric states.[1] Since then a wealth of structural detail on the conformational changes and interactions associated with proteins reactions have been delineated.

Conventional crystallographic methods however are not able to dissect protein mechanisms which typically occur in milliseconds to second owing to the time taken to collect the diffraction data, typically days. However the arrival of synchrotron radiation, new data collection techniques such as Laue diffraction and image plates coupled with insight into regulating enzyme reactions has made it possible to undertake time-resolved studies in some selected systems.

PROTEIN STRUCTURE AND MOBILITY FROM X-RAY ANALYSIS

The X-ray analysis of proteins at high resolution reveals that generally their residues have a higher apparent mobility with rms amplitudes of vibrations between .5 - 1.0 Å and even more. The surrounding network of water molecules is even more mobile - except where there are specific H bonds to well defined protein atoms. This mobility of the protein surface is an indication of protein flexibility and potential for conformational storage.

The structural changes associated with protein function such as ligand binding or catalysis vary enormously in scope. Their description by X-ray analysis depends on the different structures being either separately crystallisable or achievable in a stable form within the parent crystal structure. While the structural determination of the conformational states (or more) in a protein provides a framework within which the mechanism occurs there are however often major questions still unanswered by this approach. For example intermediate structures have not usually been characterised by this approach and thus the pathway of the rearrangement can remain obscure. Spectroscopy can provide vital clues in these studies, but here again the complexity of the systems often presents intractable difficulties.

THE CRYSTALLOGRAPHIC ANALYSIS OF STRUCTURAL PATHWAYS

The increasing capacity of X-ray analysis to delineate the structural details in a protein mechanism is best illustrated by examples, first from conventional studies and then from the recently-applied Laue technique.

Haemoglobin. The haemoglobin molecule exists in two defined states, the so called T state in which the ligand affinity is low, and the R state in which the ligand affinity is high.[2] The two states are in rapid equilibrium and differ principally in their quaternary organisation. In the absence of oxygen the molecule deoxy haemoglobin is almost entirely in the T state. In the presence of oxygen, oxyhaemoglobin, the molecule is almost entirely in the R state. Inspection of the oxygen binding sites in the T and R state (deoxy and oxyhaemoglobin) allowed Perutz to identify the low affinity with the 'out of planeness' of the FeII from the haem. The chemical and steric factors that contributed to the low affinity (the so-called tension) could however only be guessed at. To identify these the structure of the liganded T state was needed, but owing to the dynamics of the system, liganded haemoglobin, rapidly switches to the R state, precluding T state crystallisation.

The answer to this difficulty came with a new crystal form of T state haemoglobin which was robust enough to prevent switching to the R state when the molecule was liganded.[3] In these T state crystals, which diffract to high resolution, the stereochemistry at the α and β haems and the oxygen's binding site could be analysed in detail. It emerged that the origins of low affinity in the α subunit lay with the distorted octahedral FeII coordination associated with a folded haem and with the stretching of a H bond between the ND of the present his (F8) and the carbonyl O of the peptide of valine F4. In the β subunit the origins of low affinity arose from quite different behaviour. The haem was seen to be planar to the FeII coordination sphere satisfactorily octahedral when liganded by water.

The low affinity was clearly associated with the blocking of the oxygen binding site by E11 valine, as Max Perutz had earlier recognised.[1,2] Finally, the arrival of ligand in the α and β subunits was seen to be associated with structural changes that were generally in the same direction as those that occur in the T (deoxy) → R (oxy) transition.

Lipase. The lipase enzyme is activated by a non-polar polar interface.[4] It has always been assumed that activation would be accompanied by a significant structural change either exposing the catalytic residues or orienting them into catalytically competent stereochemical arrangement.

The X-ray analysis of the enzyme from a fungal source[5] and from the human pancreas[6] showed that the enzyme's inactivity in water was explained by the burial of its catalytic residues (asp.his.ser). Inspection of the enzyme structure revealed that the active site was buried by a lid made up by a small helix attached to the body of the molecule by two extended peptide strands. Such a structure by a simple movement stabilised by a non-polar environment should easily reveal the active site.

Attempts to crystallise the activated structure from non-polar/polar conditions failed. When however a covalently bound substrate analogue was attached it did prove possible to crystallise the enzyme in the activated state. The crystals were extremely fragile and diffracted poorly but nonetheless data were collected to 3Å spacing. The structure of the complex showed that, as expected, the helix lid had simply rolled away from the active site. In doing this an extensive surface of non-polar residues made up from the environs of the catalytic site and from the previously buried residues of the helix lid, is exposed.

From these two structures we propose that the positions of the helix lid are in equilibrium. In polar and aqueous conditions the helix buried the catalytic site and its non-polar conditions the helix is stabilised in the conformation that exposes its non-polar residues and buried its polar side chains. Since a tryptophan is involved in this rearrangement, fluorescence experiments might provide a measure of the rate of this structural change and make it possible to determine the various factors that govern the enzyme's kinetics.

Time-resolved experiments. The analysis of the transient events in crystals has been made possible by using the very intense X-rays generated by synchrotron radiation. In these experiments diffraction data are collected by conventional monochromatic methods in 10-40 minutes and by the polychromatic Laue method in seconds or less. Providing the reaction can be initiated in the crystal coherently so that all cells are undergoing the same structural changes at the same time, it is possible to dissect in detail the stereochemistry of a protein mechanism.

Two time-resolved studies have been carried out, the first on the enzyme phosphorylase and the second on the ras p21 protein

The initiation and rate of reaction of the phosphorylase were controlled by an activation molecule which made it possible to collect monochromatic data sets on coherently reacting enzymes well within the reaction time.[7] These experiments revealed beautifully the accumulation of phosphate on the substrate during the course of the reaction and the concomitant adjustments of the surrounding protein.[8] The polychromatic Laue technique has recently advanced these experiments very impressively. Some with Laue photographs data sets of ca. 2.5Å spacing can be collected in seconds.

The time-resolved experiments on the oncogene product p21 revealed the structural changes associated with GTP hydrolysis when it converts from its 'active' to inactive states.[9] The half life of the p21 GTP complex is about 40 minutes making the faster Laue technique essential. The initial complex with p21 was with caged ATP activatable by laser light. The analysis of the electron density 4 and 14 minutes after activation revealed the process of structural change in the ATP and ADP conversion and in the associated protein.

CONCLUSIONS

The functional behaviour of proteins is becoming increasingly accessible to X-ray analysis. There are four principal reasons for this. The first is that protein crystals can be very tolerant of limited (though functionally significant) changes. In protein crystals the solvent volume is often 40-60% and the protein molecules are thus mostly bathed in ligand. Thus, apart from the specific contacts, structural changes in ligand binding can often be accommodated within the crystal. This makes protein crystals splendid laboratories which would incidentally justify much more research into designing lattices for specific function. Secondly crystals of these liganded or reacted proteins are frequently small or fragile and in the past their study has been a major obstacle. The development of greatly improved X-ray data collection devices and the availability of synchrotron radiation has meant that useful high resolution data can usually be obtained from such poor and precious crystals. Thirdly the time needed to determine and, crucially important, to refine protein crystal structures is in many cases reducing to months.

Fourthly, the Laue method has made it possible to dissect the processes of protein mechanisms in stereochemical detail and has greatly enlarged the scope of time-resolved methods. It now falls to the biochemist and chemist to solve the subtle problems of initiating reactions coherently in the crystal; once this can be achieved protein structural behaviour will be analysed in new and fundamental ways.

REFERENCES

1.   M.F. Perutz, Nature, 1970, 228, 726-739.

2.   M.F. Perutz, Am. Rev. Biochem., 1979, 48, 327.

3.   R. Liddington et al., Nature, 1988, 331, 725-728.

4.   W.A. Pieterson et al., Biochemistry, 1974, 13, 1445-1460.

5.   R.L. Brady et al., Nature, 1990, 343, 767-770.

6.   F. Winkler et al., Nature, 1990, 343, 771-774.

7.   J. Hadju et al., EMBO, 1987, 6, 539-546.

8.   J. Hadju et al., Nature, 1987, 329, 178-181.

9.   I. Schlichting et al., Nature, 1990, 345, 309-315.

ULTRAFAST INFRARED SPECTROSCOPIC STUDIES OF PROTEIN DYNAMICS

R. M. Hochstrasser

Department of Chemistry, University of Pennsylvania
Philadelphia, PA   19104-6323

Introduction:

Recent advances and demonstrations of transient IR methods in studies of solution phase and protein dynamics (1-10) have established that this approach is a most promising one that can lead to definitive structural interpretations of the dynamics. In this lecture an account will be given of various approaches we have used to carry out infrared spectroscopy with ultrafast time resolution.   Optically pumped samples can be probed in the IR with time resolution down to a few hundred femtosecond either by gating CW carbon monoxide lasers (1,2) or by generating IR pulses at high rep rate using a parametric and regenerative amplifier.   We have also accomplished tunable pump: probe IR spectroscopy with low energy pulses at high rep rate (3,4).   These methods are applicable to a wide range of questions in biological dynamics.   The capability of both pumping and probing with continuously tunable infrared pulses has opened up many new possibilities for studying dynamics, including vibrational energy transport and bond orientational relaxation.

One of the strongest aspects of transient IR spectroscopy is its potential for studying protein dynamics, independently of the chromophores that trigger the changes.   An example of this is found with bacteriorhodopsin (11) which shows a significant change in its IR spectrum in the amide I region a few ns after isomerisation of the chromophore. This work will be presented as a Poster by R. Diller.

This talk will focus on recent results on the photodissociation of hemoglobin bound to CO and NO. Interesting new effects have been discovered that relate to geometrical structure and heme pocket fluctuations, cooling of the protein, multiphoton processes and structural relaxations occurring after photolysis.

Geometrical Structure Fluctuations: Previous measurements of IR/Optical polarization spectra of hemes (12,13) have assumed that the heme is rigidly held inside the protein so that $\langle P_2[\hat{z}(o) \cdot \hat{z}(t)]\rangle$ is influenced only by overall motion of the protein ($\hat{z}$ is the heme-plane normal). Recent improvements in IR methods allow the extent of any wobbling of the heme in the pocket to be evaluated. For a wobble frequency ω, corresponding to harmonic motions about two mutually perpendicular axes in the plane of the heme, a heme moment of inertia I ($\sigma = K_B T/I\omega^2$ in the equation below) and effective diffusion coefficient D for the heme rotation in the heme pocket and D_p for the protein, the value of $\langle P_2(\hat{z}(o) \cdot \hat{z}(t))\rangle$ is given by:

$$\langle P_2[\hat{z}(o) \cdot \hat{z}(t)]\rangle = \exp(-6D_p t) \, [1 + 3 \exp \{-4\sigma \, (1 - \exp \, [-Dt/\sigma])\}]$$

The wobbling is found, by fitting to the above equation, to have an amplitude of less than $7°$. In ambient solution the heme is fixed rigidly inside the protein with a geometry similar to that found in the crystal. The experiment shown in Figure 1 is the first such measurement (14). The HbCO is photolyzed with a 200 fs pulse. The amount of bleaching at the bound carbonyl absorption is measured at 1951 cm^{-1} using polarized infrared light before the protein has time to rotate significantly. The bleaching signals for the optical pump polarized parallel and perpendicular to the IR yield the anisotropy which corresponds to the correlation function given above. The observed value is less than the ideal value of −0.2 (for α = 0°, or C≡O perpendicular to the heme plane), because the equilibrium bond angle is 18° and is _not_ a result of motion of the heme relative to the protein.

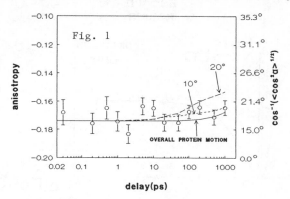

Fig. 1

Multiphoton Absorption and Cooling: Optical pumping of HbCO with fs pulses causes changes in the IR absorption for Hb, HbCO (6) and HbNO. These changes are not due to the FeCO or FeNO absorption. A subpicosecond IR absorption spike is observed followed by a 19 ps time constant increase in transmission. We suggest now that the spike is due to coherent coupling of the IR and visible beams (two-photon absorption). The slow increase in transmission was attributed to a change in the D_2O spectrum that follows the optical heating of the heme (6). Recent improvements in the sensitivity have permitted both these processes to be considerably better defined. The slow change in transmission fits well to an exponential process with time constant 19 ps and does not recover on the picosecond timescale. Experiments are underway to prove whether this effect is a result of the deligation influencing the IR spectrum of D_2O.

Ligand and Protein Dynamics

Our results with HbCO and on-going experiments with HbNO have suggested that the diatomic quickly moves to a location about 4.5A° from

the iron, or about 4A° above the heme plane. From the Morse potential
(Fig. 2) for Fe-CO motion we define an escape radius at 3.5A°, where the

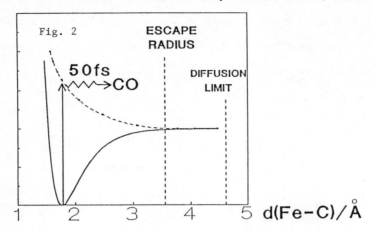

attraction is balanced by kT. The diffusion limit is the distance that
CO would diffuse in the time for the observed spectrum to stabilize.

Acknowledgement: This research was supported by NSF and NIH.

1.   R.M. Hochstrasser, P.A. Anfinrud, R. Diller, C. Han, M. Iannone, T. Lian and B. Locke in Ultrafast Phenomena VII (eds. C.B. Harris, E.P. Ippen, G.A. Mourou and A.H. Zewail) Springer-Verlag, Berlin-Heidelberg, 1990, p 429.
2.   P.A. Anfinrud, C-H Han, T. Lian and R.M. Hochstrasser, J. Phys. Chem, 1991, 95, 574.
3.   M. Iannone, B.R. Cowen, R. Diller, S. Maiti and R.M. Hochstrasser, Optics Letters, in press.
4.   J. Owrutsky, M. Li, Y.R. Kim and R.M. Hochstrasser, "Vibrational energy relaxation for N_3 in H_2O".
5.   P.A. Anfinrud, C. Han and R.M. Hochstrasser, Proc. Natl. Acad. Sci., 1989, 86, 8387.
6.   T.M.Jedju and L. Rothberg, Applied Optics 27, 615 (1988).
7.   L. Wang, X. Zhu and K.G. Spears, J. Am. Chem. Soc. 110, 8695 (1988).
8.   R.B. Dyer, K.A. Peterson, P.O. Stoutland and W.H. Woodruff, Biophys. J. 59, 472a (1991).
9.   M.K. Hong, E. Shyamsunder and R.H. Austin, "Time-Resolved Infrared Studies of Molecular Diffusion in Myoglobin" (preprint).
10.   H. Graener, G. Seifert and A. Laubereau, Chem. Phys. Lett. 172, 435 (1990); H. Graener, T.Q. Ye and A. Laubereau, J. Chem. Phys. 91, 1043 (1989).
11.   R. Diller, M. Iannone and R.M. Hochstrasser, Biophys. J. (in press).
12.   J.N. Moore, P.A. Hansen and R.M. Hochstrasser, Proc. Natl. Acad. Sci. 85,5062-5066 (1988).
13.   P.A. Hansen, J.N. Moore and R.M. Hochstrasser, Chem. Phys. 131, 49 (1989).
14.   T. Lian, R.B. Locke and R.M. Hochstrasser, "Femtosecond Motions of the Heme in HbCO" (submitted).

SPECTROSCOPIC INVESTIGATIONS OF THE PRIMARY PROCESSES OF THE SOLAR-TO-ELECTRIC ENERGY CONVERSION BY THE OTHER PHOTOSYNTHETIC SYSTEM IN NATURE

M. A. El-Sayed

Department of Chemistry and Biochemistry
University of California
Los Angeles, California 90024

INTRODUCTION

The retinylidene protein bacteriorhodopsin (bR), the other photosynthetic system besides chlorophyll, is a protein pigment found in the purple membrane (PM) of Halobacterium halobium.[1,2]   A remarkable feature of PM is its two-dimensional crystalline structure of P3 symmetry with trimers of bR molecules located at the points of the hexagonal lattice.[3]   Light adapted bR contains an all-trans-retinal, which is covalently bound via a protonated Schiff base (PSB) linkage to the protein.[1,4]   Upon the absorption of visible light, bR proceeds through the photochemical cycle:[5]

$$bR_{568} \xrightarrow[0.5ps]{h\nu} J_{625} \xrightarrow{5ps} K_{610} \xrightarrow{2\mu s} L_{550} \xrightarrow[60\mu s]{-H^+} M_{412} \xrightarrow[ms]{+H^+} O_{640} \rightarrow bR_{568}$$

As a result, the PSB is deprotonated during the $L_{550} \rightarrow M_{412}$ step, leading to a proton-pumping process that increases the proton concentration of the outside surface of the membrane.[6]   The created proton gradient across the membrane is then used to transform ADP into ATP in the final step of the photosynthesis of bR.[7]
   bR normally contains bound Ca^{2+} and Mg^{2+}.[8,9] Acidification or removal of metal cations from bR produces the blue form of bR which is incapable of forming the M_{412} intermediate but allows the formation of the K_{610} and L_{550} intermediates.[10,11] Thus metal cations are somehow required for the deprotonation of the PSB and thus for the proton pumping process.

QUESTIONS RAISED

a. Is there an antenna system in bR?

The biphasic CD spectrum of bR was described in terms of an exciton model that couples the retinal trimers of bR in the PM. It is found that an interaction energy of

~200 cm^{-1} could fit the observed biphasic spectrum.[12] This predicts an excitation energy randomization time of hundreds of femtoseconds.

We have used laser photoselection techniques[13] to determine the orientation of the transition moment of the K daughter relation to that of the absorbing parent bR molecule. The results showed that the orientation of the moment does not change in the time between the excitation of the parent and the formation of the K daughter. Since the transition moment of the excited $\pi \rightarrow \pi*$ transition is fixed along the retinal conjugated axis, one concludes that the excitation and the isomerization processes (to form the K intermediate) occur on the same retinal molecule. This suggests that energy transfer is not observed in the time between excitation and isomerization. If indeed excitation transfer occurs in ~100 fs, the above results suggest that isomerization must occur in a time much faster than 100 fs.

Using ultrafast spectroscopic techniques, Kaiser et al.,[14] have recently shown that changes in the optical absorption of the retinal in bR occurs in 400 fs. When rotation around the $C_{13}-C_{14}$ bond is hindered by chemical "anchors" of CH_2 groups, such optical changes are not observed.[14] These authors concluded that isomerization around the $C_{13}-C_{14}$ bond must occur in bR in 400 fs.

We have used our previously developed resonance Raman technique[15] to record the Raman spectrum of femtosecond transients in the retinal conformation-sensitive vibration region of the spectrum.[16] In this technique, bR sample flows rapidly across tightly focused low energy (few nJ/pulse), high rep. rate (80 MHz), ultrashort (700 fs) laser pulses so that every pulse causes photolysis and gives Raman scattered radiation. The sample exposed to one pulse is flown out of the laser focus before the next pulse arrives. By recording the Raman scattered radiation at low and high laser intensities (by using a spectrometer with an OMA), the transient spectrum of species formed during the laser pulse width can be extracted by subtraction techniques. The results are shown in Fig. 1. These results suggest that a new band appears at 1195 cm^{-1}, which has been previously observed on the picosecond time scale and assigned to the distorted 13-cis retinal.[17] Furthermore, the observed C=C stretching frequencies for the transients on this time scale are found to correlate with the optical absorption maximum observed by Mathies et al.[18] and was assigned to isomerization. This strongly suggests the previous conclusion that the observed optical changes of the retinal absorption on the 400 fs time scale correspond to retinal isomerization. If this is the case, then our linear dichroism results[13] suggest that energy transfer between different retinals, if present, must be an order of magnitude slower than isomerization, i.e., on picosecond time scale. According to the uncertainty principle, this corresponds to coupling energy of a few to tens of cm^{-1}. This energy leads to a small exciton

splitting incapable of giving a biphasic CD spectrum.
This strongly suggests that the biphasic nature of the CD
spectrum of bR results from heterogeneity. Distortion of
the retinal system in the bR trimers is blamed for the
origin of its CD. The different sense of distortion
around the retinal axis could be the cause of the
different signs of the CD bands.[19] The observed monophasic
CD spectrum as well as the small rotational strength of
reduced bR,[20] in which the retinal is planar, support this
notion.

b. What is the Involvement of Metal Cations?

Removing the 4-5 Ca^{2+} or the Mg^{2+} strongly bound to bR
is found to inhibit the photocycle of the L → M step,
i.e., the deprotonation of the PSB.[10,11] Are those metal
cations involved directly (by interaction with the PSB
allowing it to deprotonate) or indirectly (by controlling
the deprotonation reaction path of the protein)? We need
to determine the location of these metal cations. Time
resolved spectroscopy of Eu^{3+}-regenerated bR is used to
determine the number and environment of the different
sites (hydrophobic or hydrophylic) of these metal cations.
The results of these studies will be discussed in terms of
the binding constants of metal cations to bR.

REFERENCES

1.   Oesterhelt, D. & Stoeckenius, W. (1971) Nature
     (London) New Biol. 233, 149-152.
2.   Stoeckenius, W. & Bogomolni, R. A. (1982) Annu.
     Rev. Bio-chem. 52, 587-619.
3.   Unwin, P.N.T. & Henderson, R. (1975) J. Mol.
     Biol. 94, 425-440.
4.   Bridgen, J. & Walker, I.D.(1976) Biochemistry
     15, 792-798.
5.   Lozier, R., Bogomolni, R.A. Stoeckenius, W.
     (1975) Bio-phys. J. 15, 215-278.
6.   Dencher, N. & Wilms, M. (1975) Biophys. Struc.
     Mech. 1, 259-271.
7.   Stoeckenius, W. (1980) Acc. Chem. Res. 13, 337-
     344.
8.   Kimura, Y., Ikegami, A. & Stoeckenius, W. (1984)
     Photochem. Photobiol. 40, 641-646.
9.   Chang, C.-H., Chen, J.-G., Govindjee, R. &
     Ebrey, T. (1985) Proc. Natl. Acad. Sci. USA 82,
     396-400.
10. Kobayashi, T., Ohtani, H., Iwai, J., Ikegami, A.
     & Uchiki, H. (1983) FEBS Lett. 162, 197-200.
11. Chronister, E. L., Corcoran, T. C., Song, L. &
     El-Sayed, M. A. (1986) Proc. Natl. Acad. Sci.
     USA 83, 8580-8584.
12. Ebrey, T., et al., (1977) J. Mol. Biol. 112,
     377.
13. El-Sayed, M.A., Karvaly, B. & Fukumoto, J.M.

(1981) <u>Proc. Natl. Acad. Sci.</u> **78**, (12), 7512.

14. Polland, J.J., et al., (1986) <u>Biophys. J.</u> **49**, 651.
15. El-Sayed, M.A., (1985) <u>Pure and Appl. Chem.</u> **57**, 187.
16. van den Berg, Jang, D.-J., Hitting, H.C., & El-Sayed, M.A., (1990) <u>J. Phys. Chem.</u> **58**, 135.
17. Hsieh, C.-L., Nagumo, M., Nicol, M. & El-Sayed, M.A. (1981) <u>J. Phys. Chem.</u> **85**, 2714.
18. Mathies, R., et al., (1988) <u>Science</u> **240**, **777-779**.
19. El-Sayed, M.A., Lin, C. T. & Mason, W. R. (1989) <u>Proc. Natl. Acad. Sci. USA</u> **86**, 5376-5379.
20. Wu, S. & El-Sayed, M.A. <u>Biophys. J.</u> in press.

<u>Figure 1</u> The resonance Raman Spectra of retinal in bR in the fingerprint isomerization-sensitive region on the 700 fs time scale. <u>A:</u> at low laser intensity; <u>B:</u> at high laser intensity, <u>C:</u> is (B - A) the difference. The fact that there is a difference suggests a change in the retinal conformation in a time ≤700 fs.

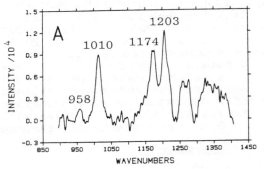

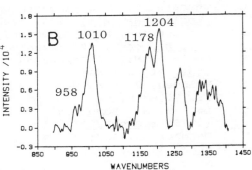

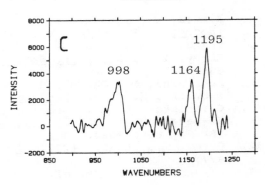

TRANSIENT RAMAN AND TRANSIENT ABSORPTION SPECTROSCOPIES OF PHOTOSYNTHETIC PIGMENTS

Y. Koyama

Faculty of Science, Kwansei Gakuin University
Uegahara, Nishinomiya 662
Japan

Carotenoids.

Roles of Carotenoids. Carotenoids have dual functions of light-harvesting and photo-protection.[1-3] In the case of purple photosynthetic bacteria, natural selection of carotenoid configurations has been found in relation to the above functions; the all-trans configuration is selected by the light-harvesting complex (LHC) for the light-harvesting function, while the 15-cis configuration is selcted by the reaction center (RC) for the photo-protective function.[4-9] In relation to the light-harvesting function in the LHC, the S_1 state of carotenoid which is involved in singlet energy transfer to bacteriochlorophyll has been an important issue. In relation to the photo-protective function, the mechanism of energy dissipation by the T_1 state carotenoid in the RC is also an important issue to be investigated.

The S_1 State. Information concerning the S_1 state of carotenoids has been obtained by picosecond Raman spectroscopy. We detected an S_1 Raman line of a carotenoid, β-carotene, by transient Raman spectroscopy.[10] The frequency was as high as 1777 cm^{-1} (for all-trans-β-carotene in benzene solution), and it was ascribed to an a_g type C=C stretching vibration in the $2A_g$ state which is vibronically coupled with that in the S_0 (A_g) state. It means that the S_1 state probed by Raman spectroscopy is actually the optically forbidden $2A_g$ state which resides below the optically allowed B_u state. The result provided a definite support for the proposal of Thrash et al.[11] on the basis of the Raman excitation profile. Later, the above Raman line was detected for a variety of carotenoids. Transient Raman spectroscopy of carotenoids bound to the chromatophore[12] and thylakoid[13] membranes detected the particular Raman line with lower intensity and in the lower frequency region when compared to the carotenoids free in solution; the lower intensity (the lower frequency) indicates shorter lifetime (weakened vibronic coupling) upon binding of the carotenoids to the apoprotein. More importantly, it ensures that the $2A_g$ state is still the S_1 state which is involved in the singlet energy transfer.

Another piece of information obtained by picosecond Raman spectroscopy is concerned with intersystem crossing (ISC) of carotenoids, the efficiency of which has been believed to be negligible.[14] Detection of T_1 Raman lines of β-carotene[15] and β-apo-8'-carotenal[16] by picosecond pulse train(s) strongly suggested the presence of ISC. Transient Raman and absorption spectroscopies of isomeric β-carotene evidenced (1) that the quantum yield of ISC is on the order of 10^{-3} and (2) that no isomerization takes place in the S_1 state.[17]

The T_1 state. In order to answer the intriguing question, why the 15-cis configuration has been selected by the RC for the photo-protective function, we compared the T_1 species generated through sensitization from isomeric β-carotene. Time-resolved Raman spectroscopy showed (1) that all-trans, 7-cis, 9-cis and 13-cis generate their own T_1 species, and (2) that 15-cis generates the "all-trans" T_1 instead of the "15-cis" T_1.[18] Time-resolved absorption spectroscopy supported the above conclusion;[19] the T_1 species generated from all-trans and 15-cis showed the same T_n <- T_1 absorption and the T_1 lifetime as each other. HPLC analysis of triplet-sensitized isomerization showed that the quantum yield of isomerization from 15-cis to all-trans is as high as 0.98, while a PPP-CI calculation of a model polyene rationalized the efficient isomerization via the T_1 state. The above results support the idea of extremely efficient isomerization from the "15-cis" T_1 to the "all-trans" T_1, which is expected to play a most important role in the mechanism of dissipating the triplet energy.[8,9]

Chlorophylls.

Roles of chlorophylls. Chlorophylls have multiple functions in photosynthesis: (1) Singlet energy transfer between chlorophylls through the dipole mechanism. (2) Triplet energy transfer between chlorophylls and from a chlorophyll to a carotenoid through the exchange mechanism, where the T_1 state structure of the donor and the overlap of the LUMOs and HOMOs between the adjacent molecules should play an important role. (3) Charge separation at the special pair chlorophylls, where the S_1 state structure of one of them and the overlap of the LUMOs between them should play an important role. Transient Raman spectroscopy is the only means, at present, to reveal the excited-state structures of chlorophylls.

The T_1 and S_1 states. We recorded the Raman spectra of chlorophyll a in the T_1 state by using nanosecond pulses.[20] Changes in Raman frequencies upon T_1 <- S_0 excitation indicated increase in the C_b-C_b bond order and decrease in the C_a-C_b, C_a-C_m and C_a-N bond orders, both

being predicted by an ab initio calculation of ethyl
chlorophyllide a.[21] The T_1 Raman spectra of bacteriochlorophyll a indicated changes in the bond orders in the
macrocycle upon triplet excitation, i.e. decrease
(increase) in the bond order of a bond having a double-
bond (single-bond) nature.[22] Effect of axial coordination of the solvent molecule(s) was also examined for T_1
bacteriochlorophyll a.[23] The ring-breathing mode in the
$1600 - 1570 cm^{-1}$ region was found to function as a
coordination-marker Raman line, but its frequency change
was not discrete but continuous.

We recorded the S_1 Raman spectrum of bacteriochlorophyll a by using picosecond pulses. Changes in the bond
orders upon $S_1 <- S_0$ excitation were similar, but a
little more enhanced, when compared to those upon $T_1 <-$
S_0 excitation described above.

Retinoids.

The T_1 and S_1 States. As an initial step for the
understanding of the isomerization property of the
retinylidene chromophore in bacteriorhodopsin, we
examined the excited states of retinal (C20 aldehyde)
free in solution. Nanosecond transient Raman spectroscopy
of isomeric retinal showed (1) that all-trans and 13-cis
generate the "all-trans" and "13-cis" T_1, but (2) that 7-
cis, 9-cis and 11-cis generate the "all-trans" T_1.[24]
Picosecond time-resolved absorption spectroscopy detected
the process of isomerization of each "cis" T_1 into the
"all-trans" T_1.[25,26] Time-resolved absorption spectroscopy of retinylideneacetaldehyde (C22 aldehyde) having a
longer polyene chain detected the S_1 species, the process
of intersystem crossing, and the resultant T_1 species.[27]
No isomerization was detected in the S_1 state, but the
same type of isomerization from each "cis" T_1 to the
"all-trans" T_1 as described above was detected.
Comparison of the T_1 Raman spectra and of the isomerization properties upon sensitized- and direct-photoexcita-
tion (examined by HPLC) among β-ionylideneacetaldehyde
(C15 aldehyde), β-ionylidenecrotonaldehyde (C17 aldehyde)
and the above C20 and C22 aldehydes showed (1) that T_1
potential has minima at each cis position and at the all-
trans position for all the aldehydes, but (2) that the
cis minima are as stable as the trans minimum for C15
aldehyde, while cis minima are far less stable than the
trans minimum for C17, C20 and C22 aldehydes.[28] Spectral
comparison indicated also the presence of what we call
"the triplet-excited region" which triggers isomerization.[29]

REFERENCES

1.   P. Mathis and C.C. Schenck, 'Carotenoid Chemistry and
     Biochemistry', G. Britton and T.W. Goodwin eds., Pergamon,
     Oxford, 1982, p.339.
2.   R.J. Cogdell and H.A. Frank, Biochim. Biophys. Acta, 1987, 895,

63.

3.   Y. Koyama, <u>J. Photochem. Photobiol.</u>, in press.

4.   M. Lutz, J. Kleo and F. Reiss-Husson, <u>Biochem. Biophys. Res. Commn.</u>, 1976, <u>69</u>, 711.

5.   Y. Koyama, T. Takii, K. Saiki and K. Tsukida, <u>Photobiochem. Photobiophys.</u>, 1983, <u>5</u>, 139.

6.   M. Lutz, W. Szponarski, G. Berger, B. Robert and J.-M. Neumann, <u>Biochim. Biophys. Acta</u>, 1987, <u>894</u>, 423.

7.   Y. Koyama, M. Kanaji and T. Shimamura, <u>Photochem. Photobiol.</u>, 1988, <u>48</u>, 107.

8.   Y. Koyama, I. Takatsuka, M. Kanaji, K. Tomimoto, M. Kito, T. Shimamura, J. Yamashita, K. Saiki and K. Tsukida, <u>Photochem. Photobiol.</u>, 1990, <u>51</u>, 119.

9.   Y. Koyama, 'Carotenoids: Chemistry and Biology', N.I. Krinsky, M.M. Mathews-Roth and R.F. Taylor eds., Plenum, New York, 1990, p. 207.

10.   H. Hashimoto and Y. Koyama, <u>Chem. Phys. Lett.</u>, 1989, <u>154</u>, 321.

11.   R.J. Thrash, H.L.-B. Fang and G.E. Leroi, <u>J. Chem. Phys.</u>, 1977, <u>67</u>, 5930.

12.   M. Kuki, H. Hashimoto and Y, Koyama, <u>Chem. Phys. Lett.</u>, 1990, <u>165</u>, 417.

13.   H. Hashimoto and Y. Koyama, <u>Biochim. Biophys. Acta</u>, 1990, <u>1017</u>, 181.

14.   R. Bensasson, E.A. Dawe, D.A. Long and E.J. Land, <u>J. Chem. Soc. Faraday Trans. I</u>, 1977, <u>73</u>, 1319.

15.   H. Hashimoto and Y, Koyama, <u>Chem. Phys. Lett.</u>, 1989, <u>163</u>, 251.

16.   H. Hashimoto and Y. Koyama, <u>Chem. Phys. Lett.</u>, 1989, <u>162</u>, 523.

17.   H. Hashimoto, Y. Koyama, Y. Hirata and N. Mataga, <u>J. Phys. Chem.</u>, in press.

18.   H. Hashimoto and Y. Koyama, <u>J. Phys. Chem.</u>, 1988, <u>92</u>, 2101.

19.   H. Hashimoto, Y. Koyama, K. Ichimura and T. Kobayashi, <u>Chem. Phys. Lett.</u>, 1989, <u>162</u>, 517.

20.   E. Nishizawa, H. Hashimoto and Y. Koyama, <u>Chem. Phys. Lett.</u>, 1989, <u>164</u>, 155.

21.   J.D. Petke, G.M. Maggiora, L. Shipman and R.E. Christoffersen, <u>Photochem. Photobiol.</u>, 1979, <u>30</u>, 203.

22.   E. Nishizawa and Y. Koyama, <u>Chem. Phys. Lett.</u>, 1990, <u>172</u>, 317.

23.   E. Nishizawa and Y. Koyama, <u>Chem. Phys. Lett.</u>, 1991, <u>176</u>, 390.

24.   H. Hamaguchi, H. Okamoto, M. Tasumi, Y. Mukai and Y. Koyama, <u>Chem. Phys. Lett.</u>, 1984, <u>107</u>, 355.

25.   H. Hirata, N. Mataga, Y. Mukai and Y. Koyama, <u>Chem. Phys. Lett.</u>, 1987, <u>134</u>, 166.

26.   Y. Mukai, Y. Koyama, Y. Hirata and N. Mataga, <u>J. Phys. Chem.</u>, 1988, <u>92</u>, 4649.

27.   Y. Hirata, N. Mataga, Y. Mukai and Y. Koyama, <u>J. Phys. Chem.</u>, 1987, <u>91</u>, 5238.

28.   Y. Mukai, H. Hashimoto and Y. Koyama, <u>J. Phys. Chem.</u>, 1990, <u>94</u>, 4042.

29.   H. Hashimoto, Y. Mukai and Y. Koyama, <u>Chem. Phys. Lett.</u>, 1988, <u>152</u>, 319.

VIBRATIONAL RAMAN OPTICAL ACTIVITY OF BIOLOGICAL MOLECULES

L. D. Barron, A. R. Gargaro, L. Hecht and Z. Q. Wen

Chemistry Department
The University
Glasgow G12 8QQ, U.K.

INTRODUCTION

Vibrational optical activity studies on chiral molecules can provide a wealth of new stereochemical information.[1] This is because a vibrational optical activity spectrum contains bands associated with every part of the molecule, each having a sign that depends on the absolute configuration of the part of the structure embraced by the corresponding normal mode and an intensity that is much more sensitive to molecular conformation than the parent vibrational band intensity. Vibrational optical activity in typical chiral molecules in the liquid phase was first observed using the Raman optical activity (ROA) technique, which measures a small difference in the Raman-scattered intensity in right- and left-circularly polarized incident light.[2] Until recently, lack of sensitivity has restricted ROA studies to favourable samples such as neat liquids,[3] with the complementary technique of vibrational circular dichroism (VCD) finding more application in detailed stereochemical studies.[4,5] However, a major advance in instrumentation based on the use of a backscattering geometry[6] together with a cooled CCD detector[7] has now rendered most biological molecules in aqueous solution accessible to ROA studies.

Vibrational optical activity is an ideal form of spectroscopy for biological molecules since most are chiral and solution conformation is central to biological function; indeed VCD has been applied sucessfully in this area for several years.[5] For the same reasons that have led conventional Raman spectroscopy to find many applications in biology (water is a good solvent for Raman studies, the complete vibrational spectrum is accessible, resonance enhancement enables sites of biological function to be probed directly, etc.), ROA is expected to become a powerful new biochemical spectroscopy. This is born out by our preliminary studies of samples such as peptides and proteins[8] and carbohydrates.[9,10] Here we present the first results from a new ROA instrument constructed in Glasgow and optimized for biological samples which is now providing spectra of greatly improved quality.

THEORY

ROA originates in interference between light waves scattered *via* the polarizability and optical activity tensors of the molecule.[1] The basic ROA measurement is $I^R - I^L$, where I^R and I^L are the Raman-scattered intensities in right and left circularly polarized incident light. An appropriate quantity for comparing theory and experiment is a dimensionless circular intensity difference (CID) which, for the particular case of backscattering from a chiral molecule composed entirely of idealized axially-symmetric bonds, takes the form[6]

$$\Delta = (I^R - I^L) / (I^R + I^L) = 64\beta(G')^2 / 2c[45\alpha^2 + 7\beta(\alpha^2)]$$

where α and $\beta(\alpha^2)$ are the isotropic and anisotropic invariants of the Raman transition polarizability tensor, and $\beta(G')^2$ is the anisotropic invariant of the product of the polarizability with the electric dipole—magnetic dipole optical activity. It has been shown, both theoretically and experimentally, that a given ROA signal—to—noise ratio is achieved almost one order of magnitude faster in backscattering than in the previously used right-angle scattering geometry.[6]

EXPERIMENTAL

ROA instruments usually utilize an electro-optic modulator to switch the polarization state of an incident argon—ion laser beam between right and left circular, with the detection of the Raman light synchronized with the modulation so that the circular intensity sum and difference signals can be extracted after a suitable number of modulation cycles.

The new Glasgow multichannel ROA instrument is built around a high luminosity single-grating spectrograph (Jobin Yvon HR 250S) with a holographic edge filter (Physical Optics Corporation)[11] to suppress stray light from the Rayleigh line, and a cooled CCD camera (Wright Instruments AT1) as detector. This instrument is currently roughly five times faster than our previous multichannel ROA instrument, with a further factor of five to come when our CCD is upgraded shortly to a backthinned version (which has a much higher quantum efficiency). Our backscattering optical system is based on the usual 'mirror with a hole' idea with two crucial additional components: a calcite Lyot depolarizer and a lens, both with central holes, which together ensure that the backscattered cone of Raman light is depolarized and collimated before it strikes the mirror.[6,12]

The quality and reliability of the ROA spectra provided by this new instrument is illustrated by the almost perfect mirror—image spectra of the two enantiomers of alanine in H_2O shown in Figure 1. *Ab initio* computations of the ROA spectrum of alanine carried out by Polavarapu agree remarkably well with the experimental spectrum in the lower-frequency region,[13] suggesting that it should eventually be possible to interpret ROA spectra of biological molecules at a fundamental level.

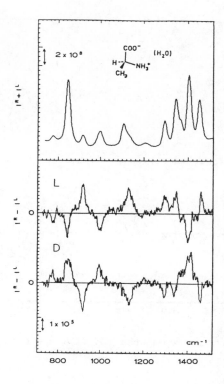

Figure 1. The backscattered Raman ($I^R + I^L$) and ROA ($I^R - I^L$) spectra of L-alanine (upper) and D-alanine (lower) in H_2O using 488.0 nm excitation.

Figure 2 shows the first ever "complete" vibrational optical activity spectrum of a protein, the ROA spectrum of lysozyme in water. The features in the amide III region at 1250—1340 cm^{-1} have been reported previously:[7] they originate in coupled C_α—H and N—H deformations of the peptide backbone and appear to be sensitive to the details of the secondary conformation. New features that might also be sensitive to the secondary conformation include the splendid ROA couplet covering the 1600 to 1700 cm^{-1} region and apparantly centred on the amide I band, a broad couplet in the C—N stretch region at 1060—1140 cm^{-1}, detailed structure in the C—C stretch region at 900—1000 cm^{-1}, and a broad underlying couplet (positive at low frequency and negative at high) spanning the region from below 400 to over 800 cm^{-1} that might be associated with C=O deformations of the peptide backbone. Running throughout the spectrum are a number of reproducible smaller ROA features, often quite sharp, which are presumably associated mostly with the side groups: these features are particularly prominent in bands assigned to tryptophan residues. Hence information about both side group and secondary backbone conformation should be available from protein ROA spectra.

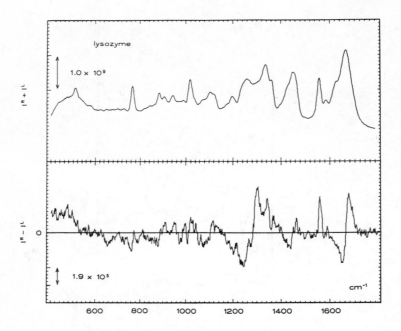

Figure 2. The backscattered Raman and ROA spectra of lysozyme in H_2O using 514.5 nm excitation.

REFERENCES

1. L. D. Barron, 'Molecular Light Scattering and Optical Activity', Cambridge University Press, Cambridge, 1982.
2. L. D. Barron, M. P. Bogaard and A. D. Buckingham, *J. Am. Chem. Soc.*, 1972, **95**, 603.
3. L. D. Barron, *Vibrational Spectra and Structure*, 1989, **17B**, 343.
4. T. B. Freedman and L. A. Nafie, *Topics in Stereochemistry*, 1987, **17**, 113.
5. T. A. Keiderling, in 'Practical Fourier Transform Spectroscopy', Academic Press, New York, 1990, p. 203.
6. L. Hecht, L. D. Barron and W. Hug, *Chem. Phys. Letts.*, 1989, **158**, 341.
7. L. D. Barron, L. Hecht, W. Hug and M. J. MacIntosh, *J. Am. Chem. Soc.*, 1989, **111**, 8731.
8. L. D. Barron, A. R. Gargaro and Z. Q. Wen, *J. Chem. Soc. Chem. Commun.*, 1990, p. 1034.
9. L. D. Barron, A. R. Gargaro, Z. Q. Wen, D. D. MacNicol and C. Butters, *Tetrahedron: Asymmetry*, 1990, **1**, 513.
10. L. D. Barron, A. R. Gargaro and Z. Q. Wen, *Carbohydr. Res.*, in press.
11. M. M. Carrabba, K. M. Spencer, C. Rich and D. Rauh, *Appl. Spectrosc.*, 1990, **44**, 1558.
12. W. Hug, in 'Raman Spectroscopy', Wiley—Heyden, Chichester, 1982, p. 3.
13. L. D. Barron, A. R. Gargaro, L. Hecht and P. L. Polavarapu, *Spectrochim. Acta,* in press.

PROTEIN CONFORMATIONAL STUDIES WITH VIBRATIONAL CIRCULAR DICHROISM

Timothy A. Keiderling*[a] and Petr Pancoska[b]

[a]Department of Chemistry,
University of Illinois at Chicago,
Box 4348, Chicago, IL 60680 USA

[b]Department of Chemical Physics,
Faculty of Mathematics and Physics,
Charles University,
Prague 2, Czechoslovakia

INTRODUCTION

Spectroscopic studies of proteins have a long history in the biochemical and biophysical sciences and have been particularly important in protein studies for preliminary, qualitative estimations of secondary structure. These have involved ultraviolet (UV) spectra involving electronic transitions between delocalized π-electron states, infrared (IR) and Raman spectra of vibrational transitions in the ground state, as well as magnetic resonance studies. Of these, optical activity measurements have unique sensitivity to structural variation due to their dependence on the three dimensional electron distribution in the molecule. In particular, circular dichroism (UVCD) of the n-π* and π-π* electronic transitions of the amide groups has become an indispensable tool for qualitative characterization of proteins in solution. However the amide transitions available in the near-UV are limited in number and are broad and overlapping. Interactions among them yield information about the polymeric backbone, but since such amide electronic excitations are relatively delocalized, the resulting UV spectral bands are often affected by environmental or local perturbations. Furthermore, the amide transitions are overlapped by several, aromatic π-π* transitions. By comparison, IR and Raman spectroscopies have inherently resolved transitions but are limited to measurement of small frequency shifts characteristic of the perturbation of bond strengths by the structure.

This set of circumstances has stimulated the development of vibrational CD (VCD) as a hybrid of these more established spectroscopic tools.[1] Unlike UVCD, VCD can be used to sense several different spectrally resolved transitions involving different localized vibrations of the molecule; and, unlike IR and Raman spectroscopies, each will have a distinct dependence on molecular stereochemistry.

Empirical correlation of spectral features with secondary structure has historically been the most profitable route for stereochemical utilization of both electronic CD and vibrational (IR and Raman) spectroscopies. Thus it seems such an approach is also optimal for VCD. The difference in the physical origins of CD measured in the two spectral regions suggests that they bear a complementary relationship that could enhance the quality and quantity of structural information derivable from either one alone and compensate for the shortcomings of each. How best to carry out such a coupled analysis is a topic of our ongoing studies. Here is reviewed our first *uniform systematic analysis* of data available from each CD, vibrational and electronic.

Experimentally, VCD spectra are routinely measured at the University of

121

Illinois on either a dispersive or an FTIR-based instrument, both described in detail elsewhere in the literature.[1] The dispersive data are generally obtained at ~10 cm^{-1} resolution by averaging several repetitive scans over the band of interest. The protein FTIR-VCD spectra are typically of lower signal-to-noise ratio (S/N) since, in aqueous solution, only narrow spectral windows are accessible.[2] Our spectra are obtained on solutions held in short path length sample cells made with CaF_2 windows. For purposes of comparison and further analyses, higher resolution and better S/N FTIR absorption spectra are obtained for multiple bands with a Digilab FTS-60, and UVCD spectra down to 180 nm are obtained with a JASCO J-600. Spectra from all these instruments are uniformly treated using SpectraCalc for data manipulation, our own factor analysis routine, and EINSIGHT for cluster analysis.

RESULTS AND DISCUSSION

In extending the application of VCD to proteins, we have obtained reproducible protein VCD spectra in the amide I' region (C=O stretch) and have profitably compared it to UVCD and IR data on the same proteins.[2] As compared to IR spectra, the sign variation inherent in VCD gives it an effective resolution advantage in differentiating between proteins. Furthermore, since the different types of secondary structure contribute on a comparable basis to VCD, it has more variation in band shape than UVCD, which is singularly dominated by the α-helical contribution. This latter effect may be a result of the importance of short-range interactions in determination of VCD band shapes.[3] For example, as shown in Figure 1, hemoglobin is in the class of proteins whose secondary structure is dominated by the α-helix, while concanavalin A is dominated by its β-sheet components. Each of these has similarly shaped UVCD with the main difference being intensity[2] and has an IR absorption band that differs only ~20 cm^{-1} in peak position. In VCD, they exhibit oppositely signed band shapes and significant frequency shifts thereby evidencing more sensitivity to the structural variation.[4] Globular proteins of similarly predominantly α-helical or β-sheet secondary structures have similar VCD bandshapes, and those with a mix of α and β components have VCD spectra resembling a linear combination of these two forms.[2,4]

The basis for our first systematic method of empirical VCD analysis[4] is the principal component method of factor analysis whereby the spectra are reduced to relatively small vectors of coefficients that serve to characterize the spectral shapes.[5] Our initial results show that 20 protein amide I' VCD spectra can be fit to their error limit with linear combinations of six orthogonal subspectra. The first subspectrum represents the most common elements of the experimental spectra used in the analysis. The second represents the major deviations from the first. Each successive one becomes less significant, with those beyond the sixth representing the noise contributions.

The relative contribution of these subspectra, which are *determined directly from experimental VCD without any prior assumptions,* to a protein VCD spectrum can be used to determine aspects of its secondary structure. Selected coefficients of the subspectra can be correlated at a statistically significant level with α-helical, β-sheet, bend and "other" contributions to the secondary structure as determined from the x-ray structures.[6] At this time, turns cannot be reliably determined from this data. Our selective regression analysis avoids the problems inherent in the total back transformation of CD data into secondary structure components as is commonly used in other analyses.[7,8] It is similar to, but more restricted than, a recent analysis of FTIR spectra.[9] We have carried out exactly the same analysis on remeasured UVCD data to show that the information content of both types of spectra is qualitatively similar but that the VCD does better quantitatively, particularly for β-sheet determination.[10] For those proteins whose spectra are well-grouped with the 'known' ones in our initial study, as judged by cluster analysis,[11] adequate predictions of secondary structure contents can be made.

We have found it most useful to characterize the crystal structure results using the Kabsch-Sander algorithm[6] for generating fractional contributions to the secondary structure. A 5-vector characterization of the secondary structure was used encompassing the fraction of α-helix, β-sheet, bend, turn and "other" which includes all other conformational types. This classification is not unique, nor is it obvious that it is the best for any of the spectral techniques that we wish to study. Tests using the Levitt and Greer[6] "low resolution" analysis of crystal structures led to similar results.[4]

Cluster analysis can be used to isolate those proteins which have spectral behavior far from the norm based on standard crystal structure-spectra correlations. This was possible in our study of the pH dependence of the secondary structure of phosvitin, a highly charged glycophosphoprotein.[16] The VCD spectra of phosvitin evidenced considerable frequency shift and at low pH did not cluster with those of our set of 'known' proteins, thus any derivation of the secondary structures for these forms of phosvitin from the spectra are likely to have considerable error.

An alternate approach to statistical analysis of spectra is the partial least squares (PLS) method[12] which treats the 'known' proteins as a calibration set and has been recently applied to protein FTIR spectra.[13] Our regression approach gave about the same average result as the PLS one; but, since PLS involves a total back-transformation, some predictions (ex. β-sheet) were much worse than the factor analysis-regression approach. Furthermore, for the 'unknown' proteins, the regression method successes and failures were predictable via cluster analysis. For UVCD, the PLS method was unable to determine the β-sheet, bend and turn fractions with reliability.

Our factor analysis and limited regression approach is more conservative and predicts fewer fractional contributions than some others,[7,9,13] but it emphasizes those aspects of the structure to which the analysis is sensitive and evaluates the degree of confidence one should place in the results. Target transformation methods neglect other environmental contributions to the spectra or implicitly assume that they cancel. For purposes of understanding protein structure, *accuracy* is more important than precision. Accuracy is enhanced by our method in that non-relevant factors have diminished influence in our selective parameter method for interpretation of 'unknown' structures. We are now extending this systematic method of analysis to spectral techniques beyond CD,[14] to VCD of other vibrational transitions[15] and to a broader range of proteins.

ACKNOWLEDGEMENT. We wish to thank the National Institutes of Health for support of this research (GM 30147) and NIH and NSF for instrumentation.

REFERENCES

1. T. A. Keiderling, in 'Practical Fourier Transform Infrared Spectroscopy' (Ferraro, J. R., and Krishnan, K., Eds.) p. 203-284, Academic, San Diego, 1990; *Appl. Spectr. Rev.*, 1981, *17*, 189-226; and references.
2. P. Pancoska, S. C. Yasui and T. A. Keiderling *Biochemistry* 1989, *28*, 5917-23.
3. R. K. Dukor, T. A. Keiderling, and Gut, V. *Int. J. Pept. Prot. Res.*, 1991 (in press). R. K. Dukor and T. A. Keiderling, *Biopolymers* 1991, (submitted for publication). S. C. Yasui, T. A. Keiderling, F. Formaggio, G. M. Bonora and C. Toniolo, *J. Am. Chem. Soc.,*1986, *108*, 4988-93

4. P. Pancoska, S. C. Yasui, and T. A. Keiderling, *Biochemistry* 1991 (in press)
5. E. R. Malinowski and D. G. Howery, 'Factor Analysis in Chemistry' Wiley, New York, 1980. Pancoska, P., Fric, I. and Blaha, K. *Coll. Czech. Chem. Comm.* 1979, *44*, 1296-1311; *44*, 1289-95.
6. W. Kabsch and C. Sander, *Biopolymers*, 1983, *22*, 2577-2637. M. Levitt and J. Greer, *J. Molec. Biol.* 1977, *114*, 181-293.
7. J. P. Hennessey and W. C. Johnson, *Biochemistry*, 1981, *20*, 1085-94.
8. M. Manning, *J. Pharmceut. Biomed. Anal.* 1989, 7 1103-19.
9. D. C. Lee, P. I. Haris, D. Chapman and R. C. Mitchell, *Biochemistry*, 1990, *29*, 9185-93.
10. P. Pancoska and T. A. Keiderling, *Biochemistry*, 1991, (submitted for publication)
11. D. L. Massart and L. Kaufman, 'The Interpretation of Analytical Chemical Data by the Use of Cluster Analysis, Chemical Analysis', Vol. 65, J. Wiley & Sons, New York, 1983.
12. D. M. Haaland and E. V. Thomas, *Anal. Chem.,* 1988, *60*, 1193-1202, 1202-8.
13. F. Dousseau and M. Pezolet, *Biochemistry* 1990, *29*, 8771-79.
14. P. Pancoska, L. Wang and T. A. Keiderling, (to be submitted).
15. V. P. Gupta and T. A. Keiderling, (to be submitted)
16. S. C. Yasui, P. Pancoska, R. K. Dukor, T. A. Keiderling, V. Renugopala-krishnan, M. J. Glimcher and R. C. Clark, *J. Biol. Chem.* 1990, *265*, 3780-88.

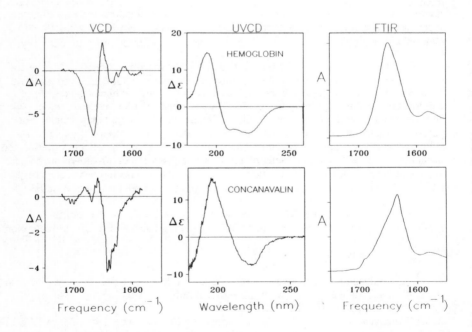

Figure 1. Comparison of VCD, UVCD and FTIR spectra of hemoglobin and concanavalin A. The vibrational spectra are for the amide I' only and are normalized to a peak absorbance of $A_{max} = 1.0$. The units of VCD are presented as $\Delta A * 10^5$.

FEMTOSECOND DYNAMICS OF ß-CAROTENE SOLUTIONS

S.Califano[1], P.Foggi[1], V.F.Kamalov[2]*, R.Righini[1], R.Torre[1]

[1]European Laboratory for Nonlinear Spectroscopy
University of Florence
Largo E.Fermi, 2, Florence 50125 Italy

[2]Photochemistry Department
Institute of Chemical Physics
Academy of Sciences USSR
Kosygin str., 4, Moscow 117977 USSR

INTRODUCTION

Depolarized Rayleigh wing scattering is well known as a method for study molecular dynamics of liquids and solutions. Time-domain picture of such processes was resolved due to progress in femtosecond laser pulses generation. Temporal evolution of low frequency intramolecular vibrational coordinates and ultrafast decay component attributed to librational motion were studied for simple liquids and solutions recently by optical Kerr effect (OKE) [1].

The purpose of this study is the creation of anisotropy of ß-carotene solutions by OKE to study dynamical behavior of ß-carotene at the femto and picosecond time scale. Interactions of optical chromophores with the close environment can be studied on the base of technique developed.

EXPERIMENTAL

The anisotropy created in ß-carotene solutions by powerful pump pulse were measured by transmittance of probe pulse passed through the crossed polarizers. The angle between linearly polarized pump and probe was $45°$. Durations of pump and probe pulses were 100 fs. Wavelength of pump was 605 nm, probe was tuned close to ß-carotene electronic transition around 500 nm. Dependence of anisotropy created on the time delay between pump and probe pulses were measured for ß-carotene n-dodecane solutions with concentration 100 μM.

RESULTS AND DISCUSSION

At low pump intensity the anisotropy created during the laser pulses decays exponentially with the time 8 ps reflects the creation of the gap in the orientational distribution of molecules in ground electronic state due to one-photon absorption of pump beam, and subsequent decay of anisotropy due to repopulation of ground state with the time 8 ps.

The contributions from nonlinear optical processes were increased with the pump intensity. The signal of anisotropy detected at high pump intensities caused by birefringence generated in the ß-carotene solutions due to OKE. High level of the signal observed was originated due to intrinsic anisotropy of ß-carotene, high value of ß-carotene hyperpolarizability [2], and its resonant enhancement due to resonance of probe beam with the electronic transition. Three components in the anisotropy rise and subsequent decay were observed at high pump intensities. The first one is the instantaneous response due to pure electronic hyperpolarizability of ß-carotene. The second response lay in the femtosecond time scale with characteristic time 250 fs for dodecane solution. The last component was slow decay with the time 200 ps for dodecane.

Slow component in decay with 200 ps was attributed to rotational diffusion process, and rotational diffusion rate dependence on the viscosity was measured. Femtosecond response with 250 fs decay time can be assign to interaction induced (I-I) decay of created anisotropic distribution of ß-carotene due to interaction of ß-carotene with surrounding dodecane molecules. The I-I effects lead to anisotropy decay through the intermolecular motions, librations, that change the relative position and orientation of ß-carotene molecules.

Also fast ß-carotene intramolecular conformational dynamics can cause decay of anisotropy created and molecular dynamics of chromophores with conjugated chains will be discussed.

REFERENCES

1. D.McMorrow, W.T.Lotshaw, G.A.Kenney-Wallace, IEEE J.Quantum Electron., 1988, QE-24, 443.
2. J.P.Hermann, D.Ricard, J.Ducuing, Appl.Phys.Lett., 1973, 23, 178.

CONFORMATIONAL DYNAMICS IN BACTERIORHODOPSIN

Carey K. Johnson

Department of Chemistry
University of Kansas
Lawrence, KS 66045 U.S.A.

INTRODUCTION

Bacteriorhodopsin (BR) is a photoactivated membrane protein that serves to pump protons across the cell membrane via a cyclic process that involves isomerization, structural relaxation, a series of proton transfers, and protein conformational changes. The questions one wishes to answer about the function of the protein are: how does the protein control the reaction path and proton-transfer sequence? how do protein and chromophore interact? and what role, if any, do protein conformational changes play?

In order to answer these questions, one needs to probe the dynamics of BR on a wide range of time scales, from picoseconds to milliseconds. We have undertaken time-resolved optical pump-probe experiments focussing on protein dynamics from the picosecond to millisecond time scales, using a flexible pump-probe laser system capable of spanning these time regimes in a single experiment.[1] Rates were measured by transient absorption experiments, and conformational motion was probed by polarization sensitive methods.

ORIENTATIONAL MOTION

Excitation of BR with a polarized pump pulse at about 570 nm selects an orientation of the chromophore to initiate the photocycle. The result is an anisotropic population of photocycle intermediates, and an orientational hole in the BR ground state. These anisotropies persist for milliseconds due to the photocycling time and slow rotational diffusion of membrane particles. Anisotropy decay on shorter time scales is related either to restricted internal motion within the protein or motion of BR monomers within the membrane. We detected anisotropy and its decay by time-resolved polarization spectroscopy and time-resolved linear dichroism spectroscopy.

Time-Resolved Polarization Spectroscopy

In time-resolved polarization spectroscopy the intensity of the probe pulse transmitted through the sample with polarization perpendicular to the incident probe polarization is detected following excitation with a pump pulse oriented at 45° with respect to the probe polarization. The time dependence of the signal is determined by both the anisotropy and the population of the various states present, and hence is sensitive to anisotropy decay due to orientational motion. A feature of these experiments is the possibility of interferences between the signal induced by the anisotropic ground-state hole and the signal induced by the anisotropic intermediate-state populations.[2]

Time-resolved polarization experiments for BR over times scales from picoseconds through hundreds of microseconds[2] reveal restricted orientational motion on the time scale of the transition from the K to L intermediates. These scans also display the effect of interference between ground-state and intermediate-state anisotropies. In picosecond scans, evidence of possible orientational motion with a 50-100 ps correlation time is observed.

Time-Resolved Linear Dichroism

In order to investigate the possibility of internal motion in BR further, we have used time-resolved linear dichroism (TRLD) spectroscopy.[3] These experiments allow direct comparison of the dichroism signal, which is sensitive to orientational motion, to transient absorption (TA) scans which are not sensitive to orientational motion. Clear differences are observed in the time and wavelength dependence of TRLD and TA scans. Our analysis[3] leads us to the conclusion that a protein conformational change involving orientational motion of the chromophore occurs in the L intermediate immediately following the K to L transition. We place an upper limit of 30 ° on the reorientation of the chromophore. This motion may be related to the proposed T to C protein conformational change in BR.[4] Recent TRLD experiments have focused on orientational motion in later stages of the BR photocycle.

TEMPERATURE-INDUCED TRANSITION

Transient-absorption scans on the microsecond time scale have been used to measure the K to L transition rate as a function of temperature both in the native purple membrane, and reconstituted in micelles. In the low temperature regime, Arrhenius behavior is observed, with an activation energy of 55 kJ per mole. The activation energies in the native purple membrane and in micelles are nearly equal, although the transition rate is slightly faster in micelles. At higher temperatures, both the purple membrane and micelles deviate from Arrhenius behavior in the temperature regime of the reversible "pretransition" that has been observed calorimetrically in both media.[5] These results show the effect of a thermally induced protein conformational change on the kinetics of one of the early steps in the BR photocycle.

ACKNOWLEDGEMENTS
I thank Dr. Wan Chaozhi, Jun Qian, and Sarah Mounter for their vital contributions to these experiments, and Dr. Christie Brouillette for micellar BR samples and helpful discussions. This work was supported by NIH FIRST Award GM 40071, and by the University of Kansas General Research Fund.

REFERENCES
1. C.K. Johnson, J.M. Bostick, S.A. Mounter, K.L. Ratzlaff, and D.E. Schloemer, *Rev. Sci. Instrum.* 1988, *59*, 2375.
2. C. Wan, J. Qian, and C.K. Johnson, *J. Phys. Chem.* 1990, *94*, 8417.
3. C. Wan, J. Qian, and C.K. Johnson, *Biochemistry* 1991, *30*, 394.
4. S.P.A. Fodor, J.B. Ames, E.M.M. van den Berg, W. Stoeckenius, J. Lugtenburg, and R. Mathies, *Biochemistry* 1988, *27*, 7097.
5. C.G. Brouillette, D.D. Muccio, and T.K. Finney, *Biochemistry*, 1987, *26*, 7431; C.G. Brouillette, R.B. McMichens, L.J. Stern, and H.G. Khorana, *Proteins*, 1989, *5*, 38.

ULTRAFAST INFRARED STUDY OF BACTERIORHODOPSIN

R. Diller, M. Iannone, B. Cowen, S. Maiti, R. Bogomolni[#] and R. M. Hochstrasser[*]

Department of Chemistry, University of Pennsylvania
Philadelphia, PA   19104-6323
[#] Department of Chemistry, University of California
Santa Cruz, CA   95064

The development of time resolved vibrational spectroscopic methods enables the study of dynamical procecesses in biological molecules on a microscopic level.   In the case of Bacteriorhodopsin (BR), the interaction between the chromophore and the protein during the photocycle is of particular interest for the understanding of the proton pump mechanism.   Fourier transform infrared (FTIR) and resonance Raman (RR) spectroscopy have revealed a great deal of information about the involvement of several aminoacid residues (1,2,3) and the kinetics of the chromophore (4,5) on various timescales.   However, the nanosecond and subnanosecond states so far have only been studied by RR and low temperature FTIR spectroscopy, where the intermediate protein states are frozen.

We have developed a high repetition rate infrared laser system for the application to the BR photocycle at ambient temperatures. Recorded 100 ps, 1 ns and 14 ns after the pump pulse at 550 nm (25 ps FWHM), infrared difference spectra are obtained between 1700 cm^{-1} and 1560 cm^{-1}.   The spectra show a strong bleach in the Amide I region, including the C=NH stretch vibration of BR570 at 1537 cm^{-1}. Positive bands at 1622, 1610 and 1580 cm^{-1} are assigned to the K state.   The risetime of both the negative and the positive signals are not resolved, reflecting the fast photoisomerization process.

Significant spectral changes appear in the Amide I region between 1 ns and 14 ns.   Changes of the chromophore spectra on the same time scale have been reported earlier (6,7).

Comparison with low temperature FTIR spectra of the BR-K transition (2) reveals overall similarities but significant differences at about 1570 cm^{-1}, indicating protein flexibility.

The presented method allows recording of the spectral changes in the infrared on the picosecond timescale.   Measuring the vibrational anisotropy of certain infrared modes will permit the study of the ultrafast dynamics of both the chromophore and the protein at ambient temperature.

REFERENCES

1. K. Bagley, G. Dollinger, L. Eisenstein, A.K. Singh and L. Zimanyi, PNAS, 1982, 79, 4972.
2. K. Gerwert and F. Siebert, EMBO J., 1986, 5, 805.
3. M.S. Braiman, T. Mogi, T. Marti, L.J. Stern, G.H. Khorana and K.J. Rothschild, Biochemistry, 1988, 27, 8518.
4. M. Stockburger, T. Alshuth, D. Oesterhelt and W. Gaertner in Spectroscopy of Biological Systems (eds. R.J.H. Clark and R. E. Hester) Wiley & Sons, London, 1986, p 483.
5. S.P.A. Fodor, J.B. Ames, R. Gebhard, E.M.M. van den Berg, W. Stoeckenius and R.A. Mathies, Biochem., 1988, 27, 7097.
6. S.J. Milder and D.S. Kliger, Biophys. J., 1988, 53, 465.
7. D. Stern and R. Mathies in Time Resolved Vibrational Spectroscopy (eds. A. Laubereau and M. Stockburger) Springer Berlin, 1985, p 250.

INFRARED (VIBRATIONAL) CD OF PEPTIDES IN THE AMIDE I AND AMIDE III SPECTRAL REGIONS

S.Birke, C.Farrell, O.Lee, I.Agbaje, G.Roberts[+] and M.Diem

Department of Chemistry and Biochemistry, City University of New York, Hunter College, New York, USA
[+]Department of Biophysics, University of Michigan, Ann Arbor, MI USA

INTRODUCTION

Infrared or Vibrational Circular Dichroism (VCD) has been used extensively to study the solution conformation of peptides[1] and proteins[2]. To date, nearly all these efforts have been carried out in the amide I spectral region, which is due mostly to the C=O stretching vibration. In this contribution, we report efforts to observe and interpret amide III VCD signals of a number of peptides.

EXPERIMENTAL ASPECTS

All VCD spectra shown here were recorded on one of two dispersive VCD spectrometers constructed in our laboratory, and described elsewhere.[3] Sample concentrations are typically about 40 mg/mL for the amide I and about 150 mg/mL for the amide III spectra. H_2O and DMSO-d_6 were used for solvents in the amide III region, and D_2O and DMSO in the amide I region.

RESULTS AND DISCUSSION

We have shown *via* detailed deuteration studies[4,5] that the amide III mode is a highly coupled vibration consisting of the N-H and the methine C-H deformation coordinates. This coupling is non-polar, and differs from the coupling responsible for the exciton VCD features observed in the amide I spectral region. Nevertheless, the coupling is sensitive to the geometry between the C-H and the N-H deformation coordinates.

In Figure 1 are shown the amide I and amide III VCD spectra of poly-L-lysine in a conformation which is generally referred to as a "random coil". The amide I VCD, as well as vibrational studies, have suggested that this conformation is a left-handed helical structure, which has distinct order on the interaction distance responsible for exciton interactions in VCD.

We have found that a positive/negative or positive/negative/negative signal is indicative of a left-handed structure.

131

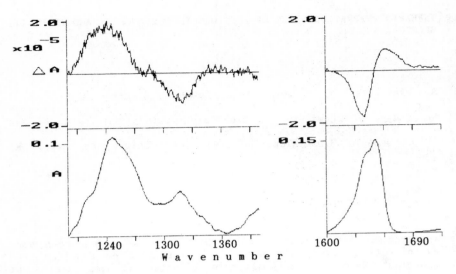

Fig.1 VCD (top) and IR absorption spectra of poly-L-Lys
in the Amide III and Amide I Regions

A positive/negative/negative signal was also observed in a very
small peptide, (L-Ala)$_3$, which was shown to exist in a left-handed
structure by the analysis of the amide I VCD spectra.[5] A mostly
negative signal is indicative of an α-helix. This was established
by comparing the observed amide I and amide III spectral patterns
for peptides for which the secondary structure can be altered by
changing the acidity of the solvent.

It appears, however, that the VCD changes in the amide III
region do not always agree with the changes in the amide I VCD
features. At the acid induced transition point from left- to right
handed poly-L-Tyr, we observe that the amide I region might still
indicate a left handed structure, while the amide III already
indicates a conformational change. This might be due to the fact
that the coupling distances of these vibrational modes is drastically different. Similar arguments were used by Keiderling[1] to
explain apparently different conformations deduced by CD and amide
I VCD spectra.

REFERENCES

1. T.A.Keiderling, *Nature*, **1986** 322, 851, and references cited
   herein
2. T.A.Keiderling, in *'Practical Fourier Transform Infrared Spectroscopy'*, **1990** Academic Press, NY, pp 203-283
3. M.Diem, G.M.Roberts, O.Lee, and A.Barlow, *Appl.Spectrosc.* **1988**
   42, 20
4. M.R.Oboodi, C.Alva and M.Diem, *J. Phys. Chem.* **1984**, 88, 501
5. G.M.Roberts, PhD Dissertation, City University of New York, 1991
6. O.Lee, G.M.Roberts, and M.Diem, *Biopolymers*, **1989**, 29, 1759

CONFORMATIONAL STUDY OF α-LACTALBUMIN USING VIBRATIONAL CIRCULAR DICHROISM. EVIDENCE FOR A DIFFERENCE IN CRYSTAL AND SOLUTION STRUCTURES

Marie Urbanova, Rina K. Dukor, Petr Pancoska, and Timothy A. Keiderling

Department of Chemistry,
University of Illinois at Chicago
Box 4348, Chicago, Illinois 60680 USA

α-Lactalbumin (α-Lac), a Ca^{+2} binding protein, is found in mammalian milk whey and has recently been shown to exhibit potential antitumor activity. The three-dimensional structure of baboon α-Lac has recently been solved at 1.7Å resolution.[1] It was there demonstrated that *the α-Lac crystal structure is very similar to that of hen egg white lysozyme (HEWL)*, which is highly homologous to it. This x-ray data on baboon α-Lac has been further compared to model predictions for bovine α-Lac to show that their structures are essentially the same.[2]

Recently, using resolution enhanced FTIR studies on aqueous solutions of these two proteins, Prestrelski et al.[3] concluded that the conformations are indeed alike, even though the envelope of the absorption spectra was significantly different. Their conclusion was based on the calculation of the relative areas of the individual peaks, but two out of seven peaks had a questionable assignment (there are two peaks assigned to an α-helix and a low frequency band assigned to a 3_{10}-helix).

Our solution studies on these proteins were done using vibrational circular dichroism (VCD). This technique combines the conformational sensitivity of an optical activity measurement (CD) and the resolution enhancement of vibrational region. We have demonstrated that VCD leads to an enhanced sensitivity to protein conformation in solution.[4]

The VCD spectra of bovine α–Lac (with or without Ca^{+2}) and HEWL in D_2O or pH 7.6 buffer solution are in fact quite different (Fig. 1). To understand these differences, the conformations of these two proteins were studied as a function of pH and added propanol concentration. With increase in propanol concentration for bovine α-Lac a transformation from the spectrum shown in Fig. 1, which is representative of a substantial contribution from extended structures, to a spectrum typical of helical forms is observed. In 30% propanol, the VCD spectrum of α-Lac is very much like that of lysozyme in buffer. On the contrary, addition of propanol has no effect on the VCD spectrum of lysozyme.

α-Lac is known to undergo two types of transition at acidic pH.[5] However, no changes in the VCD spectrum are observed in the pH region of 11-4.5. Some changes are seen at pH=1 but they probably correspond to a protein in a denatured state (α-Lac exists in a partially unfolded conformation between pH 2 and 4). The crystals for X-ray study of lysozyme were grown at pH=4.7, thus to eliminate pH effects we measured VCD spectra of lysozyme at different pH's. No changes in VCD spectra are observed in the range of pH=3-7.

The differences observed in the VCD spectra of bovine α-lactalbumin and hen egg white lysozyme and solvent sensitivity of the α-Lac spectra suggest that the *crystal structure* may represent a conformation different from that seen in *aqueous solution.*

References:

1. Acharya, K. R., Stuart, D. I., Walker, N. P. C., Lewis, M., and Phillips, D. C., *J. Mol. Biol.* 1989, *208*, 99-127
2. Acharya, K. R., Stuart, D. I., Phillips, D. C., and Scheraga, H. A., *J. Prot. Chem.*, 1990, *9*, 549-563
3. Prestrelski, S. J., Byler, D. M., and Thompson, M. P., *Int. J. Pept. Prot. Res.*, 1990, in print
4. Pancoska, P., Yasui, S., and Keiderling, *Biochemistry*, 1989, *28*, 5917-23
5. Kuwajima, K., Nitta, K., and Sugai, S., *Biochim. Biophys. Acta*, 1980, *623*, 389-401

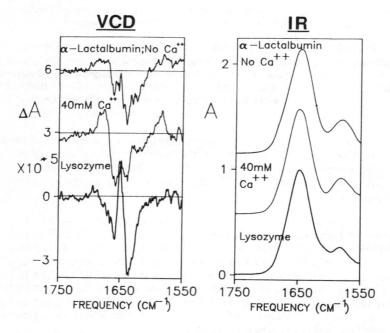

Figure 1. Comparison of VCD and IR spectra of bovine α-lactalbumin containing no Ca^{+2}, 40mM Ca^{+2} in imidazole buffer (in D_2O) and hen egg white lysozyme in phosphate buffer (in D_2O), pH=7. The absorption spectra are normalized to a peak absorbance of A_{max}=1.0 for ease of comparison.

INFRARED (VIBRATIONAL) CD OF DNA MODELS

M.Gulotta, W.Zhong, D.J.Goss, H.Votavova[+] and M.Diem

Department of Chemistry and Biochemistry, City University of
New York, Hunter College, New York, USA
[+]Institute for Organic Chemistry and Biochemistry, Czechoslovak
Academy of Science, Prague, Czechoslovakia

INTRODUCTION

Infrared Vibrational CD (VCD) has been observed for a number of
polynucleotides in buffered aqueous solution in the carbonyl
stretching vibrations of the nucleotide bases (1550-1750 cm^{-1}), as
well as the phosphate backbone stretching region (1050-1150 cm^{-1}).
Both these regions are sensitive toward conformational changes of
the DNA models. The VCD features in the C=O stretching vibrations
have been reproduced computationally, using a simple exciton formalism.

EXPERIMENTAL ASPECTS

All VCD spectra shown here were recorded on one of two dispersive
VCD spectrometers described elsewhere.[1] Sample concentrations were
typically about 10 mg/mL for the polymers discussed below. H_2O was
used for the 1100 - 1300 cm^{-1} region, and D_2O in the carbonyl
stretching region. A cacodylate buffer was used in both cases.

RESULTS AND DISCUSSION

The observed VCD features of similar
DNA models, such as poly(dG-dC)·
poly(dG-dC) and poly(dC)·poly(dG) in
the B-conformation show rather different features,[2] cf. Figure 1. The
observed spectral differences are
reproduced well by VCD intensity
calculations, based on fiber diffraction geometries.[2] The VCD spectra are
relatively insensitive to temperature
changes, and even DNA models which
are known to undergo distinct melting, such as poly(dA-dT)·poly(dA-dT),
have shown only minimal variations in
the VCD spectra as a function of
temperature.

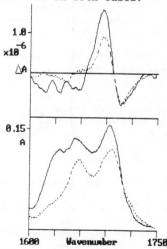

Figure 1. Observed VCD (top) and IR (bottom)
spectra of poly(dG-dC)$_2$ and poly(dG)·poly(dC)

The VCD spectrum of poly(dG-dC)·poly(dG-dC) changes at high ionic strength of the solvent, Figure 2. The VCD and absorption spectra of the Z form are shifted toward lower wavenumber, and the VCD intensities are inverted. Both these observations are reproduced computationally. The spectral changes due to the Z form occur at lower NaCl concentration if bivalent metal ions are present, but the VCD features induced in this way are different from those induced by NaCl alone.

B-form polymers containing AT sequences have been studied as well.[2] Poly(dA-dT)·poly(dA-dT) needs to be studied at much lower concentration, since aggregation occurs. Since A has no carbonyl group, and T has two, the spectra are more difficult to reproduce computationally. In addition, the observed VCD features of both poly(dA-dT)·poly(dA-dT) and poly(dA)·poly(dT) are more complicated than those of the GC polymers.

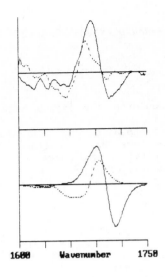

Figure 2. Observed (top) and computed (bottom) VCD spectra of B (solid) and Z (dashed) forms DNA

Small double stranded oligonucleotides, containing AT and GC regions have been studied as well. One of these is d(CGCGAATTCGCG). The VCD spectra appear as a superposition of poly(dG-dC)·poly(dG-dC) and poly(dA)·poly(dT) spectra, and are clearly due to a B-conformation. The molecule maintains this conformation even at very high ionic strength, as suggested by other techniques as well. X-ray structural parameters for the carbonyl groups, and transition dipole moments from monomeric species, were used to compute the C=O stretching VCD spectra within the exciton formalism.

The DNA models studied to date show good agreement between observed VCD spectra and computed spectral features. This indicates that the formalism used to compute VCD, the extended coupled oscillator model[3], works well for DNA. Furthermore, it appears that average structural features in solution are well represented by crystallographic data.

REFERENCES

1. M.Diem, G.M.Roberts, O.Lee, and A.Barlow, *Appl.Spectrosc.* **1988** <u>42</u>, 20
2. W.Zhong, M.Gulotta,D.J.Goss, and M.Diem, *Biochemistry* **1990** <u>29</u>, 7485
3. M.Gulotta, D.J.Goss, and M.Diem, *Biopolymers*, **1989** <u>28</u>, 2047

CONFORMATIONAL PHASE TRANSITIONS (A-B AND B-Z) OF DNA AND MODELS USING VIBRATIONAL CIRCULAR DICHROISM

Lijiang Wang, Ligang Yang and Timothy A. Keiderling

Department of Chemistry,
University of Illinois at Chicago,
Box 4348, Chicago, IL 60680 USA

INTRODUCTION

It is known that DNA can adopt several double-helical conformations whose relative stability depends upon environmental conditions such as salt concentration, extent of dehydration, relative humidity and base sequence. B-type DNA is the common form under aqueous conditions, and A-DNA is found under low humidity conditions and is conformationally like the duplex RNA structure. Both are double helical forms with a right handed screw sense. These two major right-handed forms are differentiated by sugar puckering, base tilt relative to the helix axis, distance of base pairs from the center of the helix and relative dimensions of the major and minor grooves. A left-handed helical form, called Z-DNA, is quite different in many respects from the A- and B-forms and can be induced under high salt or high alcohol content conditions.

These conformations of DNA have been studied with a multitude of spectro-scopic techniques. Recently we[1-3] and Diem's group[4] have reported vibrational circular dichroism (VCD) spectra for RNAs and DNAs in aqueous solution. The VCD spectra of the B- to Z-form DNA transition for poly(dG-dC) and related oligomers have been reported.[2,4] VCD spectra of both forms of DNA give rise to large, but oppositely signed couplet-shaped VCD spectra in the base and C=O stretching region as well as to weaker signals associated with the phosphate transitions. This distinct difference in the VCD spectra measured for the two conformational forms appears to be due to the difference in handedness of the twist of their respective polymeric backbones.

A related question of interest to us is characterization of the VCD changes that are associated with the transition from the A- to B-form DNA and evaluation of its information content. Solution phase VCD of A- and B-form DNAs from calf thymus, chicken blood and several model nucleic acids, all measured in the C=O stretching region, have been measured. Due to interference from the alcohol solvent (required to form the A-form in solution), comparable data has not been obtained in the PO_2^- stretching region. In a parallel study, we have attempted to measure the VCD of DNA films at varying humidities; but, to date, no consistent results have been obtained.

RESULTS AND DISCUSSION

A VCD study of the effect of adding TFE (trifluoroethanol) to an aqueous calf thymus DNA solution resulted in little difference in either IR or VCD band shape over the range of 0% to 50% TFE added. Some differences are apparent at TFE concentrations over 70%, which indicates that a phase transition takes place during the last stages of the process (Figure 1). UVCD spectra confirmed the formation of A-form DNA in these high percentage TFE samples. The A-form DNA VCD resembles that of B-form DNA but is sharper, as is the IR absorption, with both components of the VCD couplet gaining in intensity.

Considering that A-DNA closely resembles RNA in terms of their structural parameters, we have compared the VCD of natural DNAs to that of natural, partially "A-form-like" t-RNA. t-RNA has a smaller VCD signal than does A-form DNA, when normalized to the absorbance area, and loses the weaker positive band at ~1640 cm^{-1} seen in both natural DNA forms. A parallel comparison can be made with synthetic DNA and RNA. The VCD of poly(dG-dC) [B-form] as compared to that of poly(rI)·poly(rC) [A-like form of RNA] (Fig. 2) has the same type of relationship as that of B-form to A-form DNA (Fig. 1) except that in the RNA the main VCD band is higher in energy than that of the DNA. Similar studies were made using poly(dA-dT) and resulted in evidence for a complex equilibrium between double, triple and single helices. In 80% TFE/D$_2$O, poly(dG-dC) does not show A-form VCD but instead gives rise to a left-handed Z-form with a VCD sign pattern inverted and frequencies lowered with respect to those of the right-handed A- & B-forms.

Acknowledgement. This work was supported by the National Institutes of Health (GM30147).

REFERENCES

1. A. Annamalai and T. A. Keiderling, *J. Am. Chem. Soc.* 109, 3125 (1987).
2. T. A. Keiderling, S. C. Yasui, P. Pancoska, R. K. Dukor, and L. Yang *SPIE Reports* 1057, 7 (1989).
3. L. Yang and T. A. Keiderling, *Biopolymers*, to be submitted.
4. M. Diem, D. J. Goss, and M. Gulotta, *Biopolymers*, 28, 2047(1989).
5. W. Zhong, M. Gulotta, D. J. Goss and M. Diem *Biochmistry* (in press).

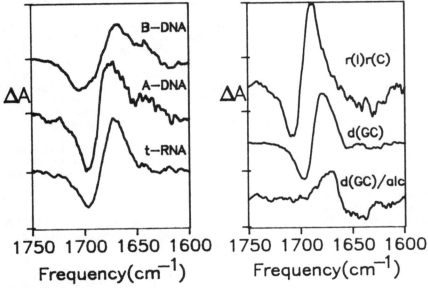

Figure 1. Comparison of B-form and A-form DNA VCD spectra from calf thymus DNA (top)B-form, (middle) A-form, and (bottom) t-RNA. VCD are normalized to a standard area for the IR absorbance. Each division corresponds to $\Delta A = 1.25 \times 10^{-4}$.

Figure 2. Comparison of the VCD for (top) a synthetic RNA, poly(rI)·poly(rC), with (middle) a synthetic DNA in the B-form and (bottom) in the Z-form (85% TFE) poly(dG-dC). Spectra are normalized to a constant absorbance area. Spectra are separated by $\Delta A = 1.5 \times 10^{-4}$.

VIBRATIONAL CIRCULAR DICHROISM IS A SENSITIVE PROBE STUDY OF HEME PROTEIN-LIGAND INTERACTIONS

SANFORD A. ASHER, RICHARD W. BORMETT, PETER J. LARKIN, WILLIAM G. GUSTAFSON, Dept. of Chem., Univ. of Pittsburgh, Pittsburgh, PA; N. RAGUNATHAN, TERESA B. FREEDMAN, LAURENCE A. NAFIE, Dept. of Chem., Syracuse Univ., Syracuse, NY; NAI-TENG YU, School of Chem., Georgia Institute of Technology, Atlanta, GA; KLAUS GERSONDE, Abteilung Phys. Chemie, Rheinische-Westfalische Technishe Hochschule, D-5100, Aachen, Fed. Rep. of Germany; ROBERT W. NOBLE, Dept. of Med. and Biochem., State Univ. of New York, Veterans Admin. Hosp., Buffalo, NY; BARRY A. SPRINGER, STEPHEN G. SLIGAR, Dept. of Biochem., Univ. of Illinois at Urbana-Champaign, Urbana, IL; SRIRAM BALASUBRAMANIAN, STEVEN G. BOXER, Dept. of Chem., Stanford Univ., Stanford, CA

The molecular basis of the control mechanism determining ligand affinity in heme proteins is often dependent on weak interactions between the ligand and the protein residues in the heme pocket. We demonstrate that vibrational circular dichroism (VCD) of the azide asymmetric stretch of the azide ligand bound to a heme protein is sensitive to the energetically small interactions between the ligand and distal globin amino acid residues E11 and E7. The VCD intensities are reported as anisotropy ratios g, where $g = \Delta A/A_T$, with $\Delta A = A_L - A_R$ and $A_T = \frac{1}{2}(A_L + A_R)$, total absorbance.

We examined the VCD of the asymmetric stretch of covalently bound azide (ca. 2020 cm^{-1}) in detail for the first time using a number of evolutionarily diverse and mutagenic heme proteins. These heme proteins include horse, sperm whale and elephant myoglobins, Chironomus thummi thummi monomeric hemoglobin III, carp and human hemoglobin and mutants of sperm whale (Gly E7) and human (Asn E11) myoglobin. The VCD of sperm whale and horse myoglobin are identical with an anisotropy ratio, g, of ca. 9.7 x 10^{-4}. The carp and human hemoglobin VCD intensities are reduced by ca. 15% compared to the horse Mb. The elephant Mb which has the distal histidine replaced by a glutamine E7 exhibits a VCD intensity that is ca. 40% less than the horse Mb. The mutant sperm whale Mb which substitutes the E7 histidine with glycine has no observable VCD and has an azide ligand binding affinity that is smaller by an order of magnitude. The same result was obtained for mutant human Mb which has the distal valine E11 replaced by asparagine.

VCD is extremely sensitive to the interations between the azide ligand the E7 and E11 residues of heme proteins. These experiments clearly indicate that these interactions cannot be described simply by the steric interactions in the heme pocket but must include hydrogen bonding or van der Waals interactions between the lone electron pairs on the azide ligand and the E7 and E11 residues.

The mutant Mb results also indicate that VCD is sensitive to the same energetic interactions that affect the kinetics of ligand binding. Figure 1 shows the FTIR and VCD spectra of various heme proteins while Figure 2 presents a model for the VCD mechanism.

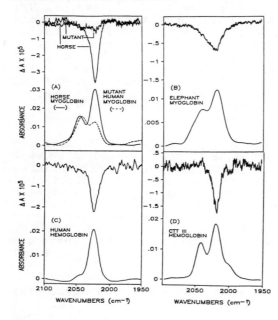

Figure 1. IR absorption (8 cm^{-1} resolution) spectra and VCD (6 cm^{-1} resolution) spectra of: (A) 11.5 mM horse MbN$_3$ and 7.1 mM mutant (Asn E11) human MbN$_3$ with 2.7 mM unbound N$_3^-$; (B) 6.0 mM Elephant MbN$_3$; (C) 9.0 mM human HbN$_3$; (D) 6.0 mM CTT III HbN$_3$. Protein samples were buffered at *ca.* pH 7 with 0.01 M phosphate buffer adjusted to an ionic strength of 0.7 with KCl. All spectra were measured in a 26μm CaF$_2$ cell and normalized to obtain a low spin band intensity comparable to horse Mb, for ease of comparison.

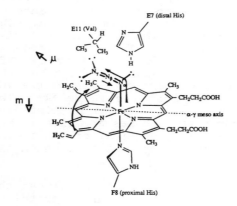

Figure 2. VCD mechanism showing the heme and bound azide ligand interactions with the Val E11 and His E7 in the distal heme pocket. Large arrows show the direction of positive electronic current flow resulting from the N$_\beta$ motion (small arrow). Approximate directions of the electric (μ) and magnetic (m) dipole transition moments are indicated.

140

TRANSIENT ELECTRIC BIREFRINGENCE MEASUREMENTS ON NATIVE AND RE-CONSTITUTED α-CRYSTALLIN

Bart van Haeringen*, Don Eden[&], Rienk van Grondelle and Michael Bloemendal[#]

Dept. of Biophysics and [#] Dept. of General and Analytical Chemistry, Free University, De Boelelaan 1081, 1081 HV Amsterdam (The Netherlands)
[&]Dept. of Chemistry and Biochemistry, San Francisco State University, San Francisco (USA)

INTRODUCTION

On the basis of size, three classes of mammalian water-soluble eyelens proteins can be distinguished. One of these, α-crystallin has been studied extensively with a variety of physio-chemical techniques. This has not resulted however in a clear understanding and a general agreement on its three-dimensional structure. This presentation tries to contribute to the, sometimes intense[1], discussion which is being held on the subject[2,3].

α-Crystallin exists as an assembly of two types of subunits, αA and αB, each of which has a molecular weight of about 20 kDa. Dissociation and subsequent reassociation of the subunits of α-crystallin isolated at room temperature or lower (α_c; 800 kDa) results in an α-crystallin aggregate of considerable smaller apparent molecular weight (α_r; 320 kDa). Isolation at 37 °C yields a particle of comparable or smaller size $(\alpha_m)^{3,4,5}$.

Thomson et al. suggest that α_c and α_r are artefactual aggregates of α_m[5]. Tardieu et al. assume a three-layer tetrahedral model for α_c, where α_r and α_m represent the inner two layers[2]. Particles with an inducible and/or permanent electric dipolemoment can be oriented in solution by a square pulse of an electric field. For optically anisotropic particles this leads to a difference in the refractive indices of the solution parallel and perpendicular to the direction of orientation. The orientation and disorientation can be monitored by the time-resolved birefringence of the solution, which is very sensitive to molecular dimensions[6]. Hence, transient electric birefringence (TEB) measurement is a powerful technique to study rotational motions of macromolecules in solution.

We performed TEB-measurements on bovine α-crystallin obtained from Fast Protein Liquid Chromatography (FPLC®, Pharmacia). The total peak, subfractions of it and reconstituted α-crystallin have been studied. The results are compared with those obtained by linear dichroism (LD) spectroscopy on compressed gels[7].

RESULTS

In all cases two distinct relaxation times were found when the field free decay of the birefringence was fitted to a continuous distribution of exponentials. From a comparence of rise and decay of the birefringence the relative contribution of permanent and induced electric dipole (P/Q) to the orientation mechanism could be determined. Normalized LD-spectra for the different fractions (not shown) are very similar in shape, but differ in intensity. Results obtained by TEB-and LD-measurements are summarized in Table 1.

Table 1. Relaxation times τ_i , values for P/Q and the reduced LD, $(\Delta A/3A)_{280}$. τ_i describe the field free decay of the birefringence and are corrected to 20 °C. The pools indicate the different fractions from the FPLC-peak; pool 1 contains protein with highest molecular mass, pool 5 consists of protein with lowest mass, α_c and α_r are complete FPLC-peaks.

sample	τ_1 (μs)	τ_2(μs)	τ_1/τ_2	P/Q	$(\Delta A/3A)_{280}$
pool 1	4.8	1.5	3.2	0.43	0.0062
pool 2	3.8	1.1	3.4	0.15	0.0035
pool 3	3.8	1.0	4.0	0.03	0.0030
pool 4	-.-	-.-	-.-	-.-	0.0028
pool 5	3.0	0.8	3.6	0.00	-.-
α_c	3.9	1.2	3.2	-.-	-.-
α_r	3.2	0.6	5.3	0.00	-.-

DISCUSSION

Two interpretations, which can not be distinguished by TEB-measurements alone, are possible. Firstly the α-crystallin population may consist of asymmetric ellipsoids with the ratio a/b over b/c (a, b and c being the semi axes of the ellipsoid) reflected by τ_1/τ_2. In that case protein from pool 1 would be the most spherical particle However the deviation of P/Q from zero for pool 1 indicates electric asymmetry. It is questionable whether this should be regarded independently from a possible spatial asymmetry. This is stressed by the LD-results. The similar shape and different intensities of the spectra suggest that the protein from pool 1 deviates more from a spherical particle than those from pools 2, 3 and 4.

The two relaxation times found can also be explained by a mixture of two ellipsoids of revolution with different rotational diffusion coefficients. This means that the two modes of the distribution correspond to two differently shaped particles. Both particles possess a size distribution, as reflected by the dependence of the relaxation times on elution volume. A bimodal distribution of particles with equal symmetry is not very probable, since the ratio of relaxation times corresponds directly to the volume ratio. FPLC should be able to seperate particles with volumes differing by a factor 3 to 4. Furthermore all pools show a considerable amount of linear dichroism, indicating that at least one of the relaxation times belongs to a non-spherical particle.

We therefore conclude that the α-crystallin population is bimodal, and that at least one of the modes consists of non-spherical particles.

REFERENCES

1. R.C. Augusteyn and A. Stevens, 'Abstracts of the 9th Int. Congress of Eye Research', Helsinki, 1990, 263.
2. A. Tardieu, D. Laporte, P. Licinio, B. Krop and M. Delaye, J. Mol. Biol., 1986, 192, 711.
3. J.F. Koretz and R.C. Augusteyn, Cur. Eye Res., 1988, 7, 25.
4. J.A. Thomson and R.C. Augusteyn, J. Biol. Chem., 1984, 259, 4339.
5. J.A. Thomson, J.A. and R.C. Augusteyn, Exp. Eye Res., 1983, 37, 367.
6. E. Fredericq and C. Houssier. 'Electric Dichroism and Electric Birefringence', Clarendon Press, Oxford, 1973.
7. M. Bloemendal, H. van Amerongen, H. Bloemendal and R. van Grondelle, Eur. J. Biochem., 1989, 184, 427.

CHARACTERISTICS OF THE CU-BOUND STATE OF α-LACTALBUMIN

H. Van Dael[*], E. Tieghem, F. Van Cauwelaert

Interdisciplinair Research Centrum

K.U.Leuven Campus Kortrijk
B-8500 Kortrijk (Belgium)

In the studies on the binding of metal ions to bovine α-lactalbumin, traditionally most attention was paid to the Ca^{2+}-binding site. Since also a distinct zinc-binding site was found, the binding of Cu^{2+} at this zinc site was proven[1]. In a previous study,[2] we were able to determine the groups involved in Cu^{2+} binding. At low concentration, Cu^{2+} preferentially coordinates with His 68 while at higher concentration the nitrogens of the N-terminal amine and the deprotonated amide groups become the principal agents for Cu^{2+} binding. The conformational state of the copper-protein complex mostly was described[3,4] as "apo-like" i.e. similar but not identical to the apo-state. By circular dichroism experiments, we will prove the existence of a typical Cu-bound state which conformation significantly differs from both the Ca- and the apo-state.

As in the near-UV region an important ligand to metal charge-transfer band overlaps with the aromatic band of the protein, a subtraction method was developed in order to determine the netto effect of Cu^{2+} ions on the tertiary structure of the protein. The results (fig.1) show that the complete change in tertiary structure is already induced by the primary binding of Cu^{2+} ions to the His group ($0 < [Cu^{2+}] < 3$) and further Cu^{2+} binding has no additional effect on the conformation.

Observation of the secondary structure by CD experiments at 220 nm demonstrates that the binding of up to three Cu^{2+} ions is accompanied by an important reduction of secondary structure (fig.2). Decalcification of native bovine α-lactalbumin on the other hand does not give noticeable changes in secondary structure. The Cu-bound state thus cannot be described as "apo-like". By GdnHCl titration, this intermediate Cu-bound state has already lost a substantial part of its helix structure so that the temperature unfolding curves do not show the typical sigmoidal helix to random coil transition.

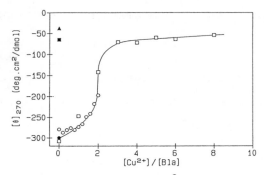

Fig.1 : The net effect of Cu^{2+} on the tertiary structure
of BLA,derived from the subtraction of the results in 7 M
GdnHCl from the results in Hepes buffer. The filled symbols
are taken from ref.5 for the native (circle),the apo
(square) and the acid (triangle) state.

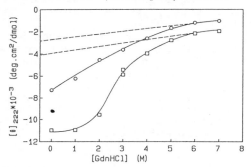

Fig.2 : Changes in the secondary structure by GdnHCl
titration of BLA without (□) and with M.R. = 6 Cu^{2+} (o). The
filled circle is the result on disulfide reduced BLA from
ref.6.

REFERENCES

1. K. Murakami and L.J. Berliner,Biochemistry, 1983, 22,
   3370
2. E. Tieghem,H. Van Dael,F. Van Cauwelaert,J.
   Inorg.Biochem., 1991, to appear
3. E.A. Permyakov, L.A. Morozova,L.P. Kalinichenko and V.Y.
   Derezhkov, Biophys.Chem., 1988, 32, 37
4. G. Musci and L.J. Berliner,Biochemistry, 1985, 24, 3852
5. T. Segawa and S. Sugai,J.Biochem., 1983, 93, 1321
6. M. Ikeguchi and S. Sugai,Int.J.Peptide Protein Res.,
   1989, 33,289

IV

PHOSPHOLIPIDS AND MEMBRANES

CARBOHYDRATES

RHODOPSINS

POLARIZED ATR-IR: THE ORIENTATION OF PHOSPHOLIPID MEMBRANES FROM [13]C=O LABELED LIPIDS

Henry H. Mantsch* and Wigand Hübner

Institute for Molecular Sciences, National Research Council of Canada, Ottawa, K1A-0R6, CANADA (HHM) and Institute of Physical Chemistry, University of Freiburg, D-7800 Freiburg, GERMANY (WH).

INTRODUCTION

Since the structure of biological membranes is a direct consequence of the nature of the constituent lipid molecules, a knowledge of the molecular conformations of these amphiphilic molecules is a prerequisite for understanding their functional properties. Much effort has been devoted to the question of phospholipid conformation, most of it dealing with phosphatidylcholines (PCs) and phosphatidylethanolamines (PEs), the major mammalian membrane phospholipids. What is known today about lipid conformation is largely based on the X-ray single crystal structures of dimyristoyl PC and dilauroyl PE [1,2]. It provides a picture of these lipids as having the glycerol moiety perpendicular to the bilayer plane with a fully extended all-trans acyl chain connected to the sn-1 glycerol carbon, while the sn-2 chain is oriented along the membrane plane until a 90° bend allows both chains to become parallel to each other. Neutron diffraction experiments of specifically deuterium labeled phospholipids showed that this molecular arrangement also exists in the hydrated state of the bilayer [3], while ^{2}H-NMR studies of PCs demonstrated a conformational non-equivalence of the first segments of the sn-1 and sn-2 acyl chains [4]. However, a unique structural picture of phospholipid bilayers had to be reconsidered when the X-ray structure of dilauroyl phosphatidic acid revealed a reversed alignment of the initial segments of the two acyl chains [5], namely a 90° bend next to the C_2 atom of the sn-1 chain, and a fully extended all-trans sn-2 chain, starting from the $\check{C}_2$-glycerol backbone. Furthermore, ^{13}C-NMR studies of DPPC liposomes ^{13}C enriched at the carbonyl position of the sn-2 chain [6], and a study of macroscopically aligned egg yolk PC [7], were interpreted as indicating that the sn-2 carbonyl bond direction is close to that of the magic angle (54.7°), while the sn-1 carbonyl bond direction is approximately perpendicular to the rotation axis of the lipid chains (or the bilayer normal). Such an orientation of the sn-2 carbonyl group differs from the values of 61 and 80° obtained for the two lipid molecules in the unit cell of crystalline DMPC dihydrate [1,2]. Thus, it is no surprise that the molecular conformation of phospholipids in bilayers is still under active investigation, and molecular spectroscopic techniques, such as ATR-IR, can make useful contributions to this issue.

147

EXPERIMENTAL

$^{13}C=O$ labeled phospholipids were synthesized according to standard procedures using fatty acids 99% ^{13}C enriched in the carboxyl carbon atom. Polarized ATR measurements were performed by use of an overhead ATR unit, a parallelogram-shaped germanium ATR crystal (50x20x2 mm, face angle 45°) and a grid polarizer on KRS5. Oriented films were prepared by spreading a stock solution of lipid in chloroform on a germanium plate and stroking the solution with a Teflon bar along the crystal until complete evaporation of the solvent. A film thickness between 2.5-4 μm was obtained, which is much greater than the penetration depth of the evanescent wave (0.2 to 0.8 μm).

RESULTS AND DISCUSSION

Polarized infrared spectroscopy

It is possible to measure three distinct parameters of an infrared absorption band, (i) its frequency, (ii) its intensity and (iii) its state of polarization, i. e. the direction of the transition moment with respect to some fixed axis of the molecule or crystal. If the electric vector **E** of the radiation makes an angle θ with the direction of the transition moment **M**, the absorption is proportional to $(ME \cos\theta)^2$, hence maximum absorption occurs when the electric vector and the transition moment are parallel and is zero when the two are perpendicular. Polarized infrared spectroscopy has been used in the past with nonlabeled lipid bilayers [8-10]. We have resorted to phospholipids with specifically ^{13}C labeled carbonyl groups which allows the assignment of separate vibrational bands to the sn-1 and sn-2 ester group, a prerequisite for the assessment of orientational differences near the glycerol backbone.

Macroscopically aligned multilayers of 2(1-^{13}C)DMPC

A first step is to determine the direction of the transition moment of the acyl chains in these oriented macromolecular assemblies. This is illustrated in Fig. 1 which shows polarized ATR-IR spectra in the region of the CH_2 wagging band progression. The seven CH_2 wagging bands of the palmitate chain (at 1198, 1220, 1242, 1265, 1287, 1310 and 1331 cm^{-1}), have the transition moment along the direction of the all-trans methylene chain and thus exhibit a very high dichroic ratio

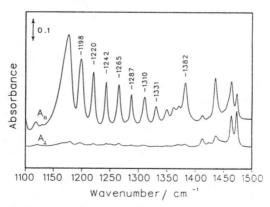

Fig. 1. ATR-IR spectra of macroscopically oriented multilayers of methyl palmitate, recorded with parallel and perpendicular polarized light relative to the plane of incidence.

148

of R>30. R represents the intensity ratio of bands recorded with parallel polarized infrared light to that recorded with perpendicular polarized light. The high z-polarization indicates that the acyl chains are oriented perpendicular to the xy-plane of the ATR crystal. The transition moment obtained for the alkyl CH_3 deformation band at 1382 cm^{-1} is also z-polarized. The xy-polarization of the CH_2 scissoring bands at 1462 and 1472 cm^{-1} and that of the αCH_2 scissoring band at 1410 cm^{-1} (R<1) confirm the perpendicular orientation of the alkyl chains relative to the crystal surface, as the transition moment of the CH_2 scissoring vibration is located in the plane spanned by the two C-H bonds of the CH_2 groups [11].

Fig. 2 (left hand panel) shows the infrared spectra of 2(1-^{13}C)DMPC in the carbonyl stretching region, recorded with parallel and perpendicular polarized light. The two ester C=O and CO-O single bond stretching vibrations can be assigned unambiguously. In the dry film, the 1740 cm^{-1} band comes from the unlabeled sn-1 chain and the band at 1695 cm^{-1} from the ^{13}C=O labeled sn-2 chain. The relative intensities of the sn-1 ^{12}C=O and sn-2 ^{13}C=O bands are very similar in the solid and hydrated films. In the dry film (Fig. 2a) the dichroic ratio of the sn-1 ^{12}C=O group is 1.4, leading to an angle of 64° relative to the bilayer plane (angle β), while the dichroic ratio of the sn-2 ^{13}C=O group is 1.33, leading to an angle β of 65.5°. In the gel state the tilt angles are 64° for the sn-1 group and 66° for the sn-2 C=O group; in the liquid crystalline state they are 62° for the sn-1 group and 66° for the sn-2 C=O ester group.

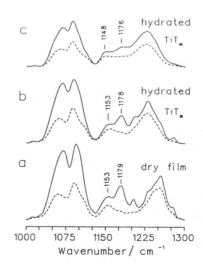

Fig. 2. Polarized ATR-IR spectra in the region of the C=O stretching bands (left hand panel) and in the region of the CO-O stretching bands (right hand panel) of macroscopically oriented 2(1-^{13}C)DMPC films in the solid state (a), in the gel state (b) and in the liquid crystalline state (c). Solid traces represent spectra recorded with parallel polarized light, broken traces spectra recorded with perpendicular polarized light.

Further information on the structural arrangement of the polar headgroup can be extracted from infrared spectra in the region 1000-1300 cm^{-1} (see Fig. 2, right hand panel). In the dry state, the antisymmetric PO_2^- stretching vibration is found at 1255 cm^{-1}, in the hydrated state it shifts to 1229 cm^{-1}. The dichroic ratio of this vibration is ~1.1 both in the dry state and in the gel state, resulting in an average tilt angle of $\beta = 72°$. In the liquid crystalline phase, R increases to 1.3 ($\beta = 66°$). Thus, the transition moment of this vibration (directed along the connecting line of the two non-esterified phosphate oxygens) lies preferentially in the bilayer plane. An important feature in Fig. 2 is the dichroic behavior of the ester CO-O single bonds. The CO-O stretching band at 1179 cm^{-1}, which originates from the sn-1 CH_2CO-O fragment, shows strong z-polarization, resulting in an average tilt angle of this transition moment of ~30° relative to the bilayer normal, the same tilt angle as that of the acyl chains. Above T_m the maximum of the sn-1 CO-O stretching band shifts from 1179 to 1175 cm^{-1}; the dichroic ratio of the band changes from 7 to ~4, and the tilt angle β increases to ~40°.

In contrast to the high z-polarization of the sn-1 CO-O single bond vibration, the ester sn-2 $^{13}CO-O$ single bond vibration of 2(1-^{13}C)DMPC (which occurs at 1153 cm^{-1}) exhibits a R value of 1.5, leading to a tilt angle relative to the bilayer normal of $\beta = 62°$. The dichroic ratio of the sn-2 CO-O single bond vibration does not change significantly above T_m (R = 1.55).

CONCLUDING REMARKS

As shown, the use of specifically ^{13}C=O labeled phospholipids allows the direct measurement of the orientation of individual sn-1 and sn-2 C=O and CO-O bonds in hydrated diacyl PCs. The results presented herein demonstrate that in PC multilayers the ester carbonyl double bonds are both aligned in the membrane plane with a tilt angle β relative to the bilayer normal greater than 60°. There are no significant differences between the orientation of the sn-1 and sn-2 C=O double bonds in the dry and the hydrated state. On the other hand, the transition moments of the two ester CO-O single bond vibrations are oriented very differently within the bilayer of PCs, the CO-O segment of the sn-1 chain being oriented along the all-trans methylene chain, whereas the same segment of the sn-2 chain is oriented in the bilayer plane. At the time of this writing we are investigating the orientation of ester groups in other lipids, in particular phosphatic acids and phosphatidylserines, and the results will be reported at the conference.

REFERENCES

1. R. H. Pearson and I. Pascher, *Nature*, 1979, **281,** 499-501.
2. P B. Hitchcock, R. Mason, K. M. Thomas and G. G. Shipley, *Proc. Nat. Acad. Sci. USA*, 1974, **71,** 3036-3040.
3. G. Büldt and J. Seelig, *Biochemistry*, 1980, **19,** 6170-6175.
4. A. Seelig and J. Seelig, *Biochim. Biophys. Acta*, 1975, **406,** 1-5.
5. H. Hauser, I. Pascher and S. Sundell, 1988, *Biochemistry* 27, 9166-9174.
6. R. J. Wittebord, C. F. Schmidt and R. G. Griffin, *Biochemistry*,1981, **20,** 4223-4228.
7. V. L. B. Braach-Maskvytis, and B. A. Cornell, *Biophysical J.,* 1988, **53,** 839-843.
8. U. P. Fringeli, 1977, *Z. Naturforsch.* **32c,** 20-45.
9. H. Akutsu, M. Ikematsu and Y. Kyogoku, *Chem. Phys. Lipids*,1981, **28,** 149-158.
10. E. Okamura, J. Umemura and T. Takenaka, *Biochim. Biophys. Acta*, 1990, **1025,** 94-98.
11. W. Hübner and H. H. Mantsch, *Biophysical J.*, 1991, **59,** 1375-1383.

QUANTITATIVE DETERMINATION OF CONFORMATIONAL DISORDER IN BIOLOGICAL MEMBRANES BY FT-IR

R. Mendelsohn

Department of Chemistry
Rutgers University, Newark College of Arts and Sciences
Newark, New Jersey, USA 07102

INTRODUCTION

The fluid mosaic model proposed two decades ago by Singer and Nicolson[1] provides a suitable framework for investigation of structure-function relationships in biological membranes. Yet the origins at the molecular level of many fundamental phospholipid-dependent membrane phenomena such as fusion, lipid transverse asymmetry, domain formation, and the lipid control of membrane protein function, remain obscure. The plethora of biophysical techniques has thus been applied to the problem of accurately defining the structure and dynamics of phospholipids and their binary mixtures with cholesterol and intrinsic membrane proteins. Yet, for reasons detailed elsewhere[2], a complete molecular description remains elusive.

The goal of the current experiments is to acquire a complete profile of conformational disorder (defined as the extent and type of one, two and three bond acyl chain configurations) as a function of depth in the bilayer. The detailed understanding of these trans-gauche isomerizations, the fastest motions available to the acyl chains, should then provide clues to the origins of slower motions (seen in NMR or EPR spectroscopies) available to the phospholipids.

Toward the above goal, we have at various stages of development, three FT-IR experiments, as follows:

CD$_2$ ROCKING MODES TO PROBE TRANS-GAUCHE ISOMERIZATION

The first measurement utilizes the CD$_2$ rocking vibrations of specifically deuterated phospholipids as quantitative markers of the fraction of gauche bonds at particular positions in DPPC and DPPC/cholesterol bilayers. The rocking mode appears at 622 cm^{-1} for $\underline{tt}$ conformations around a central CD$_2$ group and at 646 or 652 cm^{-1} for $\underline{gt}$ forms, the slight shifts in the latter arising from changed configurations further along the chain (Table I).

Table I Conformation-Dependence of the Rocking Frequencies in Isolated and Consecutive CD_2 Groups

Isolated CD_2 Group		Consecutive CD_2 Groups	
ν (cm^{-1})	Conformation[a]	ν (cm^{-1})	Conformation
622	<u>tt</u>	562	ttt
652	<u>tg</u>t	575	gtt
652	g<u>tg</u>'	589	ttt($\overline{CHDCD_2}$) + gtg'
646	<u>tg</u>g	600	gtg
646	g'<u>tg</u>g	640	ggt

[a] The underlined bond pair surrounds the central C atom of a $-CH_2CD_2CH_2-$ sequence in the acyl chain.

To apply this experiment to phospholipid bilayer membranes, we have synthesized a series of specifically deuterated derivatives of DPPC. The extent of conformational order has been determined above and below the gel→liquid crystal phase transition temperature (T_m = 41 °C) in the presence and absence of cholesterol. Results are shown in Figure 1 and 2.

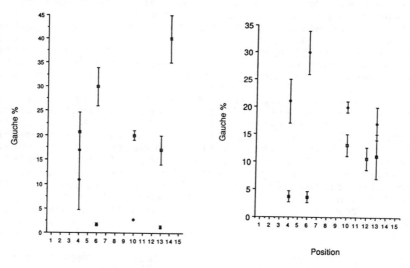

Figure 1. Disorder in DPPC (i) 50° (open squares) (ii) 25° (filled diamonds) The point at the 14-15 bond position comes from CH_2 wagging modes

Figure 2. Disorder in DPPC/Cholesterol (2:1 mol:mol) at 50° (i) DPPC (open squares) (ii) DPPC/cholesterol (filled diamonds)

Conformational order in the gel phase is very high (~4% gauche) at all positions except the 4(4') location, where 10% gauche forms occur (Figure 1). This averages to <0.5 gauche bonds/chain. The higher value at the 4(4') position may reflect the static bend in chain 2 in the crystal structure. Above T_m, conformational order is constant (about 20%, except for an anomaly at position 6) from positions 4-13, with a sudden increase around the $C_{14}-C_{15}$ bond near the center of the bilayer, the latter measured from CH_2 wagging modes. We note that the maximum disorder allowed (from the rotational isomeric state model) in isotropic liquid alkanes is about 37%. Thus, significant conformational constraints are evident due to the bilayer structure.

The strong depth-dependent effects of cholesterol on DPPC conformational disordering above T_m are remarkable (Figure 2). The current data place on more quantitative grounds the widely accepted structural model for cholesterol insertion into lipid bilayers. Conformational disorder is most highly constrained (by factors of 6-9) for those regions of the phospholipid located at depths which match the location of the rigid sterol nucleus. At greater depths in the bilayer (positions 10-16), the conformational constraints are relieved and the extent of disordering is reduced only by about factors of 1.5-2 compared with the cholesterol-free systems.

A more general result from the above data is that mammalian membranes (which contain substantial levels of cholesterol) are more conformationally ordered than is generally appreciated. The rotational isomeric state model predicts about 5 gauche bonds for a C_{16} chain at $41^{\circ}C$, whereas according to the current experiments, 3.6-4.2 gauche bonds/chain are noted for DPPC, and only about 1.5 gauche bonds/chain are present at 33 mol % cholesterol.

CH_2 WAGGING MODES TO PROBE DOUBLE GAUCHE AND END-GAUCHE CONFIGURATIONS

The second measurement uses the CH_2 wagging modes of fully proteated chains to measure overall levels of specific conformer types .Three wagging modes at 1368, 1353 and 1341 cm^{-1} arise from localized (gtg'(kink) + gtg), double gauche (gg) and end-gauche sequences respectively. Intensities (extinction coefficients) for quantitative analysis in each case are calibrated from alkane spectra. Our results are summarized in Table II. Three trends are evident. First, the number of double gauche forms/chain in bilayers is 1/3-1/2 that in alkanes. This reduction appears to be the main source of conformational ordering seen in bilayers compared with free alkanes.

Table II Conformational Disordering in L_α and H_{II} Phases

Molecule	Phase	gg/chain	eg/chain	(kink+gtg)/chain
$C_{16}H_{34}$	liq.	1.1	0.6	1.2
DPPC	L_α	0.4	0.45	1.0
DPPE	L_α	0.2	0.15	1.0
POPE	L_α	0.2	0.05	0.8
POPE	H_{II}	0.4	0.1	1.0
DEPE	L_α	0.3	0.2	1.1
DEPE	H_{II}	0.4	0.25	1.2

Second, small increases in the number of double gauche are noted during the bilayer→H_{II} interconversion in both POPE and DEPE, indicating increased conformational flexibility in that phase. Finally, the main source of conformational disordering in all systems is the formation of (kink+gtg) states (approximately 1 per chain).

COUPLED CD_2 ROCKING MODES TO DISTINGUISH KINK FROM SINGLE GAUCHE CONFIGURATIONS

The third measurement, at preliminary stages , utilizes coupled rocking modes of two adjacent CD_2 groups within each acyl chain. Under these conditions[2] , the rocking modes couple and produce normal modes sensitive to the local geometry around the $-C(H_2)-C(D_2)-C(D_2)-C(H_2)-$ skeleton as summarized in Table I. The purpose of this experiment is to determine quantitatively the number of kink vs. single gauche (gtt) conformers as a function of depth in the bilayer. The feasibility of the experiment is demonstrated for 7,7,7'7',8,8,8',8'-d_8 DPPC. Below T_m ,the most intense rocking bands in the spectrum appear at 561 cm^{-1} (ttt sequences), 589 cm^{-1} (arising from ttt modes of partially deuterated species overlapped with a small contribution from full deuterated kink sequences), and 600 cm^{-1} (suggested to arise from gtg forms). Above T_m , a band at 574 cm^{-1} arising from gtt sequences, emerges from the high frequency wing of the 561 cm^{-1} peak. The intensity near 589 cm^{-1} is enhanced substantially, and presumably arises from increasing kink components of the band. Quantitative aspects must await better S/N ratios.

REFERENCES

1. S.J. Singer and G. Nicolson, Science, 1972, 175, 720.
2. R.Mendelsohn, M.A. Davies, J.W. Brauner, H.F. Schuster and R.A. Dluhy, Biochemistry, 1989, 28, 8934.

MOLECULAR REORGANIZATIONS AND DOMAIN STRUCTURES IN BIOLOGICAL MEMBRANES

Ira W. Levin
Laboratory of Chemical Physics
National Institute of Diabetes and Digestive and Kidney Diseases
National Institutes of Health
Bethesda, MD 20892
USA

INTRODUCTION

An understanding at the molecular level of the physiological and pharmacological processes occurring at biological interfaces requires a detailed architectural view of cellular membranes. Toward this end, refinements of the fluid mosaic membrane model, in which an amphipathic bilayer acts as a matrix for expressing the activities of both peripheral and integral membrane components, has dictated for about two decades the general picture for these organized, highly selective cellular barriers. Recently, however, the suggestion and demonstration of explicit domain structures within the bilayer has provided a basis for better understanding the means by which the various lipid-lipid and lipid-protein interactions required for membrane function are modulated. Our strategy for characterizing domain properties in membranes is to apply the noninvasive techniques of Raman spectroscopy to probe the intra- and intermolecular dynamics defining the lipid matrices   of reconstituted model membrane assemblies.

In the present discussion we emphasize three different types of model bilayer systems exhibiting specific membrane domain behavior. For two classes of domain structures we induce the lateral segregation of separate lipid species into distinct bilayer regions. Thus, in the first class of structures we utilize the distinctively different properties of ether- and acyl-linked lipids with respect to chain interdigitation in aqueous phases to elicit domain formation. To illustrate a second class of bilayer domains we prepare a membrane by dispersing in aqueous media symmetric and asymmetric chain lipids. We take advantage of the different packing characteristics of asymmetric chain lipids to control the final bilayer thickness. Finally, we examine systems which exhibit novel microdomain packing arrangements of a single lipid class, namely, lipid molecules whose sn-1 and sn-2 chain positions reflect completely saturated and highly unsaturated chain moieties, respectively.

MIXED DHPC AND DPPC BILAYER DISPERSIONS

The first liposomal system to be considered consists of a binary mixture of dihexadecylphosphatidylcholine (DHPC), an ether-linked phospholipid which forms a completely interdigitated-chain bilayer in aqueous dispersions, and a conventional bilayer prepared from dipalmitoylphosphatidylcholine containing perdeuterated chains (DPPC-d_{62}). Temperature profiles derived from spectral data in the C-H and C-D stretching mode regions indicate the formation of three characteristic lipid types for the mixed system; namely, separate domains of bulk DHPC and DPPC-d_{62} and interfacial, or boundary region, lipids between the separate lipid pools. Figure 1 demonstrates domain formation with DHPC and DPPC lipids. For the mixed bilayers the lipid dynamics of DPPC are invariant to the constitution of the lipid assembly, while the more flexible and mobile ether-linked chains of DHPC allow compositionally dependent variations in the various inter- and intramolecular properties of these laterally segregated domains.[1]

INTERDIGITATED ASYMMETRIC CHAIN LIPIDS

A second domain-forming system exhibits two types of bilayer interdigitation on the inclusion of the asymmetric chain lipid C(18)C(10)PC in a binary mixture with dimyristoylphosphatidylcholine (DMPC), a lipid containing symmetric C(14) acyl chains. The lateral phase separation which occurs in the gel phase arises from the different packing characteristics of the C(18)C(10)PC molecule; one gel phase domain contains the C(18)C(10)PC species packed with 3 chains per headgroup, while the second gel phase domain exhibits a 2 chain per headgroup arrangement for this molecule. Domain formation is driven by two opposing forces. That is, one packing motif for the bilayer is based upon a matching of the DMPC bilayer thickness by the C(18)C(10)PC species, while a second packing arrangement is derived from a matrix ordering effect which results from the loss of the bilayer midplane when the C(18)C(10)PC species packs with 2 chains per headgroup. It is of interest that the vibrational data can explain in molecular terms the phase diagram determined calorimetrically for this novel binary eutectic system.[2]

UNSATURATED BILAYER LIPIDS

A final example of microdomain heterogeneity within a bilayer system concerns systems more nearly resembling mammalian membranes in that the lipid molecules are comprised of highly unsaturated sn-2 acyl chains and completely saturated sn-1 chains. This type of assembly is found, for example, in the retinal rod outer segment disk membrane. The behavior of the various Raman spectral parameters for the lipid gel phase reflects the formation of lateral clusters whose packing properties maximize the van der Waals interactions between sn-1 chains.[3] We suggest that this microdomain heterogeneity may provide a mechanism by which organisms control bilayer physical properties responsible for optimizing the membrane functions associated with integral proteins.

Multilamellar dispersions of the highly unsaturated lipids were composed of POPC (1-palmitoyl-2-oleoylphosphatidylcholine), PAPC (1-palmitoyl-2-arachidonoyl phosphatidylcholine) and PDPC (1-palmitoyl-2-docosahexaenoyl-phosphatidylcholine), lipid systems whose unsaturated chains contain one, four and six double bonds, respectively. The spectral characteristics of the saturated sn-1 chains were determined in two ways. First, liquid phase spectra of arachidonic and docosahexaenoic acid, respectively, were subtracted from the 2900 cm^{-1} C-H stretching mode regions of the PAPC and PDPC bilayers before constructing temperature profiles. The temperature curves were derived from peak height intensity ratios reflecting intrachain and interchain order/disorder characteristics. The second approach was to perdeuterate the sn-1 chains and to observe directly spectra in the 2100 cm^{-1} C-D stretching mode region, that is, spectra free from subtraction procedures. The resulting temperature profiles for the unsaturated PAPC and PDPC systems display both an abnormal scatter in the gel to liquid crystalline phase transition curves and a breadth in the phase transition regions that strongly suggest relatively noncooperative melting processes involving small gel phase clusters. Temperature profiles monitoring intrachain trans-gauche isomerization exhibit well-behaved characteristics in contrast to the chaotic behavior for parameters measuring pure interchain order/disorder effects. Introduction of cholesterol into various membrane assemblies significantly reduces the phase transition cooperativities for POPC and PDPC, but does not affect the breadth of the phase transition curve of the PAPC multilayers. Even more interestingly, cholesterol reduces the dynamic range over which the lateral chain-chain order/disorder parameter varies at any given temperature. The microdomain packing characteristics of these highly unsaturated bilayer lipids imply that changes in the acyl chain free volume within the hydrophobic region of the matrix act to modulate the conformational changes of integral membrane proteins. These bilayer microdomains are illustrated in figure 2.

ACKNOWLEDGMENT
These various studies were performed in collaboration with Drs. E. N. Lewis, B. J. Litman, J. L. Slater and M. T. Devlin.

REFERENCES
1.   M. T. Devlin and I. W. Levin, Biochemistry, 1989, 28, 8912,
2.   J. L. Slater, C. Huang and I. W. Levin, Biophys. J., 1990, 59, 502a.
3.   B. J. Litman, E. N. Lewis and I. W. Levin, Biochemistry, 1991, 30, 313.

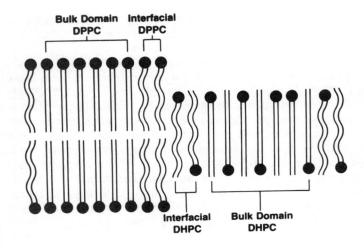

Figure 1. Packing arrangement of DPPC:DHPC
bilayers into laterally segregated species

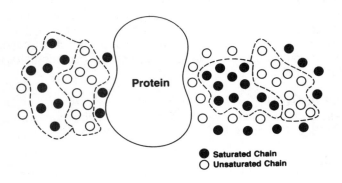

Figure 2. Highly unsaturated bilayer microclusters

FT-IR Studies of Bacteriorhodopsin: Structural and Functional Aspects

K. Fahmy[1], M. Engelhard[2], H. Sigrist[3] and F. Siebert[4]
(1) Institut für Biophysik und Strahlenbiologie der Universität, D-7800 Freiburg, F.R.G.
(2) MPI für Ernährungsphysiologie, D-4600 Dortmund, F.R.G.
(3) Institut für Biochemie der Universität, CH-3012 Bern, Switzerland
(4) MPI für Biophysik, D-6000 Frankfurt a.M. 70, F.R.G.

Much progress has recently been made in the elucidation of the structure of bacteriorhodopsin, the light-driven proton pump located in the plasma menbrane of the bacterium Halobacterium halobium. Using cryo-electron-microscopy, R. Henderson was able to propose a first structure at molecular level.[1] However, to deduce the structure, it was necessary to rely on a number of results obtained by other techniques. In addition, helix D appears to be the least certain part of the model. Therefore, additional information is desirable to test the validity of the proposed structure. In a recent paper we have shown that important structural information is obtained by the application of polarized FT-IR difference spectroscopy to bacteriorhodopsin.[2] With the so-called dichroism method, the sample, and thus the purple membrane, is tilted with respect to the infrared beam, providing information about the angle between the membrane's normal and infrared transition moments. With photoselection experiments, an anisotropy is produced by illuminating the sample with polarized visible light. Hereby, the angle between infrared transition moments and the visible transition moment can be determined.

Earlier FT-IR investigations have suggested that internal aspartic acids take part in the proton pathway.[3] The role was emphasized by the study of bacteriorhodopsin modifications produced by site-directed mutagenesis.[4,5] FT-IR difference spectroscopy has proved to be an essential tool to characterize these mutants.[6,7] However, in some cases the mutation altered the system to such a degree as to render a comparison with the wild type difficult. Therefore, we used another approach to study the role of aspartic acids in the pumping process. Incorporating isotopically labelled amino acids into bacteriorhodopsin allows bands to be assigned in the FT-IR difference spectra to a specific type of amino acid but it is not possible to discriminate between different amino acids of the same type. The ideal method would be "site-directed labelling", i.e. labelling of only a single amino acid in the polypeptide chain. This method is practically not feasible as yet,

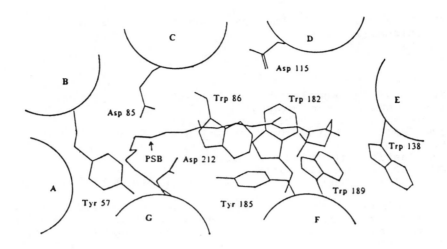

Tyr185-HOOP	Tyr185-CO	Trp86	Trp	Asp85	Asp96	Asp115
θ	90	90	41	37	?	41	55
Φ	90	<8	62	~45	>75	70	68

Fig. 1 Model of the bacteriorhodopsin structure as derived by Henderson. Shown are the amino acids in the neighbourhood of the retinal chromophore. View is from the cytoplasmatic side onto the membrane. For certain modes of selected amino acids, the table gives the direction of the infrared transition moments with respect to the membrane's normal (θ) and the retinal axis (Φ).

and the technique we employed represents a first limited approach. As has been shown by one of us (H.S.)[8], it is possible to reconstitute bacteriorhodopsin functionally from two fragments obtained by proteolytic cleavage with V8 protease. The fragments comprise helices A-E (V_1) and F-G (V_2). If one fragment contains labelled aspartic acid(s), it is possible to discriminate between asp 85, 96, 115 on the large part, and asp 212 on the small fragment.

Fig. 1 and the table show a summary of the polarized FT-IR investigations. They fit nicely into the structure proposed by Henderson. The following features should be mentioned: The plane of tyr185 which is located on helix F is oriented perpendicularly to the mebrane plane and parallel to the average retinal plane, and the direction of the long axis is approx. parallel to the membrane plane. The OH group of this tyrosine can form an hydrogen bond with asp212, facilitating thereby the protonation of the Schiff base. The normals of the planes of Trp86 and of another Trp form large angles both with respect to the membrane normal and with respect to the retinal axis. The tryptophans form part

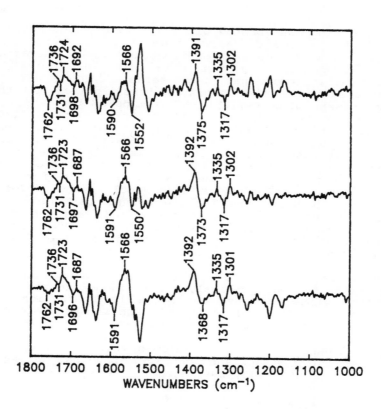

Fig. 2 Subtraction of the BR-M difference spectra of bacteriorhodopsin reconstituted from the two tryptic fragments V_1 and V_2. The subtraction constant in the upper trace is too large (positive fingerprint bands) and in the lower trace too small (negative fingerprint bands).

of the rigid retinal binding pocket. From measurements of a ^{13}C-asp labelled mutant, in which asp85 is changed into a glutamic acid we concluded tentatively that Asp85 is the putative proton acceptor for the deprotonation of the Schiff base (to be published). Its bisection points towards the Schiff base and forms an angle of approx. 90° with the retinal axis. The data obtained for the intermediates indicate that, whereas the aromatic residues remain in a rather fixed position, the carboxyl groups exhibit greater flexibility.

Fig. 2 shows the effect of labelling fragment 1 and fragment 2 on the BR-M difference spectra of bacterio-rhodopsin reconstituted from the two V8 fragments. In M, the Schiff base is deprotonated and the question arises if asp85 actually becomes protonated with the formation of M. Another uncertainty is the functional role of asp212, especially whether it undergoes protonation changes during the photocycle. To clearly demonstrate the effects of the different labellings in Fig.2, the spectrum in which V_2 was labelled has been subtracted from the spectrum in which V_1 was labelled. The subtraction constant has been selected in a way to cancel the fingerprint bands of the retinal in the region between 1300 and 1150 cm^{-1} (middle trace).It can be concluded that the spectral features between 1700 and 1800 cm^{-1}, which reflect protonated carboxyl groups, are only due to aspartic acids located on V_1. The differential band observed at 1392 cm^{-1} is caused by a deprotonated aspartic acid. From this, it can be inferred that asp85 is the proton acceptor for the Schiff base. Asp212 does apparently not undergo protonation changes. The feature around 1317 cm^{-1} can probably be assigned to this group and reflects mere environmental alterations, such as changes in hydrogen bonding accompanying the deprotonation of the Schiff base and the retinal isomerization.

Together with our FT-IR results obtained on the orientation of the retinal,[2] its isomerization and remaining twists, a model can be derived in which the protonation changes of internal asp(s) are closely linked to the retinal molecular changes.

1. R. Henderson, J.M. Baldwin, T.A. Ceska, F. Zemlin, E. Beckmann and K.H. Downing, J. Mol. Biol., 1990, **213**, 899.
2. K. Fahmy, M.F. Großjean, F. Siebert and P. Tavan, J. Mol. Struct., 1989, **214**, 257.
3. M. Engelhard, K. Gerwert, B. Hess, W. Kreutz and F. Siebert, Biochemistry, 1985, **24**, 400.
4. L.J. Stern, P.L. Ahl, T. Marti, T. Mogi, M. Dunach, S. Berkowitz, K.J. Rothschild and H.G. Khorana, Biochemistry, 1989, **28**, 10035.
5. H.J. Butt, K. Fendler, E. Bamberg, J. Tittor and D. Oesterhelt, EMBO Journal, 1989, **8**, 1657.
6. M.S. Braiman, T. Mogi, T. Marti, L.J. Stern, H.G. Khorana and K.J. Rothschild, Biochemistry, 1988, **27**, 8516.
7. K. Gerwert, B. Hess, J. Soppa and D. Oesterhelt, Proc. Natl. Acad. Sci. USA, 1989, **86**, 4943.
8. H. Sigrist, R.H. Wenger, E. Kisling and M. Wüthrich, Eur. J. Biochem., 1988, **177**, 125.

STRUCTURE DETERMINATION OF SINGLE CRYSTALS OF LIPIDS-RELATED COMPOUNDS BY THREE DIMENSIONAL POLARIZATION ANALYSIS OF IR SPECTROSCOPY

F. Kaneko and M. Kobayashi

Department of Macromolecular Science
Faculty of Science
Osaka University
Toyonaka, Osaka 560, Japan

INTRODUCTION

Lipids and related long chain compounds crystallize into various crystalline phases, depending on environmental conditions as well as thermal histories, and show many kinds of solid state phase transitions. The knowledge of the crystal structures and the mechanism of phase transitions is fundamentally important for understanding the character of these compounds. However, X-ray works on these compounds often result in a deadlock because of the difficulty in obtaining fairly good single crystals. To conquer this problem, we developed two IR techniques in transmission and ATR spectroscopy using single crystals or oriented specimens for three dimensional structural analysis. These are widely applicable for the study on polymorphism.

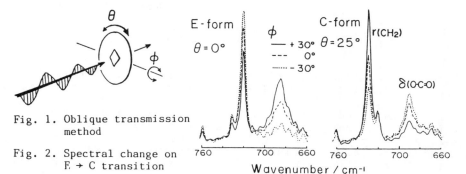

Fig. 1. Oblique transmission method

Fig. 2. Spectral change on E → C transition

OBLIQUE TRANSMISSION METHOD

Measurement of the magnitude and the direction of the hydrocarbon chain inclination and their changes on phase transitions is very important in the structural study of lipid compounds. The oblique transmission method (Fig. 1) is powerful for this purpose. Three dimensional arrangement of various functional groups can be determined unambiguously from intensity change accompanied with variation of the two sample setting angles of θ and φ.

Stearic Acid E → C Phase Transition

The E form of stearic acid transforms to the C form at about 42°C. During this transition, the external shape as well as the optical transparency of the starting single crystals were kept almost unchanged. We followed the rearrangement process of the hydrocarbon chains by the oblique transmission method, as shown in Fig. 2. At the normal incidence(φ=0), the maximum intensity position of the $r(CH_2)$ and $\delta(OCO)$ rotated by $\theta=25^{\circ}$, while the order of the intensity change with change in φ was inverted. This means that the chain inclination direction rotates by 155° around the c* axis as shown in Fig. 3. It is clarified from the result of several tens of specimens that there are two types of molecular displacement; the chain tilting direction is rotated by $150-180^{\circ}$ (case I) or $60-80^{\circ}$ (case II). In case of the double-layered polytype Orth II, the polymorphic transition from E to C is accompanied with the polytypic transition from Orth II to Mon through the two different molecular displacements occurring alternately along the layer stacking direction (Fig. 4).

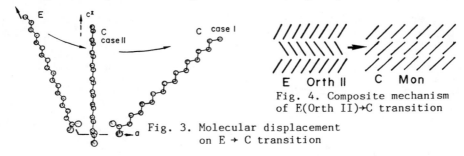

Fig. 4. Composite mechanism of E(Orth II)→C transition

Fig. 3. Molecular displacement on E → C transition

POLARIZED IR ATR TECHNIQUE USING INTERMEDIATE POLARIZATION BETWEEN P AND S

It is found that the polarized ATR technique using various polarization between p and s settings (Fig. 5) is available for cases difficult to be measured by the transmission method. Polarized IR ray generates an elliptical electric field in specimens. With the incidence around the critical angle, the long axis of the elliptical electric field inclines from the z to the y axis with changing the polarization from p to s. Therefore, the electric field of the evanescent wave can be directed to any direction by adjusting the sample orientation and the polarization. We applied this technique to the crystallization process and found that a stacking fault between Mon and Orth II polytypes often occurs in stearic acid E and B forms.

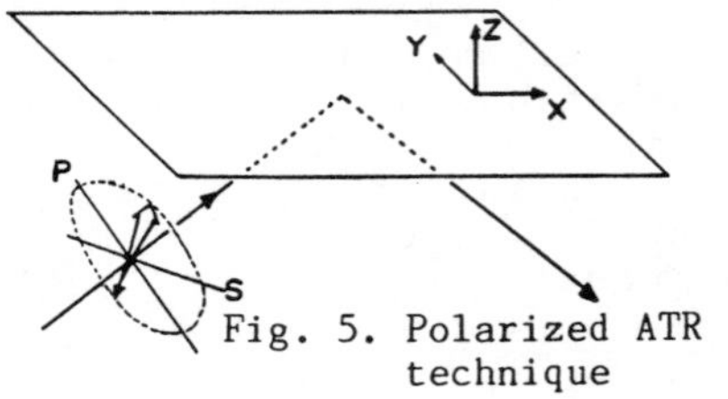

Fig. 5. Polarized ATR technique

VALINOMYCIN AND ITS INTERACTIONS WITH IONS IN ORGANIC SOLVENTS AND LIPID BILAYERS.

Michael Jackson[*] and Henry H. Mantsch

Steacie Institute for Molecular Sciences, National Research Council Canada, Ottawa, Ontario, K1A OR6, CANADA

The structure of the dodecadepsipeptide antibiotic valinomycin has been examined in a range of environments by FT-IR spectroscopy. In solvents of low polarity such as chloroform IR spectra are consistent with the bracelet structure proposed on the basis of NMR spectroscopic data [1]. A single narrow amide I absorption is seen at 1659 cm^{-1}, consistent with the presence of β-turns, together with a single band arising from non-hydrogen bonded ester $C=O$ groups (Fig 1a). Formation of a valinomycin-K^+ complex resulted in a shift to lower frequencies of the amide I and ester $C=O$ bands, indicating formation of a more compact structure with stronger amide hydrogen bonds and a decreased electron density in the ester groups (Fig. 1b). Identical results were obtained upon complexation with NH_4^+ (Fig. 1c) suggesting that valinomycin is able to efficiently bind and transport NH_4^+.

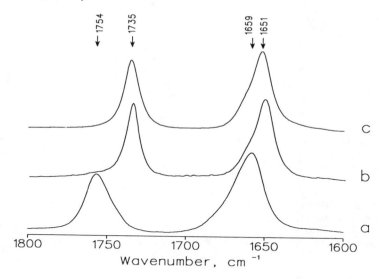

Figure 1. FT-IR spectra of valinomycin in chloroform (a) and in the presence of K^+ (b) and NH_4^+ (c).

In solvents containing only hydrogen bond acceptors ($CHCl_3$/DMSO 3:1 or pure DMSO) we find evidence of significant disruption of the internal hydrogen bonding scheme of the peptide, indicated by the appearance of an absorption due to free amide $C=O$ groups at 1666 cm^{-1}. At high concentrations of valinomycin (10 mg/ml) significant aggregation was apparent as inferred from the presence of a band at 1690 cm^{-1}, the intensity of which was concentration dependant. This band we attribute to the formation of very weak intermolecular hydrogen bonds.

Spectra of valinomycin recorded immediately upon incorporation into DHPC liposomes were found to be identical to those obtained in chloroform (Fig. 2a) indicating that valinomycin adopts a similar structure in the two environments, in agreement with NMR spectroscopic data [2]. However, prolonged incubation of the sample resulted in a decrease in the frequency of the amide and ester band maxima, which suggests significant interaction between the peptide and solvent penetrating the bilayer (Fig. 2b) . As in chloroform solution, complexation with K$^+$ resulted in spectral changes indicating strengthening of hydrogen bonds involving amide carbonyls and a decrease in electron density of the ester groups (Fig. 3a). This suggests that the structural rearrangements upon binding are identical in the two systems. Complexation is incomplete, as judged from the presence of residual absorption at 1753 and 1656 cm^{-1}. This can be related to the low rate of K$^+$ penetration into the bilayer. Complexation with NH$_4$$^+$ was again apparent in DHPC bilayers (Fig.3b).

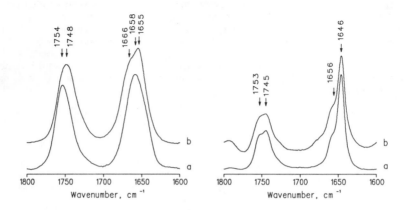

Figure 2 (left panel). FT-IR spectra of valinomycin in DHPC liposomes immediately after preparation (a) and after 24 hrs at 55°C (b). Figure 3 (right panel). Valinomycin in DHPC liposomes in the presence of K$^+$ (a) and NH$_4$$^+$ (b).

In summary, FT-IR spectroscopy provides evidence in support of the bracelet structure of valinomycin proposed on the basis of NMR spectroscopy. We also demonstrate the existence of this structure in lipid bilayers and the formation of a valinomycin-K$^+$ complex identical to that formed in organic solvents. In addition we find evidence for formation of a valinomycin-NH$_4$$^+$ complex identical in structure to the K$^+$ complex in both organic solvents and lipid bilayers, a finding not previously reported.

REFERENCES
1. Grell, E. and Funck, T., Eur. J. Biochem, 1973, 34, 415
2. Feigenson, G.W. and Meers, P.R., Nature, 1980, 283, 313

FT-IR and Calorimetric Studies of Ion Binding to Negatively Charged Phospholipids

J. Tuchtenhagen and A. Blume[*]

Fachbereich Chemie der Universität Kaiserslautern, Erwin-Schrödinger-Str. D-6750 Kaiserslautern, Federal Republic of Germany

The two phospholipids DMPA and DMPG were examined with respect to their protonation/deprotonation behaviour as well as to their ability to bind Calcium ions.

The following calorimetric titrations were carried out over a wide temperature range that included the phase transition temperatures of the different species:

$$DMPA^- + OH^- \rightleftharpoons DMPA^{2-} + H_2O \qquad (I)$$
(pH 7.0, T_m = 23 °C)   (pH 12.0, T_m = 52 °C)

$$DMPA^- + Ca^{2+} \rightleftharpoons \text{"DMPA–Ca"} \qquad (II)$$

$$DMPG^- + H^+ \rightleftharpoons DMPG \qquad (III)$$
(pH 8.5, T_m = 24 °C)   (pH 3.0, T_m = 42 °C)

$$DMPG^- + Ca^{2+} \rightleftharpoons \text{"DMPG–Ca"} \qquad (IV)$$

A ΔC_p of approx. +100 cal mol^{-1}K^{-1}, calculated from the temperature dependence of the dissociation enthalpy of DMPA$^-$, can only be explained by a drastic increase in "hydrophobic hydration" going along with the dissociation process. The protonation of DMPG$^-$ and the binding of Ca^{2+} to the negatively charged lipids all have opposite effects. The decrease in surface charge results in a dehydration of the polar/apolar interface. The simulation of the calorimetric titration curves of the DMPA-system under the assumptions of the Gouy-Chapman-theory gave an intrinsic pK value of 7.3.

FT-IR-spectroscopy in dispersion and on ATR crystals was applied to confirm the conclusions derived from the calorimetric data. At the phase transition temperature the band maximum of the antisymmetric methylene stretching vibration is shifted to higher wavenumbers (ca. 3-5 cm^{-1}) due to the increase in conformational disorder (gauche conformers). Likewise, the polymethylene chains of the higher negatively charged species (DMPA^{2-}, DMPG$^-$) absorb at higher wavenumbers than in their protonated state (DMPA$^-$, DMPG). The electrostatic repulsion among their headgroups gives rise to free volume in the hydrophobic core of the membrane which allows for higher conformational mobility and for "hydrophobic hydration". Contrarily, the methylene groups of the Calcium complexes absorb at even lower wavenumbers than the uncomplexed   lipids.

The deconvolution of the asymmetric carbonyl bands of DMPA and DMPG clearly reveals the presence of non-hydrated and hydrated carbonyl groups **(1)**. The simulation of these bands using band-shape-functions of the Gauss-Lorentz-type allow an estimate of the ratio of the number of non-hydrated to hydrated carbonyl groups (see **Fig. 1**).

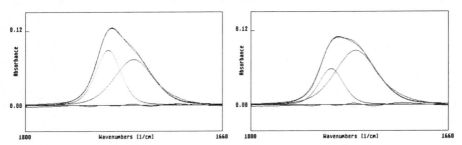

Fig. 1: Experimental and simulated spectra of DMPA⁻ in the gel phase (left) and in the liquid crystalline phase (right). Observe the change in band area for the composing bands during the phase transition. The almost straight line at the bottom is the difference between the measured and the simulated spectrum.

The following table summarizes the   results obtained for DMPA⁻, DMPA²⁻ and DMPG⁻:

Table I: Ratio of the number of non-hydrated to hydrated carbonyl groups.

	Gel phase (T [°C])		Liquid crystalline phase (T [°C])	
DMPA⁻	0.7	(50.6)	0.4	(54.7)
DMPA²⁻	0.3	(21.5)	0.4 *	(25.5)
DMPG⁻	0.4	(20.5)	0.3	(26.5)

* In this case a third band at even lower wavenumber appeared suggesting hydrogen bonding of two water molecules to one carbonyl group.

The carbonyl bands of the Calcium complexes are highly asymmetric and cannot be easily simulated using Gauss-Lorentz-functions. At least three bands appear upon deconvolution. Calcium seems to induce a very rigid hydration sphere that might account for the distortion of the carbonyl bands. The spectroscopic data strongly support our conclusions derived from the calorimetric measurements. The simulation of IR-bands proves to be very useful in the determination of the degree of "hydrophobic hydration".

Reference

Blume, A., Hübner, W., Messner, G. (1988) *Biochemistry 27*, 8239.

MOLECULAR INTERACTION BETWEEN THE LOCAL ANESTHETIC TETRACAINE AND MODEL PHOSPHOLIPIDIC MEMBRANES: A THERMOTROPIC FT-IR STUDY

J. Villaverde* and A. Morros

Unitat Biofísica, Dep. Bioquímica i Biologia Molecular
Facultat de Medicina, Universitat Autònoma de Barcelona.
08193 Bellaterra (Barcelona), Spain.

An important involvement of the phospholipidic membrane on the molecular mechanism of nerve blocking by local anesthetics has been proposed for many years[1]. We have performed a FT-IR study of the effect of the local anesthetic tetracaine (TTC) on the thermotropic behaviour and conformational properties of model phosphatidylcholine (DPPC) membranes. Hydrated membranes were equilibrated with known amounts of tetracaine, and the anesthetic binding to the lipid bilayer has been monitored by UV spectrophotometry. Different samples have been prepared at pH 5.5 and 9.5, in order to test the cationic and neutral forms of the anesthetic. The FT-IR spectra have been recorded at a wide range of temperatures, both below and above the value of the gel-to-liquid transition of the phospholipid. Two kinds of information have been obtained from these experiments:

First, modification of the thermotropic behavior of the lipid has been monitored by a variety of parameters on the lipid infrared spectrum. The main transition (gel- to-liquid) can be followed[2] by the position and width of the V (CH$_2$) band. As can be seen in Table 1, both neutral or cationic tetracaine cause similar shifts towards lower temperatures. In the case of neutral tetracaine, those shifts were paralleled by an important broadening of the transition. Gel-to-liquid conversion for the highest TTC/DPPC ratio studied, needs about 7°C to be completed, instead of 1°C as for pure DPPC . On the other hand, no loss of cooperativity was apparent for charged tetracaine. Furthermore the pre-transition, which appears in pure DPPC near 35°C, has been investigated[3] through the V (CH$_2$) bandwidth and the difference absorbance of the δ (CH$_2$) band. Even for the lowest TTC/DPPC ratios, no pre-transition was detected in the presence of either charged or neutral anesthetic.

Likewise, the FT-IR spectrum provides information about the conformational state of the lipid. Deconvolution of the ester C=O stretching band resolved two main components. In fully hydrated DPPC those two components have been assigned[4] to the non-hydrogen-bonded *sn*-1 C=O group and the hydrogen-bonded *sn*-2 C=O group. In the presence of neutral tetracaine, the relative peak height of the low frequency component (*sn*-2) showed an important decrease. This event suggests a significant perturbation of the hydrogen bonding of this group and possibly a conformation change in the hydrophylic-hydrophobic interface of the membrane.

Table 1 Gel-to-liquid transition temperatures of fully hydrated DPPC and DPPC-TTC at different molar ratios, derived from the wavenumber of the $\nu_s(CH_2)$ mode.

TTC / DPPC Molar ratio	Midpoint temperature T_m (°C)
Pure DPPC	42,0
(Neutral TTC)	
0,09	40,5
0,18	39,0
0,23	37,0
0,40	34,0
(Charged TTC)	
0,07	41,0
0,13	36,0
0,28	34,5

REFERENCES

1. P. Seeman, Prog.Anesthesiol., 1975, 1, 243-251.
2. H.L.Casal and H.H. Mantsch, Biochim.Biophys.Acta, 1984, 779, 381-401.
3. D.G. Cameron, H.L. Casal, H.H. Mantsch, Biochemistry, 1980, 19, 3665-3672.
4. P.T.T. Wong and H.H. Mantsch, Chem.Phys.Lipids, 1988, 46, 213.

INTERACTIONS BETWEEN LOCAL ANESTHETICS AND ERYTHROCYTE GHOST AND SONICATED EGG-PHOSPHATIDYLCHOLINE MEMBRANES AS STUDIED BY NOESY AND ROESY METHODS IN [1]H-NUCLEAR MAGNETIC RESONANCE SPECTROSCOPY

Y. Kuroda*, M. Wakita, Y. Fujiwara[#], and T. Nakagawa

Faculty of Pharmaceutical Sciences, Kyoto University, Sakyo-ku, Kyoto 606, Japan, and [#]Kyoto Pharmaceutical University, Yamashina-ku, Kyoto 607, Japan

Local anesthetics are known to cause their anesthetic action by affecting the function of the sodium channel in the nerve axonal membranes.[1] However, whether this action is a result of a specific anesthetic-protein interaction[2] or a non-specific perturbation of the lipid bilayer structure[3] is still unclear. Interactions between anesthetics and enzymes or membrane skeleton proteins which associate with the sodium channel proteins can also lead to anesthesia in nerves.[4] Even if which mechanism is responsible for local anesthesia, interactions with membrane lipids appear to be a most important first step in bringing about anesthesia, because it is generally accepted that the anesthetics exert their effects when present on the cytoplasmic side (i.e., inner-side) of excitable membranes, requiring outer-side to the inner-side transbilayer movement of the drug. Thus, studies on the interactions between anesthetics and the lipids have been made extensively from spectroscopic, physico- and biochemical point of view. The present work has been undertaken to inquire into the conformation of an anesthetic molecule in membranes. To investigate the conformation of the drug in a lipid matrix would lead to better understanding of the perturbation of the lipid bilayer structure by the anesthetic molecule and also lead to clear-cut concept of the anesthetic-protein interactions, if any.

Two-dimensional NOESY and ROESY methods in ^{1}H-NMR spectroscopy have been applied for obtaining the conformations of the anesthetic molecules (dibucaine and tetracaine, see Figure 1) in erythrocyte ghosts and sonicated egg phosphatidylcholine (PC) membranes. In the NOESY spectra, NOEs between protons within the drug were overwhelmed by spin diffusion[5] even at a short mixing time. This observation reduced the usefulness of the NOESY method, on the one hand; however, remarkably it facilitated to reveal signals due to the drug hidden in the broad resonances of the membranes, on the other. In the ROESY spectra, the spin diffusion phenomena were less effective; accordingly the conformations of the drugs interacting with membranes were determined by the ROESY method. The ROESY spectra were also found to be able to clearly distinguish the drugs having protons which differ in the magnitude of $T_{1\rho}$. The observed NOE data showed that dibucaine takes more than two conformations and that both dibucaine and tetracaine are present as a dimer in the membranes. These associations in dibucaine and tetracaine can exert effective perturbation on the lipid bilayer structure and/or on the function of the proteins relevant to the sodium channel. Dynamical perturbations for membranes by these anesthetics will also be discussed.

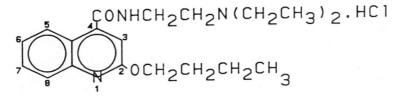

Figure 1 Structures of the hydrochlorides of dibucaine and tetracaine

REFERENCES

1. G.R. Strichartz (Editor), 'Local Anesthetics', Springer-Verlag, Berlin, 1987.
2. (a)C.D. Richards, C.A. Keightly, T.R. Hesketh, and J.C. Metcalfe, Prog. Anesthesiol., 1980, 2, 337. (b)B. Hille, Prog. Anesthesiol., 1980, 2, 1.
3. (a)A.G. Lee, Nature, 1976, 262, 545. (b)J.R. Trudell, Anesthesiology, 1977, 46, 5.
4. (a)M. Greeberg and T. Y. Tsong, J. Biol. Chem. 1982, 257, 8964. (b)M. Greenberg and T. Y. Tsong, J. Biol. Chem. 1984, 259, 13241.
5. Y. Kuroda and K. Kitamura, J. Am. Chem. Soc., 1984, 106, 1.

ERYTHROCYTE GHOST MEMBRANE HETEROGENEITY AS REVEALED BY VARIOUS FLUORESCENTLY LABELLED PHOSPHOLIPIDS: A COMPARISON STUDY BETWEEN PHASE-MODULATION AND SINGLE PHOTON COUNTING

A. J. Kungl[†], A. Hermetter[‡], F. Paltauf[‡], H. F. Kauffmann[†], and E. Prenner[‡]

[†] Institut für Physikalische Chemie, Währingerstraße 42, A-1090 Wien; and
[‡] Institut für Biochemie und Lebensmittelchemie, Petersgasse 12, A-8010 Graz; AUSTRIA

Fluorescence spectroscopy provides useful information on the physical properties (e. g. mobility, distribution, interactions) of lipids in membranes. Fluorescent phospholipids containing covalently linked labels are particularly useful in this respect, because direct spectroscopic observation of lipid molecules in membranes is possible. So far, this sort of lipid fluorophore has been applied to spectroscopic investigations of artificial membranes.

Recently we reported the first multifrequency phase fluorometry study on erythrocyte ghost membranes labelled with fluorescent phosphatidylcholine (E. Prenner et al., Proc. SPIE Vol. 1204 (1990)). Here we describe the fluorescence lifetime properties of a variety of different phopshoplipid classes in the erythrocyte ghost membrane as determined by pulse and phase fluorometry.

Erythrocyte ghosts, obtained after hypotonic treatment of whole erythrocytes, served as a model system for studies of the heterogeneous organization of biological membranes. Biomembranes contain, in general, a great variety of (phospho)lipids and proteins that are, in addition, distributed assymetrically within the bilayer.

The membrane heterogeneity should be reflected by the heterogeneity of the fluorescence lifetime properties of a suitable marker. Thus, human erythrocyte ghosts were labelled with either phosphatidylcholine, cholineplasmalogen, phosphatidylethanolamine, or shingomyelin bearing a fluorogenic DPH-propionyl residue in position sn-2 of the (glycero)phospholipid.

For the labelling procedure small unilamellar vesicles were prepared from the fluorescent phospholipids and incubated for several hours with the biomembrane at 37°C in the dark. Then the vesicles were separated by centrifugation and the remaining fluorescent ghosts were investigated.

The fluorescence decay of the labelled lipids was determined by pulse fluorometry (single photon counting) and multifrequency phase and modulation fluorometry in the time and frequency domain, respectively. The latter measurements were carried out with a commercial instrument using up to 11 different frequencies between 5 and 200 MHz. The phase angle and demodulation values were fitted to continous Lorentzian lifetime distributions using a commercial least squares fitting routine.

Single photon countin data were collected on a serial PRA 3000 transient configuration using a thyratron-gated hydrogen flash lamp. The full width at half maximum of the lamp instrumental function was typically 1.5 ns and the counts in the peak channel maximum did usually not exceed 6×10^4. Data analysis was done on a MicroVAX II.

For the evaluation of a potential fluorescence lifetime distribution of *a priori* unknown shape, a modified exponential series algorithm (Landl et al., J. Comp. Phys., in press) was used in single photon counting raw data analysis. By using terms of a series of exponentials with fixed lifetimes (usually equally, or logarithmically spaced over the time scale of fluorescence), the method allows the corresponding discrete set of amplitudes to be evaluated from the convolution integral in a linear, least-squares free-parameter optimization.

Contrary to other exponential series techniques, the algorithm used in this study is not restricted to positive amplitudes (pure decays). Also negative values (rise phase) are possible. In all cases, a rectangular distribution of initial guesses was applied in the probe function, so as not to perform preconceived bias into the computation. Generally, 70 terms were set in the analysis. When the fluorescence decays of differently labelled erythrocyte ghost membranes were investigated, the distributional width and the lifetime centers varied, depending on the phospspholipid class or subclass, indicating different organization of the respective lipids in the red blood cell membrane.

We found the phase method to be less time consuming (about 2 hours for one measurement in our case) compared to the pulse method. On the other hand, the data set is small in contrast to single photon counting. This implies that very attentive analysis and interpretation of the data is necessary to obtain reliable results, especially when distributional analysis is taken into account.

F.T. INFRARED AND N.M.R. SPECTROSCOPIC STUDIES ON MODEL BIOLOGICAL MEMBRANES

M.H. Vaughan*, R. Brown*, J. Yarwood*, R.M. Swart†.

*Department of Chemistry,   †Corporate Colloid Science Group,
Durham University,   ICI PLC,
South Road,   P.O.Box 11,
Durham,   The Heath,
DH1 3LE.   Runcorn, WA7 4QE.

INTRODUCTION

Biological membranes are highly complex systems not suitable to detailed studies of their molecular interactions. Thus our aim has been to simulate a 'model' membrane structure and study the system via ^{31}P n.m.r. and FTIR spectroscopy.

EXPERIMENTAL

Large unilamellar vesicles composed of lipids derived from egg lecithin were prepared by detergent removal through gel filtration.[1] The first 6 cm³ (the void volume) were discarded. 15 successive fractions of 1 cm³ were then collected and studied by FTIR spectroscopy. To obtain FTIR spectra which are not dominated by water absorption it is necessary to maintain the temperature, during measurement, to within ± 0.2°C. The combined fractions containing egg lecithin were also studied using ^{31}P n.m.r. spectroscopy.

RESULTS

The lipid formed small (<400 nm) vesicles (possibly unilamellar) using a 15:1 surfactant:lipid ratio. ^{31}P n.m.r data, using Mn^{2+} to detect exchange,[2] shows that integral aggregates are formed but that there may be some "leakage" over a period of a few hours. Comparison of the FTIR spectra of each fraction with cast film spectra of the detergent and lipid has shown that the lipid elutes in the first four or five fractions and the detergent (sodium cholate) in the last few. This shows that there is a clear separation of lipid and detergent.

On formation of vesicles the FTIR spectral features of the lipid show considerable narrowing compared with the spectrum of a film cast from chloroform. Typically, the full widths at half height (FWHH) are halved in the vesicle system. For example, for the CH_2 antisymmetric stretching band, the FWHH is 84 cm⁻¹ in the cast film and 29 cm⁻¹ for the vesicle system (see figure 1). This indicates a higher degree of intermolecular alignment in the vesicles and therefore a much narrower range of molecular environments.

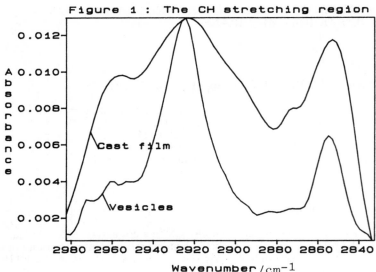

Figure 1 : The CH stretching region

However, the positions of the antisymmetric and symmetric CH_2 stretching bands in the vesicles (2924 cm^{-1} and 2851 cm^{-1} respectively) indicate that the hydrocarbon chains are disordered[3] i.e. they contain a large number of gauche conformers. This may account for vesicle penetration over a long period.

The ester band of the lipid clearly shows a doublet structure. This is caused by different conformations (Sn1 and Sn2) at the $C-\overset{\text{O}}{\overset{\|}{C}}-O$ ester group[4] and confirms the relatively disordered nature of the hydrocarbon chains in these aggregates.

The antisymmetric phosphate stretching band shows a marked shift on aggregation; from 1242 cm^{-1} in the cast film to 1230 cm^{-1} in the vesicles. This is probably due to hydration of the phosphate head group in an aqueous medium.

4   REFERENCES

1.   T.M. Allen, A.Y. Romano, H. Kercret, J.P. Segrest, Biochem. Biophys. Acta, 1980, 601, 328–342.
2.   M.J. Hope, M.B. Bally, G. Webb, P. Cullis, Biochem. Biophys. Acta, 1985, 812, 55–65.
3.   M.L. Mitchell, R.A. Dluhy, 'Recent Aspects of FT Spectroscopy', Vol. II, Mikrochimica Acta, 1987, 349.
4.   E. Mushayakarara, I.W. Levin, J. Phys. Chem., 1982, 86, 2324–2327.

INFRARED SPECTROSCOPIC STUDY OF MIXTURES OF PALMITIC ACID WITH DIPALMITOYLPHOSPHATIDYLCHOLINE USING ISOTOPIC SUBSTITUTION

José Villalaín[*] and Juan C. Gómez-Fernández
Departamento de Bioquímica y Biología Molecular, Facultad de Veterinaria, Universidad de Murcia, E-30071 MURCIA, SPAIN

INTRODUCTION

Free fatty acids have been shown to alter a variety of membrane mediated cellular functions, such as permeability, fusion, etc. Model studies employing DSC, NMR, Raman and X-ray diffraction have shown that fatty acids affect both structure and thermotropic behavior of lipid bilayers. It has been suggested that fatty acids and phospholipids form a 2:1 (mol:mol) compound stabilized by hydrogen bonds (1,2), but no direct evidence has been obtained for that.

We present in this communication our FT-IR studies with palmitic-1-13C acid (PA) and deuterated DPPC (dDPPC), showing that whereas specific interactions between different molecules of free fatty acids may occur, there are not specific interactions (such as hydrogen bonding) between free fatty acids and phospholipids.

RESULTS AND DISCUSSION

The phase transition of phospholipids can be monitored through the shift in frequency of the CH2 or CD2 stretching bands (3). This can be observed in Figure 1, where the studies of samples of different PA/dDPPC molar ratios at pH 7.4 (PA in the protonated form) , as well as pure PA and pure dDPPC, are reported. It can be observed that the transition temperatures of dDPPC and PA, separately, closely agree for each sample. This confirms the existence of good mixing between PA and dDPPC. At pH 11 (PA in the unprotonated form, characterized by the absence of the 1665 cm-1 band) the main phase transition temperatures of dDPPC and PA take place at temperatures close to that of pure dDPPC and pure PA, respectively (not shown). At this pH, a phase separation takes place, so that a phase rich in dDPPC can now be distinguished from another rich in PA. Upon PA inclusion in dDPPC, the C=O stretching band of dDPPC changes in frequency and band contour, but its two components change too little in frequency ($\approx$ 2 cm-1) to be considered an hydrogen bond effect (4). Hence, conformational variations in the interfacial part of the dDPPC molecule upon PA inclusion seems to be the cause. The inclusion of PA in dDPPC does not change the frequency of the antisymmetric and symmetric P=O stretching bands (1220 and 1088 cm-1 in pure dDPPC and at all dDPPC:PA ratios) to lower frequencies which would indicate formation of a hydrogen bond (5).

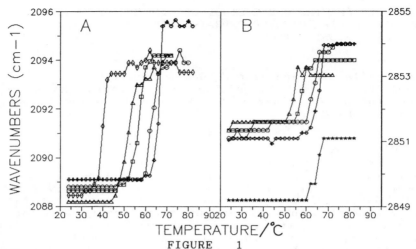

FIGURE 1

Temperature dependence of (A) CD2 and (B) CH2 symmetric stretching bands of dDPPC and PA respectively at pH 7.4 for pure PA (*), pure dDPPC (◊) and PA/dDPPC different ratios, 1/4 (Δ), 1/2 (□), 1/1 (o) and 2/1 (✦).

The other group which could be implicated in hydrogen bonding is the carboxylic group of PA itself. After deconvolution, this band has, at least, five components which are responsible for its width. They can originate from intermolecular hydrogen bonds between PA molecules or conformational variations or both. The existence of hydrogen bond bridges between different PA molecules would explain the absence of any strong effect on the dDPPC head group bands. On the contrary, deprotonation of PA would prevent the formation of hydrogen bonds between PA molecules.

Then, the characteristic properties of the 1:2 dDPPC:PA mixture do not originate from the formation of hydrogen bonding between the phospholipid and the fatty acid molecules but rather from the possible existence of Van der Waals intermolecular interactions. The tendency of free fatty acids to self associate in membranes may explain why their pKs are higher than those of other carboxylic acids. It remains to be seen whether the self-association of free fatty acids may be significant from the point of view of their biological effects.

REFERENCES
1.   Koynova, R.D., Boyanov, A.I. and Tenchov, B.G., Biochim. Biophys. Acta 903 (1987) 186-196
2.   J. M. Bogss, Biochim. Biophys. Acta 906 (1987) 353-404
3.   H.L.Casal and H.H.Mantsch, Biochim. Biophys. Acta 779 (1984) 381-401
4.   S. Fowler-Bush, H.Levin and I.W.Levin, Chem. Phys. Lipids 27 (1980) 101-111
5.   E. Mushayakarara, N. Albon and I.W.Lewin, Biochim. Biophys. Acta 686 (1982) 153-159

FTIR SPECTROSCOPY OF MICROSOMAL LIPID PEROXIDATION

T.Iwaoka and F.Tabata

Analytical and Metabolic Research Laboratories
Sankyo Co. Ltd.
1-2-58 Hiromachi, Shinagawa-ku, Tokyo Japan 140

INTRODUCTION
Spectroscopy is in principle a non-invasive technique for observing the changes of molecules and clusters in biological system. Therefore an *in situ* observation of an aqueous solution consisting of vesicles, microorganisms, or even cells would give an informative knowledge on physiological and pathological phenomena. Advances in internal reflection spectroscopy had stimulated us to explore the medium on lipid peroxidation in rat liver microsomes. The instrumentation was developed to get a reliable FTIR spectrum of the outer phase of microsomes and the monitoring of lipidperoxidation showed that the membrane protein *is* released into aqueous phase with the progress of the reaction.

EXPERIMENTAL

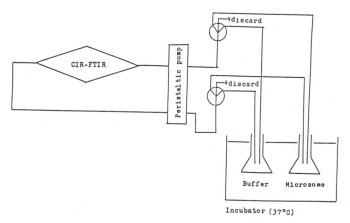

Fig. 1. ATR(CIR)-FTIR measurement system for the observation of lipid peroxidation in rat liver microsome.

Flow-through type ATR cell was made from pyrex in our laboratories so that microsomal suspension may be circulated between the cell and the incubator. Buffer solution was circulated first and then microsomal suspension was measured in the same setting of the ATR crystal in FTIR spectrometer. The subtracted spectra showed an increase of protein amide band with lipidperoxidation.

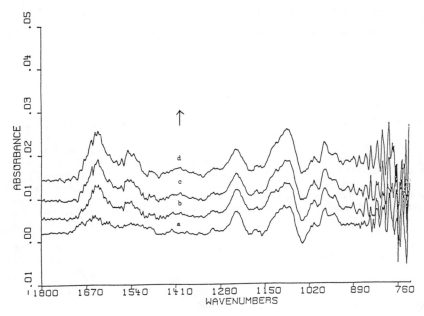

Fig.2.   FTIR spectral change of rat liver microsomal suspension after the addition of NADPH.

SUMMARY
Cylindrical internal reflection (CIR) was applied to Fourier transform infrared spectroscopy of rat liver microsomes. Spectral subtraction techniques were fully employed to eliminate the absorption of water, cofactors, and other additives to stabilize microsome. NADPH dependent lipid peroxidation was monitored by a unit of 8 min acquisition time till 125 min incubation time. Microsomal peroxidation in tris-HCl buffer caused protein release into the aqueous phase along with the progress of lipid peroxidation.

REFERENCES
1. T.Iwaoka, P.R.Griffiths, J.Kitasako and G.G.Geesey, Appl. Spectr. 1986, 40, 1062.
2. T.Iwaoka and F.Tabata, Tokyo Conf. Appl. Spectr., 1987, 41 (Japan).

CORRELATION AND MOLECULAR INTERACTIONS IN BULK AND SORPTIVE WATER

Z. Błaszczak and Z. Bochyński[*]

Laboratory of Quantum Electronics
[*]Non-Crystalline Materials Division
Institute of Physics, Adam Mickiewicz University
Grunwaldzka 6, 60-780 Poznań, POLAND

Despite intensive investigations the role of water in biological systems has not been fully understood. What we know however, is that water is not just a solvent of biological substances, it is also essential in stabilizing their native structure and in functioning of membranes as well as other organic surfaces in biological systems.

This work reports the investigations of local pure water and sorptive water in silica gel, which may be treated as a simple model of biological membranes. The employed methods of investigation were optical Kerr effect (OKE) and X-ray structural analysis (XSA). The former provides the information on molecular associates and the character of molecular interactions[1] while the latter is one of few direct methods of structural analysis of molecular and atomic liquids.[2]

A detail description of the equipment and methods of experiment as well as an analysis of the results is given elsewhere (OKE,[2] XSA[3]). The bulk water to be studied was chemically pure, twice-distilled and deionized, and the silica gels studied were of various hydration degree. For the sake of comparison we also examined quartz glass and melted quartz. The analysis of OKE results has led us to a conclusion about the presence of molecular associates in bulk water, the number of which decreases with increasing temperature. This conclusion has been confirmed by XSA investigations which also provided quantitative data. Analysis of radial functions for pure water proved that, on the average, molecules in liquid are at a distance of 2.8 Å and, at ambient temperature cannot get closer than 2.5 Å. Moreover, there are subsequent concentrations of neighbours at about 3.7 Å, 4.5 Å and 6.9 Å. The average coordination number for water molecules is about 4.5. Structural parameters obtained from XSA investigations for the conformation of sorptive water in gel only in the case of the first coordination sphere are slightly lower than for bulk water. For subsequent coordination spheres these parameters are the same as for bulk water molecules. This indicates that local orderings of bulk water, similar to the structure of ice or crystalline quartz are distorted only in the interface layer between sorptive water and silica gel. On the grounds of the obtained results the possible models of local orderings in bulk and sorptive water have been proposed and discussed in detail for the 1st, 2nd and partly 3rd coordination spheres.[4]

REFERENCES

1. Z. Błaszczak, Acta Phys. Polon. , 1987, A71, 601.
2. Z. Bochyński, Raport of the Committee of Physics Research Project 847-III, IF PAN, Warszawa, 1987, 1988.
3. Z. Błaszczak and P. Gauden, Rev. Sci. Instruments, 1987, 58, 1949 .
4. Z. Błaszczak and Z. Bochyński - in preparation.

RAMAN SPECTROSCOPIC STUDY OF THE INTERACTION BETWEEN GE-132 AND DIPALMITOYLPHOSPHATIDYLCHOLINE

Wei Wang, Laiming Li and Shiquan Xi

Changchun Institute of Applied Chemistry, Academia Sinica, Changchun 130022, P.R.China

1, 2-dipalmitoyl-sn-glycero-3-phosphocholine (DPPC) forms bilayer structures when hydrated, and has been used extensively as a model for more complex biomembranes. The interactions between the membrane and other molecules has been the subject of much discussion and many studies[1]. Raman spectroscopy provides a valuable technique for investigating changes in the structural and motional properties of lipids resulting from small intermolecular perturbation[2]. In this paper, the Raman spectra of both pure DPPC liposomes and DPPC multilayers reconstituted with β-carboxyethyl germanium sesquioxide (Ge-132) under various pH values are reported as a function of temperature from 290 to 330K. Ge-132 is an anticarcinogen. Temperature-dependent profiles constructed using both the I_{2846}/I_{2880} and I_{2930}/I_{2880} intensity ratios are shown in Fig.1 and 2 for pure DPPC and DPPC/Ge-132 dispersed in water respectively. The weight ratio is 44:1 and the pH value is 4.0 for DPPC/Ge-132 system. The profiles reflect bilayer perturbations due to the weak interaction of Ge-132. The profile for pure DPPC displays the bilayer pretransition at approx. 309K and the gel to liquid-crystalline main phase transition at approx. 315K. The width of the main phase transition ΔT is ~2K. The addition of Ge-132 suppresses the pretransition behavior and considerably broadens the main phase transition (ΔT ~10K) for reconstituted system. A qualitative measure of the relative number of gauche to trans conformers can be determined from the peak height intensity ratios, I_{1090}/I_{1126} and I_{1090}/I_{1060}. At room temperature, the height ratios I_{1090}/I_{1126} and I_{1090}/I_{1060} of pure DPPC bilayer are 0.62 and 0.49 respectively, while the ratios of reconstituted DPPC/Ge-132 multilayers are 0.80 and 0.82 respectively. This indicates that the Ge-132 increases the average number of gauche conformers in DPPC/Ge-132 system. Within investigated temperature range, the experimental result shows that a degree of rigidity of the acyl chain in the presence of Ge-132 is lower than that in pure DPPC dispersions. When the pH value of reconstituted system was adjusted to be 10.0, the Raman spectroscopic result shows that the addition of Ge-132 merely results in a few changes in bilayer structures of DPPC.

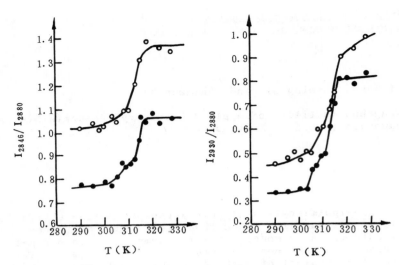

Fig. 1 I_{2846}/I_{2880} peak height intensity ratio of pure DPPC (●) and DPPC/Ge-132 (O) multilayers as a function of temperature

Fig. 2 I_{2930}/I_{2880} peak height intensity ratio of pure DPPC (●) and DPPC/Ge-132 (O) multilayers as a function of temperature

REFERENCES

1. R.N. McElhaney, Biochim. Biophys. Acta, 1986, 864, 361.
2. I.W.Levin, in Advances in Infrared and Raman Spectroscopy, Vol. 11 (R.J.H.Clark, ed.), Wiley Heyden, New York, 1984, P.1.

METAL PORPHYRINS INTERACTION WITH SPLENIC MITOCHONDRIAL MEMBRANE DESTABILIZED BY EXCESS BILIRUBIN

Ripla Beri[*] and Ramesh Chandra

Department of Chemistry
University of Delhi
Delhi - 110 007 (INDIA)

INTRODUCTION

Recent studies suggest that a number of intracellular PLA_2 exist[1] which are immunologically distinct from secretory forms[2]. Although the mechanism of action of PLA_2 has not yet been clarified, the hydrolysis of lecithin by PLA_2 is enhanced by the presence of small structural irregularities, called 'cracks' in the lipid bilayers[3]. Using rat spleen as a source, we report herein a novel form of PLA_2 and metal porphyrin effect on PLA_2 and lipid-protein interactions in mitochondrial membrane.

EXPERIMENTAL

Pregnant Wistar Strain female rats were treated with a single dose of excess bilirubin (30 μmole/Kg body weight) and the neonate rats, soon after birth were administered with a single dose of protoporphyrin-IX [PP-IX], zinc protoporphyrin [ZnPP], chromium protoporphyrin [CrPP], manganese protoporphyrin [MnPP] (40 μmole/Kg body weight) and/or metal porphyrin+bilirubin, for 2 weeks each. Control animals received equivalent volume of saline. Animals were sacrificed, spleens were perfused in situ and homogenised in 0.1M potassium phosphate buffer, pH 7.4/0.25M sucrose. Mitochondrial fractions were prepared[4] for assay of PLA_2 by using phosphatidylcholine as the standard. PLA_2 activity was calculated by measuring the phospholipid-phosphorus content[5].

RESULTS AND DISCUSSION

Prolonged excess bilirubin treatment of rats increased 25% body weight whereas PP-IX, ZnPP, CrPP and MnPP increased 7-15%. However, simultaneous administration of these metal porphyrins with bilirubin, also increased the body weight by 15-25%. Considerable increase in tissue weight is observed on simultaneous administra-

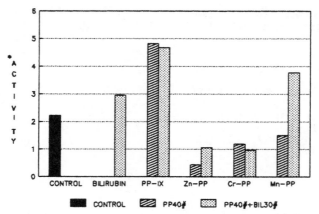

<u>Figure 1</u> Effect of metal porphyrins and bilirubin on splenic PLA$_2$ activity.
* PLA [µgm]/LPC produced/total phospholipid;# umole/Kg body weight

tion. Thus, it is confirmed that bilirubin enhances the growth rate of both the body as well as tissue weight.

The action of PLA$_2$ has been studied on mitochondrial membranes due to its effectiveness in hydrolyzing phospholipids. This enzymatic degradation of phosphatidyl choline [PC] and ethanolamine [PE] causes atleast one cytoplasmic enzyme to leak out of the cell. The results give interesting conclusions : (a) Bilirubin alone significantly increases the PLA$_2$ activity. (b) Bilirubin simultaneously with ZnPP and MnPP induces the PLA$_2$ activity, inhibited by ZnPP and MnPP when given alone. (c) Simultaneous bilirubin administration with PP-IX further elevates the PP-IX induced PLA$_2$ activity. (d) Bilirubin in conjugation with CrPP, further lowers the CrPP inhibited activity of PLA$_2$, which confirms that bilirubin has membrane repairing function only when it is labilized with CrPP. It is concluded that excess bilirubin may bring about destabilization of mitochondrial membrane, which is counteracted by CrPP and CrPP+Bilirubin administration which sterically hinders the interaction of PLA$_2$ with their substrate.

REFERENCES

1. J. Wijkander and R. Sundler, <u>FEBS Lett.</u>, 1989, <u>244</u>, 51.
2. J.G.N. deJong, H. Amesz, A.J. Aarsman, H.B.M. Lenting and H. Vanden Bosch, <u>Eur.J.Biochem.</u>, 1987, <u>164</u>, 129.
3. A.J. Slotboom, H.M. Verheij and G.H. DeHaas, <u>New Compr. Biochem.</u>, 1982, <u>4</u>, 359.
4. G.S. Drummond and A. Kappas, <u>Proc.Natl.Acad.Sci., U.S.A.</u>, 1979, <u>76</u>, 5331.
5. B.N. Ames, 'Methods Enzymology', Academic Press, New York, 1966, Vol. VIII, p.115.

INTERACTION OF INORGANIC METALS AND METAL PORPHYRINS ON HEPATIC MITOCHONDRIAL PHOSPHOLIPIDS

Ripla Beri and Ramesh Chandra*

Department of Chemistry
University of Delhi
Delhi - 110 007 (INDIA)

INTRODUCTION

Proteins and phospholipids are the major constituents of biological membranes and phospholipids predominate in membranes of non-plant origin. To study the nature of interaction between these two classes of molecules is important for understanding the structure and function of biological membranes[1]. The inner mitochondrial membrane can be regarded as a supramolecular lipoprotein complex in which the major phospholipids are phosphatidyl choline [PC], ethanolamine [PE] and cardiolipin [CL]. CL, although having the lowest abundance is highly specific to the inner mitochondrial membrane[2] and is involved in mitochondrial protein import[3] and in the control of membrane structure order[4].

EXPERIMENTAL

Female rats of Wistar Strain received subcutaneous injections (50 μmole/Kg body weight, each) of protoporphyrin IX chelates of tin [Sn^{4+}], iron [Fe^{2+}], metal salts [Co^{2+}, Ni^{2+}, Fe^{3+} and Sn^{2+}, 130 μmole/Kg body weight, each] and/or metal salts+SnPP+heme, for 2 weeks. Control animals were injected with equivalent volume of saline. Animals were sacrificed, tissues were perfused _in situ_ and homogenised for mitochondrial membrane preparation[5]. Phospholipids were extracted[6], separated on silica gel plates and phosphorus was estimated according to Ames[7].

RESULTS AND DISCUSSION

The integrity of membranes are kept intact by the phospholipids because of their amphiphatic nature. The total phospholipids in hepatic mitochondrial membrane are elevated from control by SnPP, heme, $SnCl_2$ and, also by simultaneous administration of metal salts+SnPP +heme and decreased in PP-IX, $CoCl_2$, $NiCl_2$ and $FeCl_3$

187

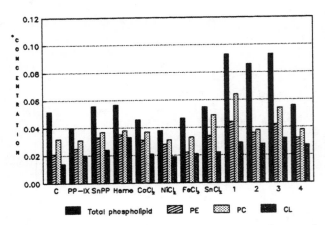

Figure 1 Effect of inorganic metals and metal porphyrins on hepatic mitochondrial phospholipids.
1 = $CoCl_2$+SnPP+heme;   2 = $NiCl_2$+SnPP+heme;   3 = $FeCl_3$+SnPP+heme;
4 = $SnCl_2$+SnPP+heme   * mg phosphorus/g tissue

(Figure 1). The metal salts and porphyrins also affect the mitochondrial hepatic PC content which is elevated by administration of metal salts and porphyrins when given alone, as well as by simultaneous administration of metal salts with SnPP+heme. But there is slight decrease in the PC content by PP-IX and $NiCl_2$ administered groups. PE and CL levels in mitochondrial membrane of liver have also shown significant increase due to administration of metal salts, porphyrins and their simultaneous administration with SnPP+heme. It is concluded that administration of metal porphyrins and metal salts affect the hepatic mitochondrial membrane and thus, induce the lipoproteinosis accompanied by increased phospholipids. These alterations noted in hepatic phospholipids may have resulted from either an increased synthesis or decreased break down of phospholipids.

REFERENCES

1. J. Dufoureq, J.F. Faucon, G. Fourche, J.L. Dasseux, M. LeMaire and T. Gulik-Krzywicki, Biochim.Biophys.Acta, 1986, 859, 33.
2. G. Daum, Biochim.Biophys.Acta, 1985, 822, 1.
3. W.J. Ou, A. Ho, M. Umeda, K. Inove and T. Omura, J.Biochem. (Tokyo), 1988, 103, 589.
4. J.S. Ellingson, T.F. Taraschi, A. Wu, R. Zimmerman and E. Rubin, Proc.Natl.Acad.Sci.,U.S.A., 1988, 85, 3353.
5. G.S. Drummond and A. Kappas, Proc.Natl.Acad.Sci., U. S. A., 1979, 76, 5331.
6. J. Folch, M. Lees and G.H. Sloane-Stanley, J.Biol.Chem., 1957, 26, 497.
7. B.N.Ames, 'Methods Enzymology', Academic Press, New York, 1966, vol. VIII, p.115.

FT–IR–Spectroscopic Studies of Ganglioside G_{M1}/Phospholipid Mixtures

E. Müller, E. Kopp and A. Blume[*]

Fachbereich Chemie der Universität Kaiserslautern, Erwin-Schrödinger-Straße
D-6750 Kaiserslautern, Federal Republic of Germany

Gangliosides, sialic acid containing glycosphingolipids, have been shown to be involved in a variety of cell surface phenomena and are particularly abundant in the central nervous system. The influence of the ganglioside G_{M1} on model membranes has been investigated using DMPC as the host lipid. Pure G_{M1} films were examined at various temperatures by FT-IR-ATR spectroscopy. The most prominent features of the spectra are the absorption bands between 1500 and 1700 cm^{-1} . This region is dominated by the amide I and amide II bands of the N-acetyl-galactosamine, the N-acetyl-neuraminic acid residues and the sphingosine base which are strongly overlapping. The additional COO$^-$ group of the sialic acid also absorbs in this part of the spectrum, complicating the band assignment.

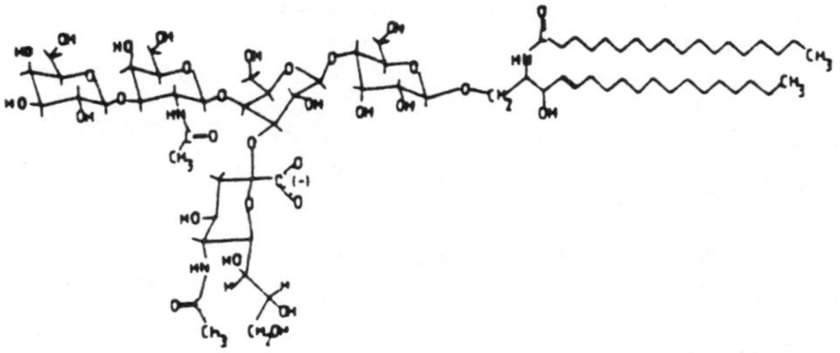

Fig. 1: Chemical structure of G_{M1}.

The deconvolved spectrum consists of at least six absorption bands (fig. 2). In order to identify these bands, the spectral region of the related compounds sphingomyelin, galactocerebroside type I and II and glucocerebrosides were compared to the G_{M1} spectrum.

The spectrum of sphingomyelin reveals two components of the amide I band at 1650 and 1630 cm^{-1} probably reflecting different types of hydrogen bonding, stronger interactions being associated with the lower frequency band.

The amide II vibrational bands for all examined lipids appear in the region between 1540 and 1550 cm^{-1}. These bands disappear if the lipid film is hydrated by D_2O because of H-D exchange.

Examination of the spectrum of the galactocerebroside type II shows the appearance of only one band at 1634 cm^{-1} while the absorption band of the amide group of the type I cerebroside is split into two components at 1633 and 1610 cm^{-1}, apparently caused by the additional OH function of the sphingosine residue.

The amide I vibrational region in the spectrum of glucocerebrosides reveals three bands at 1643, 1629 and 1611 cm^{-1} . Although glucocerebrosides build a part of the structure of G_{M1} the band at 1640 cm^{-1} is not found in the G_{M1} spectrum (see fig. 2), thus complicating the unequivocal assignment of the three amide I bands to the different amide groups in G_{M1}.

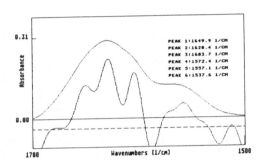

Fig. 2: The deconvolved spectrum of a G_{M1}-film (hydrated by H_2O) in the region of the amide I and amide II vibrations.

Furthermore, experiments on DMPC/G_{M1} membranes were performed to study changes in the CH_2 stretching vibrations which are sensitive to acyl chain packing. The frequencies of these absorption bands as a function of temperature reflect the phase behaviour of these mixtures. They shift to higher values at the phase transition temperature. Addition of G_{M1} to DMPC leads to increased transition temperatures with increasing ganglioside content (DMPC/G_{M1} 5:1, 5:2). This effect, indicating a stabilisation of the DMPC bilayer by G_{M1}, is accompanied by a broadening of the phase transition due to a decreased number of cooperatively melting lipid molecules.

Addition of Ca^{2+} to DMPC/G_{M1} mixtures results in a further slight shift of the transition temperature to higher values. The frequencies of the CH_2 stretching vibrations are lowered indicating tighter packing of the acyl chains. Ca^{2+} is able to bind to the sialic acid residue of G_{M1} as well as to the phosphate group of DMPC (1). The absorption region of the COO$^-$vibration (1610-1550 cm^{-1}) reveals no remarkable changes while the antisymmetric phosphate vibrational band of DMPC shows a shift to higher frequencies.

Reference

(1) McDaniel, R.V.; McLaughlin, S. (1985) *Biochim. Biophys. Acta 819*, 153-160.

CORRELATION BETWEEN FT-IR SPECTROSCOPIC AND X-RAY SMALL-ANGLE DIFFRACTION DATA AND NON-LAMELLAR STRUCTURES OF LIPOPOLYSACCHARIDES

K. Brandenburg and U. Seydel

Division of Biophysics
Forschungsinstitut Borstel
D-2061 Borstel, Germany

INTRODUCTION

Lipopolysaccharides (LPS), the endotoxins of Gram-negative bacteria, are responsible for the induction of various biological effects in mammals.[1] Beside the primary chemical structure of the single molecules the three-dimensional structures of LPS aggregates in aqueous solution and the phase states of their acyl chains (gel ß and liquid crystalline α) should be important parameters for their biological effectiveness. We have investigated the ß$\leftrightarrow\alpha$ phase transition of deep rough mutant LPS Re from *Salmonella minnesota* applying Fourier-transform infrared spectroscopy (FT-IR) and its supramolecular structures with X-ray small angle diffraction utilizing synchrotron radiation.

RESULT and DISCUSSION

In fig. 1, X-ray diffraction patterns are shown for a LPS preparation at a water concentration of 85% and at temperatures from 20 to 80 °C. In the lower temperature range 20 to 40 °C, in which the ß$\leftrightarrow\alpha$ transition takes place[2] (T_c = 32 °C), the diffraction patterns indicate the existence of non-lamellar cubic structures (main periodicity at d_Q= 8.2 nm and reflections at 4.6 nm = $1/\sqrt{3}\ d_Q$ and 3.1 nm = $1/\sqrt{7}\ d_Q$). Between 50 to 60 °C, the basic patterns allow no straightforward assignment to a definite supramolecular structure. At 70 to 80 °C, however, the LPS assembly adopts a hexagonal II (H_{II}) structure (reflections at d_H= 5.8 nm, at $1/\sqrt{3}\ d_H$ and $1/\sqrt{4}\ d_H$).

In fig. 2, typical FT-IR measurements for LPS Re in the range 960 to 925 cm^{-1}, which was shown to represent vibration(s) of the phosphate groups[3], are presented in dependence on temperature. The spectra of fig. 2 are resolution-enhanced by deconvolution of the original data and show - in the lower temperature range -distinct absorption bands which drastically change when approaching the transition into the H_{II}-phase and eventually disappear within this inverted state. Thus, the transition into the H_{II} phase is connected with a drastic reorientation of the phosphate groups of LPS.

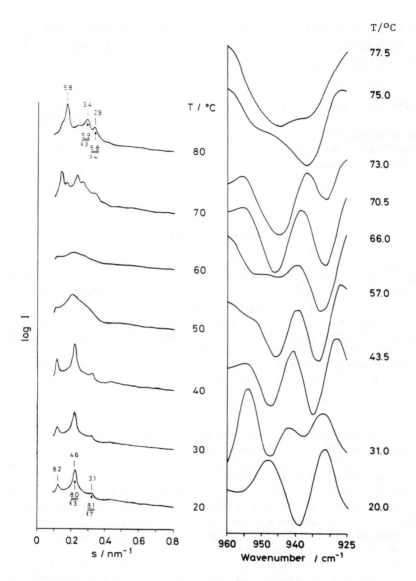

Fig.1: X-ray diffraction patterns of LPS Re in 85% water at different temperatures

Fig.2: FT-IR spectra of LPS Re in 85 % water at different temperatures

(1) E.Th. Rietschel, H.-W. Wollenweber, H. Brade, U. Zähringer, B. Lindner, U. Seydel, H. Bradaczek, G. Barnickel, H. H. Labischinski, and P. Giesbrecht in 'Handbook of Endotoxins', E.Th. Rietschel, Ed., Elsevier Amsterdam, 1984, Vol. 1, p. 187
(2) K. Brandenburg and U. Seydel, Eur. J. Biochem. 1990, 191,229
(3) K. Brandenburg and U. Seydel, Eur. Biophys. J. 1988, 16, 83

MECHANISM OF MICROGEL FORMATION OF CELLOOLIGOSACCHARIDES BY SMALL-ANGLE X-RAY SCATTERING METHOD

Yoh SANO and Takashi SASAKI

National Food Research Institute,
Tsukuba City, Ibaraki 305,
Japan

INTRODUCTION

Oligosaccharides containing sequences of D-mannose or D-glucose linked by different glycosidic linkages are widespread in nature. Oligosaccharide units occur as an inner core in both soluble and cell surface glycoproteins of yeast, fungi, higher plants and animals, and there are ample indications that they perform important biological roles, such as the stabilization of protein conformation. The structural characterization of cellooligosaccharide and maltooligosaccharide in solution are, therefore, an important base for elucidation of the spatial arrangement of oligo- and polysaccharide chains.

In the present work, the size and shape of oligosaccharides of both types in solution was determined as a function of the degree of polymerization (DP) by using the small-angle X-ray solution scattering method and the CD method.

Materials. A series of cellooligosaccharides from dimer to pentamer are prepared by hydrolysis of cellulose. The oligomers are composed of β-1, 4-linked D-glucose. All oligomers of cello- and malto- types are chromatographically pure, confirmed by high performance liquid chromatography, and are lyophylized power.

Small-Angle X-ray Scattering (SAXS). SAXS experiments were performed with the optics and detector system of SAXS, in the Photon Factory of the National Laboratory for High Energy Physics,Tsukuba,Japan. A wavelength of 0.149nm was used. The small-angle scattering was registered at 512 different angles by using the one-dimensional position sensitive proportional counter with an effective length of 200mm.

The net scattering intensities were calculated by subtracting the scattering intensities of a blank solution from those of the sample solution. Since the scattered intensity $I(Q)$ at smaller scattering angles is generally approximated in terms of an exponential function of the mean-square radius of gyration of a

193

solute, the radius of gyration Rg can be evaluated from the initial slope of the straight line by plotting the natural logarithm of I(Q) against Q^2.

As shown in Fig.1, the Guinier plots were approximated by almost straight line within smaller DP regions in cellooligosaccharide and in wide DP regions of maltooligosaccharide. In these regions, both parameters of I(0) and Rg increased almost linearly with the increase of DP. However, for higher DP regions of cellooligosaccharide, the Guinier plot curved upward in the smaller Q-ranges, as is shown with the cellopentaose in Fig.1 as an example, which indicates the presence of the microgel. The abnormal structural change by the intensive presence of microgel was also observed in cellohexaose.

CD measurements.   CD was measured by a Jasco Model J-600 spectropolarimeter.   For maltooligosaccharides, negative peak at about 190nm was observed for maltooligosaccharide of all DP regions and did not shift. However, one positive peak at about 190nm was observed for cellooligosaccharide of the smaller DP regions and this shifted to higher wavelength for the higher DP region because of the presence of microgel.

From these data of SAXS and CD, the mechanism of microgel formation of cellooligosaccharide of the higher DP regions is discussed.

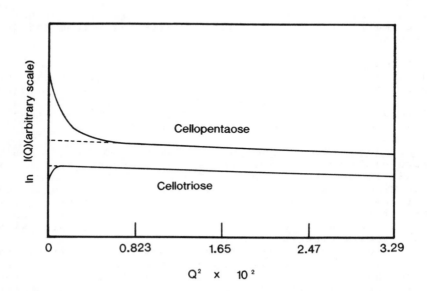

A MOLECULAR FORCE FIELD OF THE GLYCOSIDIC LINKAGE THROUGH NORMAL MODE ANALYSIS OF DISACCHARIDES.

Manuel DAUCHEZ and Gérard VERGOTEN

Laboratoire de Génie Biologique et Médical, INSERM U279,Faculté dePharmacie, Université du Droit et de la Santé de LILLE. Groupement Scientifique IBM-CNRS "Modélisation Moléculaire".

Introduction

In the field of vibrational studies of carbohydrates, the force field previously determined for both anomer molecules of glucose [1] has been applied to disaccharides. The studied molecules exhibit different positions of the glycosidic linkage between both glucose residues as trehalose dihydrate (1-1), sophorose monohydrate (1-2), laminarabiose (1-3), maltose monohydrate and cellobiose (1-4), gentiobiose (1-6), and different configurations of the glycosidic linkage as alpha for maltose (1-4) and beta for cellobiose (1-4). From a careful analysis of infrared and Raman vibrational spectra and harmonic dynamics computation in the crystalline state, the results have shown the reliability and the transferability of the previous set of parameters. Specific parameters have been obtained for each kind of glycosidic linkage.

Materials and Methods

The vibrational spectra, i.e. infrared and Raman, were recorded in the crystalline state in the 10-4000cm^{-1} range. The force field used is a modified Urey-Bradley-Shimanouchi intramolecular force field [2,3], which is combined with an intermolecular potential energy function, consisting of Van der Waals interactions, an explicit hydrogen-bond function and an electrostatic term. The partial charges were computed using the AM1 quantum mechanical procedure [4]. The initial parameters are those obtained from both anomers of glucose, used in selected combinations for each glucose residue in the disaccharide unit.

Results

The standard deviations are: trehalose dihydrate 2.15 cm^{-1} (80,17), sophorose monohydrate 3.00 cm^{-1} (71,18), laminarabiose 2.75 cm^{-1} (69,15), maltose monohydrate 3.20 cm^{-1} (61,9), cellobiose 2.45 cm^{-1} (53,16) and gentiobiose 2.75 cm^{-1} (72,11) (The number of observed frequencies and the number of modified parameters over the 190 initial parameters are given respectively in parenthesis). All the bands are refound over the spectra and particularly in the fingerprint region (800-1000 cm^{-1}) where the bands which characterized the molecules are correctly reproduced. Moreover, in the low frequency range, the usefulness of the intermolecular interactions, the electrostatic and explicit hydrogen-bond contributions, are clearly demonstrated in order to achieve a good agreement between observed and calculated frequencies.

Assuming a three-fold sinusoidal function, the rotational barriers are computed from the values of the torsional force constants. The endocyclic barriers computed for the glucose residues, constituting the disaccharide, are the same as those obtained for the corresponding anomer molecule. Minor modifications appear for the CO exocyclic barriers, involving different hydrogen-bond networks. However, it is important to point out that, for the glycosidic linkages, the barrier height is the same order for both sides of the glycosidic linkages: $V\ C_1O_1 = V\ O_1C_x' = 3.29$ kcal/mol for an alpha residue and 2.64 kcal/mol for a beta residue (x=1,2,3,4 or 6).

Conclusion

The reliability and the transferability of the set of parameters obtained for both anomers of glucose are clearly demonstrated in this work. Moreover, specific parameters for the glycosidic linkages, exhibiting different configurations and positions, are obtained and completed the set of parameters of carbohydrates. These parameters will be used in molecular dynamic simulations on glycans, oligo- and poly-saccharides.

[1] Dauchez M., Derreumaux P., Lagant P, Vergoten G. in Spectroscopy of Biological Molecules - State of the Art, Bertoluzza A., Gagnano C., Monti P. Ed.,Bologna, Italy, 259, (1989).
[2] Shimanouchi T., Pure Appl. Chem. 7(1), (1963), 131.
[3] Derreumaux P., Vergoten G., Lagant P., J. Comput. Chem., 11/5, (1990), 560.
[4] Dewar M., Zoebisch E., Healy E., Stewart J., J. Am. Chem. Soc., 107, (1985),3902.

SPECTROSCOPIC AND CONFORMATIONAL STUDIES
OF CARRAGEENANS.

M. SEKKAL[1] , **B. SOMBRET**[1] , **J. MOLLION**[3] , **G. VERGOTEN**[2] , **P. LEGRAND**[1]

1 : Laboratoire de Spectrochimie Infrarouge et Raman
(CNRS UPR A 2631 L), UST Lille Flandres Artois, Bât. C5,
59655 Villeneuve d'Ascq cedex, France.
2 : Laboratoire de Genie Biologique et Medical, Faculté de Pharmacie,
INSERM U 279, Université de droit et santé de Lille, France.
3 : Université de Tulear , Madagascar.

Carrageenans are sulphated D-galactans consisting of alternate α (1,3) and β (1,4) linked residues. They are extracted from various red algae and are widely employed in the food industry because of their ability to form thermoreversible gels.

FTIR and FT Raman spectra of some members of this family of compounds have been recorded between 4000 and 100 cm⁻¹ in the solid state. Both of the two FT spectroscopies offer a significant sensitivity advantage over dispersive instruments . Indeed more bands are observed using these techniques than by the IR dispersive systems which were used tradionally[1] in order to characterise the sulphate position on the pyranose ring as well as the kind of carrageenan. Raman scattering has only been used for the past three years[2,3] , but unlike to the IR spectroscopy this technique, as can be seen on the FT Raman spectra of kappa, iota and lambda carrageenan (Fig.b) , does not permit a different iation of all of these polysaccharides. This is certainly caused by the presence of a relatively intense band near 848 cm⁻¹. In the 800 - 850 cm⁻¹ range there is a band assigned to the pseudo-symetrical C-O-S vibration , in the kappa carrageenan it appears at 845 cm⁻¹. This band is related to a sulphatation on the C4, and is also present in the iota spectrum, but in this latter case there is another band at 805 cm⁻¹, which is assigned to the C-O-S vibration for a sulphate group on the C2 . We have observed a new band near 902 cm⁻¹, which also can be linked to the sulphatation on the C2[5] because it is also present in the spectrum of iota deviant carrageenan[4]. In the spectrum of lambda carrageenan, the width of the observed band with a maximum at 828 cm⁻¹ is caused by the three sulphatations .

Some other assignements can be made by comparing these spectra with those of oligo-saccharides. The intense band in the IR spectra of the sulphated polysaccharides between 1250 and 1240 cm⁻¹ is assigned to the O=S=O asymetric vibration. It is absent in the case of the alkali-modified porphyran (agar completely desulphated) and it becomes broader when the number of sulphate substitution groups increases from the kappa to the lambda carrgeenan. The same observation is made for the band at 1030 cm⁻¹ which is assigned[3] to the O=S=O symetrical stretching . Belton et al.[6] have related the band at 1155 cm⁻¹ to an asymetric stretching of the glycosidic linkage, but as it is also present in the galactose spectrum, we have assigned it to a ring deformation[4]. On the other hand, we related the weak band at 1125 cm⁻¹ to the asymetric stretching of the glycosidic linkage because we have observed its increasing intensity with the increase of the degree of polymerisation in some agars and kappa carrageenans. The band at 1006 cm^{-1} is assigned to an$\propto$ 1,3 glycosidic deformation. Indeed we observed it in some 1,3 disaccharides.

Conformational studies have been done using molecular mechanics techniques and an empirical force field. The initial geometry was taken from data of Arnott et al.[7] on the iota carrageenan double helix. The energy of each carrageenan (kappa, omega and mu) was minimised and the stable conformation of each of them was obtained and visualised. Except for the case of the mu carrageenan, the double helix formation is possible but the sulphate orientation with respect to the helix axis is different for each kind of the polysaccharide. On the basis of the suggestions of Arnott et al. on the gelification mechanism, where the cations have an important role in stabilising the double helix aggregats, the orientation of the sulphate groups could be the origin of the differences observed for the gelification of the different carrageenans.

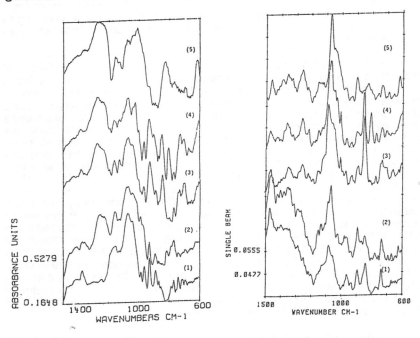

Fig.a: FTIR spectra   Fig.b: FT Raman spectra

1: porphyran alkali-modified, 2: porphyran, 3: kappa-carrageenan,
4: iota-carrageenan, 5: lambda-carrageenan.

REFERENCES:
1- D.J. Stancioff and N.F. Stanley, proc.Intl.Seaweed symp., 1969, 6, 595.
2- T. Malfait, H. Van Dael and F. Van Cauwelaert, Carbohydr. Res., 1987,163, 9.
3- T. Malfait, H. Van Dael and F. Van Cauwelaert, Int. J. Biol. Macromol., 1989, 11, 259.
4- M. Sekkal, THESE, Lille (1990).
5- J. Mollion, Andriantsiferan, M. Sekkal, Hydrobilogia, 1990, 204-205, 655.
6- P.S. Belton, R.H Wilson and D.H. Chenery, Int. Biol. Macromol, 1986, 8, 247.
7- S. Arnott, W.E. Scott, J. Chem. Soc. Perkin Trans II, 1972, 324.

NORMAL COORDINATE ANALYSIS OF MONO AND DISACCHARIDES IN THE CRYSTALLINE STATE

M. SEKKAL[1] , M. DAUCHEZ[2] , G. VERGOTEN[2] , P. LEGRAND[1]

1: Laboratoire de Spectrochimie Infrarouge et Raman
(CNRS UPR A 2631 L), UST Lille Flandres Artois, Bât. C5,
59655 Villeneuve d'Ascq cedex, France.
2: Laboratoire de Genie Biologique et Medical, Faculté de Pharmacie,
INSERM U 279, Université de droit et santé de Lille, France.

Vibrational spectra and force field calculations for alpha and beta D-galactose introduce an interesting approach, not only for our spectroscopic studies on carrageenans, but also on other galactans and polygalactans such as agars.

Previous studies concern normal modes analysis for these mono-saccharides in the case of isolated molecules[1]. In this work, an harmonic dynamics computation in the crystalline state is made, on the basis of a complete analysis of infrared and Raman spectra of both the two anomers of D-galactose and of alpha D-galactose O-D5. The method we used is a combination of a modified intramolecular Urey-Bradley-Shimanouchi potential function[2] and intermolecular potential function, which includes Van der Waals interactions by a Buckingham function[3] , electrostatic terms from a charge distribution computed by the AM_1[4] quantum-mechanical procedure, and an explicit hydrogen bond function.

The initial force field is taken from the D-glucose calculations[5] and the molecular geometry used, derives from the X-ray data of J. Ohanessian et al.[6,7] . We obtained a good agreement between calculated and observed frequencies for the case of the alpha-D-galactose (in average 3 cm^{-1} in the range below 1500 cm^{-1}) , by a fitting of force constants using the Jacobian matrix. To verify the validity of the force field obtained, we transferred it to the case of beta-D-galactose, with varying only the constants in relation with groups whose geometries are different from those of alpha-D-galactose i.e. the C_1 (anomeric carbon) and C_5 and C_6 (carbons which are concerned in the TG or GT configuration). After a fitting of the latter's constants a difference between observed and calculated vibrations is about 4 cm^{-1} in average for the beta-D-galactose. This former force field reproduces all the spectrum of the alpha-D-galactose O-D5. It is important to point out that the usefulness of the intermolecular potential function permits to obtain a perfect agreement between computed and observed vibrations below 200 cm^{-1}.

The potential energy distribution of every band is determined. The vibrational modes are very coupled to each other below 1500 cm^{-1}, some vibrations can be discussed. The C-O stretching modes are very coupled with the C-O-H bending deformations and are observed at 1100-1000 cm^{-1} , this range is very characteristic for sugars, it is constituted by intense bands in both Infrared and Raman spectra. The scissor C_6H_2 mode which appears at 1492 cm^{-1} for the alpha-D-galactose (and at 1465 cm^{-1}

for the beta-D-galactose) is represented below. We can see a motion including only the C_6 group. In other hand, the modes as those represented at 1140 cm^{-1} for the beta-D-galactose and at 762 cm^{-1} for alpha-D-galactose include almost all atoms motions. The two latters vibrations are mostly assigned to complex ring deformation. The band at 762 cm^{-1} includes ring torsional modes.

1492 cm^{-1} alpha-D-galactose   762 cm^{-1} alpha-D-galactose   1140cm^{-1} beta-D-galactose

These modes are represented after the use of the L_x matrix wich is obtained in the end of the calculations of normal modes of each molecule, this matrix contains the vibrational amplitudes of each atom at a precise frequency.

To determine the vibrational frequencies of the alpha-1,4 glycosidic linkage, the force field obtained is transferred to the case of a disacchari-de as 4-O-alpha galactopyranosyl-D-galactopyranose and fitted using the FT- Raman and FTIR spectra of this compound. And in order to establish a set of parameters which will be used in the whole potential energy function of glycans, we apprehend in the same maner as for the D-galactose, the force field calculations of alpha-L-fucose, methyl alpha-D-galactose and methyl beta-D-galactose.

REFERENCES

1- H.A. Wells Jr., and R.H. Atalla, J. Mol. Struct., 1990, 224, 385.
2- I. Shimanouchi , Pure Appl. Chem., 1963, 7(1), 131.
3- D.E. Williams, J. Chem. Phys., 1966, 45, 3770.
4- M.J.S. Dewar, E.G. Zoebisch, E.F. Healy and J.J.P. Stewart, J.A.C.S., 1985, 107; 3902.
5- M. Dauchez, THESE, Lille, 1990.
6- J. Ohanessian et H. Gillier-Pandraud, Acta. Cryst., 1976, B32, 2810.
7- F. Longchambon, J. Ohanessian, D. Avenel, A. Neuman, Acta. Cryst.1975, B31, 2623.

A SPECTROSCOPIC AND CONFORMATIONAL INVESTIGATION ON BRANCHED-CHAIN ANALOGUES OF LINEAR POLYSACCHARIDES AND ON THEIR ASSOCIATION COMPLEXES WITH Cu(II) AND Fe(III) IONS

G. Paradossi, A. Palleschi, E. Chiessi, M. Venanzi and B. Pispisa*

Dipartimento di Scienze e Tecnologie Chimiche, Universita' di Roma "Tor Vergata", 00173 Roma, and Dipartimento di Chimica, Universita' di Roma "La Sapienza", 00185 Roma (Italy).

Branched-chain analogues of linear polysaccharides have recently attracted considerable attention because both their improved solubility and increased functionality would enlarge the potentiality for specialized uses. Deoxylactit-1-yl (CHlact), 2-substituted pentanedioic acid (CHpent), and 2-substituted propanoic acid (CHprop) derivatives of chitosan (CH) were then prepared and studied in aqueous solution, over a wide range of pH, by potentiometric and spectroscopic measurements. Conformational energy calculations on CHlact and CHpent were also performed. They suggest that the former polymer is characterized by extended, conformationally mobile side-chains, whatever the protonation state of secondary amine is. Instead, both the charge state of ionizable groups and chirality of the carbon atom in the side-chain strongly affect the structural features of CHpent. When the absolute configuration of this atom is R, a six-membered chair is accomplished by an intramolecular hydrogen bond between the "distal" carboxylate and secondary amine, whatever the pH is. This makes the two proton positions in the nitrogen atom of the uncharged amine not equivalent, the S absolute configuration being that stabilized by some 3.3 kcal/mol. In contrast, when the absolute configuration of the chiral carbon atom is S, the mobility of the side-chains becomes strongly pH-dependent. Only at low pH the side-chains acquire conformational rigidity because both carboxylic groups participate at hydrogen bond interactions with the backbone. Although the network of hydrogen bonds is probably not representative of aqueous solution and may result from the lack of alternate hydrogen-bond partners, ^{1}H-NMR spectra confirm the foregoing conclusions, in that at pH $\approx$ 11 evidence of two structures with different conformational mobility is obtained, while at pH $\approx$ 2 the presence of two such structures is nomore detectable.

The field of polysaccharides-transition metal complexes is largely unexplored, and most of the work has been so far devoted to binding properties rather than to the structure of metal sites, a problem of wide interest for natural (biological) systems. We have studied the structural features of the association complexes between Cu^{2+} or Fe^{3+} ions and the aforementioned polymers as well as the parent chitosan by EPR, Mössbauer, optical and chiroptical spectra. No precipitate or opalescence, due to the hydrolysis of free metal ions, was observed, even at pH as high as 10, but Me^{n+}-CH mixtures are insoluble at pH > 6.5. According to EPR results (100 K), Cu(II) association complexes basically have a tetragonal symmetry. Visible CD spectra suggest, however, that the order of increasing departure from this geometry is Cu-CH $\approx$ Cu-CHprop < Cu-CHpent $\cong$ Cu-CHlact. Only the d-d transitions of the latter two polymers are split into three bands in CD. They were assigned to $E(\Gamma_a)$, $E(\Gamma_b)$ and B transitions (on going towards lower energies) in Cu-CHlact, and to E, B and A transitions in Cu-CHpent, the rotational strength of the last band being very small, as one could expect for a magnetically dipole forbidden transition. The distinct decrease of the symmetry around cupric ions is very likely due to the fact that both alcoholic OH groups of the conformationally mobile lactyl moiety in CHlact and the pendant carboxylates in CHpent, acting as ligands together with the donor nitrogen atom of the uncharged amine, provide a constraining environment for metal chelation.

Fe(III)-CHpent association complex at pH $\approx$ 4.5 and 6 K exhibits a sharp EPR resonance at g = 4.28, characteristic of high-spin (S = 5/2) iron(III) species with rhombic distortion. This signal is replaced by a broad band at g $\approx$ 2, as that found for all other Fe(III)-polymer mixtures, at higher pH values. This would indicate antiferromagnetic interactions between high-spin iron(III) centers, but magnetic measurements as a function of temperature are needed for a definite confirmation. Room temperature Mössbauer data of freeze-dried solutions suggest that the order of increasing distortion from idealized structures is Fe-CH < Fe-CHpent $\approx$ Fe-CHlact. The similarity in the order of increasing distortion in both types of association complexes is not surprising because stereochemical effects of polymeric ligands are predictably similar.

ORIGIN OF OPTICAL NONLINEARITY OF SUCROSE CRYSTAL AS SEEN BY OH AND CH STRETCHING VIBRATIONAL SPECTRA

M. M. Szostak and J. Giermańska

Institute of Organic and Physical Chemistry
Technical University of Wrocław
50-370 Wrocław, Poland

INTRODUCTION

Sucrose crystal generates an optical second harmonic at 1.06µm with 0.2 efficiency relative to ADP as a standard [1]. Second harmonic generation /SHG/ is a common nonlinear optical property /NOP/ of biological compounds like proteins and amino acids and it serves as a test for non-centrosymmetric structure [2].

As a result of our spectroscopic studies of aromatic nonlinear crystals, e.g. m-nitroaniline, it has been suggested that strong vibronic couplings of intramolecular charge transfer /ICT/ with some vibrations are connected with NOP's [3] and that hydrogen bonding /HB/ can cooperate [4] or compete [5] with ICT and therefore enhance or reduce NOP's.

Since any explanation of the origin of SHG in sucrose has not been given it seemed interesting to study vibrational spectra of the sucrose crystal in order to elucidate the role of hydrogen bonding /HB/ in the case of the absence of a conjugated bond system.

RESULTS AND DISCUSSION

In the paper [6] the polarized near IR spectra of the sucrose crystal in the region of the first overtones of OH and CH stretching vibrations and the unpolarized spectrum of the second overtones of these vibrations have been measured. The polarized IR and Raman spectra of the OH stretching fundamentals have been presented in the paper [7].

The assignments of the all eight OH stretching bands have been proposed by the comparison of observed dichroic ratios of the bands with those calculated from the oriented gas model and by the evaluation of force constants for seven hydrogen bonded OH groups [7].

On the basis of the observed frequencies of fundamentals and overtones the anharmonicities of the O-H..O oscillators have been estimated. It has been found that the strongest intermolecular HB /2.716 Å/ and the weakest one /2.879 Å/ are the most anharmonic [6]. These bondings are

the only ones exactly perpendicular to the direction of
the optical axis a^x [7].

According to Jeffrey and Takagi [8] the strongest
HB forms an infinite chain with possible spiral structure.
It seems that the chiral arrangement of the atoms causes
the anharmonicity of the OH oscillators and leads to the
natural optical activity.

The most striking result of these studies is the gre-
atest discrepancy from the oriented gas model in the pre-
dicting the relative intensities /dichroic ratios/ of the
NIR bands in the case of the $-CH_2$ vibration band group [6].

It makes suppose that like in the vibrational circular dichroism spectra of sugars [9] the ring current mechanism and magnetic dipole-magnetic dipole interactions
should be taken into account to explain SHG in the sucrose
crystal.

REFERENCES

1. M.J. Rosker and C.L. Tang, IEEE J.Quantum Electron.,
   1972, 18, 410.
2. M. Delfino, J.Biol.Phys., 1979, 6, 105.
3. M. M. Szostak, Chem.Phys., 1988, 121, 449.
4. J. Giermańska, M.M. Szostak and W.W. Kowala, J.Mol.
   Struct., 1990, 222, 285.
5. J. Giermańska, G. Wójcik and M.M. Szostak, J. Raman
   Spectrosc. , 1990, 21, 479.
6. M.M. Szostak and J. Giermańska, J. Mol. Struct. 1990,
   219, 95.
7. J. Giermańska and M.M. Szostak, J. Raman Spectrosc.
   1991, in press.
8. G.A. Jeffrey and S. Takagi, Acc. Chem. Res., 1978, 11,
   264.
9. M.G. Peterlini, T.B. Freedman and L.A. Nafie, J. Amer.
   Chem. Soc., 1986, 108, 1389.

TWO METHODS OF TIME-RESOLVED FTIR-SPECTROSCOPY APPLIED TO BACTERIORHODOPSIN: CONTINOUS-SCAN WITH BIDIRECTIONAL DATA ACQUISITION AND SUB-MICROSECOND STEP-SCAN TECHNIQUE

K. Nölker[1] and F. Siebert [1,2]

1) Institut für Biophysik und Strahlenbiologie der Albert-Ludwig-Universität, Albertstr. 23, D-7800 Freiburg, F.R.G.

2) Max-Planck-Institut für Biophysik Kennedyallee 70, D-6000 Frankfurt a.M. 70, F.R.G.

Time-resolved vibrational spectroscopy is a powerful tool for the study of chromoproteins (1-3). The light-driven proton pump bacteriorhodopsin (Br) has especially been the subject of many investigations. Since infrared difference spectroscopy is capable of detecting bands due to both the chromophore and the protein, time-resolved infrared spectroscopy appears to be especially suited to investigate the pumping mechanism, which involves structural changes of the chromophore and the protein. Two methods of time-resolved FTIR-spectroscopy are described.

a) The bidirectional fast-scan technique

If the relaxation time for the spectral changes is longer than the time needed for one scan of the inter-ferometer, the fast-scan technique can be employed. Previously only the forward movement of the mirror has been used for data acquisition. To shorten the time between successive interferograms, the backmovement of the mirror is also utilized in the new implementation. In this way it is possible to obtain two interferograms at a time difference of about 80 msec.

b) The step-scan technique

An interferogram $I(g(t),t)$ is a function of the optical path difference $g(t)$ in the interferometer and a function of the time-dependence of the process being studied. In the step-scan technique, the mirror of the interferometer moves stepwise from one sampling point to the next, and at each postion the time-dependence of the interferogram due to the process is measured. With the interferometer (IFS 88 series) of the Bruker company, capable of both stepwise and fast-scan movements step-scan spectroscopy is now feasable.

The step-scan technique has been described recently
(4). Two additions have now been made:
1. The 10 MHz AD-converter allows the resolution of
faster components of the process.
2. A larger buffer of 128k into which the digitized
signal is directly stored, allows even at a sampling rate
of 10 MHz the coverage of a large time period, which can
even be prolonged by switching to a slower sampling rate.
The stored signal can be further processed by transforming
the linear time-base into a quasi-logarithmic time-base
(5). In this way, the effective electronic bandwidth
matches the time constants of the process, i.e. slower
parts of the signal can be evaluated at higher precision.

With this instrument the photoreaction of Br was
measured at 278 K. The spectra were recorded at a
resolution of 4 cm^{-1}. The spectral range was limited to
below 1950 cm^{-1}. Fig. 1 shows a difference spectrum (500
nsec after the flash) revealing the typical Br-K bands
which are known from the published low-temperature FT-IR
spectra at 80 K (6). It is especially noteworthy that the
amide I bands in region between 1600 and 1700 cm^{-1}, the
fingerprint bands, and the HOOP-modes at 960 cm-1, are all
very similar to the low temperatur bands. A distinct
difference is the lower intensity of the K band at
1515 cm^{-1}, which represents the C=C mode of the retinal.

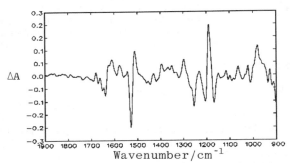

Fig. 1: Br-K difference spectrum.

1. R.J.H. Clark, R.E. Hester, 'Time-Resolved
   Spectroscopy', Advances in Spectroscopy Vol 18, 1989.
2. A. Laubereau and M. Stockburger, 'Time-Resolved
   Vibrational Spectroscopy', Springer Proceedings in
   Physics 4, 1985.
3. G.H. Atkinson, 'Time-Resolved Vibrational
   Spectroscopy', Wiley & Sons, New York, 1983.
4. W. Uhmann, A. Becker, C. Taran, F. Siebert, Applied
   Spectroscopy, 1991, in press.
5. R.H. Austin et al, Rev. Sci. Instrum., 1967, 4, 445.
6. F. Siebert and W. Mäntele, Eur. J. Biochem., 1983, 130,
   565.

HIGH RESOLUTION ^{13}C-SOLID STATE NMR OF BACTERIORHODOPSIN: CHARACTERIZATION OF [4-^{13}C]ASP RESONANCES

Günther Metz[1] ,Friedrich Siebert[1,2], Martin Engelhard[3]

[1] Institut für Biophysik und Strahlenbiologie der Universität
Albertstr.23 , D-78 Freiburg, F.R.G.

[2] Max-Planck-Institut für Biophysik
Kennedy-Allee 70, D-6 Frankfurt, F.R.G.

[3] Max-Planck-Institut für Ernährungsphysiologie
Rheinlanddamm 201, D-46 Dortmund, F.R.G.

Solid state ^{13}C nuclear magnetic resonance measurements of bacteriorhodopsin labeled with [4-^{13}C]Asp show that resonances of single amino acids can be resolved (1,2). In order to assign and characterize the resonances of specific Asp residues, three different approaches were used.

1. Determination of the chemical shift anisotropy from sideband intensities (3) provides unequivocal information about the protonation state of Asp residues. Protonation of a carboxylate results in an upfield shift of σ_{22} in the order of 30 ppm (4). Figure 1 shows the principal CSA tensor elements of the [4-^{13}C]Asp resonances. The observation of an upfield shift of σ_{22} for the resonances at 170.3 and 171.3 ppm (Asp$_{115}$, Asp$_{96}$) clearly identifies them as the only protonated internal Asp carboxyl groups detected in the BR initial state.

2. Relaxation studies and T_1-filtering allow to discriminate between resonances with different mobility. The internal Asp relax even slower than the peptid backbone carbon, whereas the Asp on the external loops show shorter T_1.

3. Resonances of aspartic acids in the close neighborhood of the chromophore are expected to be sensitive to the different isomeric compositions of the retinal in the light-adapted and dark-adapted states. Two lines due to the two internal deprotonated groups (Asp$_{85}$, Asp$_{212}$) change upon dark-light-adaptation, whereas the protonated Asp are unaffected.

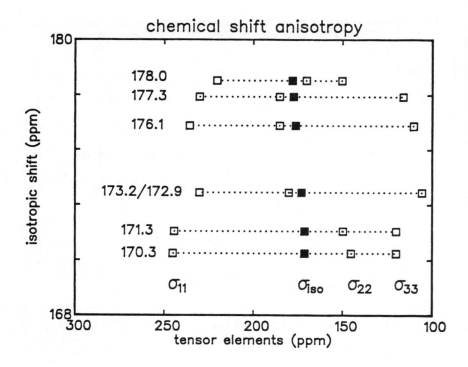

Figure 1   CSA principle tensor elements for the [4-¹³C]Asp resonances as determined by analyses of the sidebands. The values are identical for pH 7 and pH 10 as well as for room temperature, -5°C and 40°C. The errors in the tensor elements, caused by the uncertainty of peak intensity determination, are in the order of <10 ppm.

1. Engelhard, M., Gerwert, K., Hess, B., Kreutz, W. & Siebert, F. (1985) <u>Biochemistry</u> <u>24</u>, 400-407.
2. Engelhard, M., Hess, B., Emeis, D., Metz, G., Kreutz, W. & Siebert, F. (1989) <u>Biochemistry</u> <u>28</u>, 3967-3975.
3. Herzfeld, J. & Berger, A.E. (1980) <u>J. Chem. Phys.</u> <u>73</u>, 6021-6030.
4. Griffin, R.G., Pines, A., Pausak, S. & Waugh, J.S. (1975) <u>J. Chem. Phys.</u> <u>63</u>, 1267-1271.

THE PHOTOCHEMICAL CYCLE OF BACTERIORHODOPSIN STUDIED BY RESONANCE RAMAN AND OPTICAL TRANSIENT SPECTROSCOPY

C. Pusch, R. Lohrmann, M. Stockburger and W. Eisfeld

Max-Planck-Institut für biophysikalische Chemie
Am Fassberg, D-3400 Göttingen, F.R.G.

Bacteriorhodopsin (bR) is a retinal-binding protein in the purple membrane of Halobacteria where it acts as a light-driven proton pump. Excited by light bR's retinylidene chromophore, called BR-570, runs through a cyclic reaction with various intermediate states (photocycle) which under physiological conditions lasts a few milliseconds. This reaction controls bR's function as a proton pump.

BR_{570}:   Lys 216

Until now at least six different intermediate states of the chromophore could be identified by optical and resonance Raman (RR) spectroscopy. In the order of their appearance time after excitation of BR-570 they were denoted as

$$BR_{570}, \quad K_{590}, \quad L_{550}, \quad M_{412}, \quad N_{560}, \quad O_{640} \tag{1}$$

where the subscripts refer to the absorption maxima (nm) of the optical spectra.[1] It is well established that the primary step from BR to K involves all-trans to 13-cis photoisomerization of the retinal chain. All consecutive steps are thermally controlled dark reactions. During the L-to-M transition a proton is released from the nitrogen of the Schiff base (SB) group. Later in the cycle a proton is taken up from the SB nitrogen. It is thought that these reaction steps play an important role in the proton pump mechanism.

It turned out that a linear sequence of the intermediates as in (1), which had been originally proposed,[1] does not fit the kinetic and RR spectroscopic data.[2,3] Thus the intermediates L and M decay with at least two different rate constants. But on the other hand no spectroscopic evidence for the existence of two different L or M states could be found. For an understanding of this kinetic degeneracy of the chromophore we have studied the photocycle of bR by time-resolved RR experiments in the pH range 4 to 9. For the analysis of the intermediates we used their characteristic RR bands in the C=C stretching region. This allowed us to distinguish species like L-550 and N-560 whose optical spectra are strongly overlapping. Combining our results with optical kinetic data the following reaction scheme is suggested:

$\underline{\text{bR}(0)}$: (pH7, 20°C)

$$BR \leadsto K \xrightarrow{\text{5ps}} \xrightarrow{\text{2μs}} L_0 \xrightarrow{\text{40μs}} M_0(\text{cis}) \xrightarrow[\text{8ms}]{[H^+]} M_0(\text{trans}) \xrightarrow{<10\text{μs}} BR$$

$$\searrow \quad \overset{..}{\text{H}^+} \quad A_0^-$$

(2)

$\underline{\text{bR}(i)}$: (pH7, 20°C)

$$BR \leadsto K \xrightarrow{\text{5ps}} \xrightarrow{\text{2μs}} L_i \xrightarrow{>100\text{μs}} M_i \xrightarrow{<3\text{ms}} N_i(\text{cis}) \xrightarrow[\text{5ms}]{[H^+]} O(\text{trans}) \xrightarrow{\text{8ms}} BR$$

$$\searrow \quad \overset{..}{\text{H}^+} \quad A_i^-$$

In this scheme two types of parallel reaction sequences occur. The reason for this lies in the different behavior of the proton acceptor groups A_0 and A_i which are most likely due to the carboxylate side groups of internal aspartic acid residues. Whereas A_0H is deprotonated in the pH range 3 to 9, the various A_iH have different pK_a values in this range. The kinetic degeneracy of L and M thus can be explained by different acceptor/donor properties of such groups. Since A_0H is more stable than A_iH the reprotonation and re-isomerization steps occur in reversed order for the two groups. This makes just the essential difference for the two reaction sequences.

On the basis of scheme 2 in combination with complementary structural and thermodynamic data we propose a new model for proton pumping of bR. In this approach the SB proton is no longer directly involved in the pump mechanism as is usually assumed. It is rather suggested that the sequential reaction steps of the chromophore induce structural changes of the protein which control proton conduction across the membrane.

1.   R.H. Lozier, R.A. Bogomolni and W. Stoeckenius, Biophys. J., 1975, 15, 955.
2.   R. Maurer, J. Vogel and S. Schneider, Photochem. Photobiol., 1987, 46, 255.
3.   R. Diller and M. Stockburger, Biochemistry, 1988, 27, 7641.

QUANTUM YIELDS FOR THE PHOTOCHROMIC EQUILIBRIUM BETWEEN BACTERIORHODOPSIN AND ITS K INTERMEDIATE DETERMINED BY LASER–INDUCED OPTOACOUSTIC SPECTROSCOPY (LIOAS)

Mathias Rohr,[1] Wolfgang Gärtner,[2] and Silvia E. Braslavsky[1]

[1]Max–Planck–Institut für Strahlenchemie, Stiftstr. 34–36, D–4330 Mülheim/Ruhr, and [2]Institut für Biologie I (Zoologie) der Universität, Albertstr. 21a, D–7800 Freiburg, Germany

The overlap between the absorbances of bacteriorhodopsin (BR) and its early photochemical intermediates (J and K)[1] impedes the accurate determination of the quantum yield of formation of these intermediates by optical methods. Laser–induced optoacoustic spectroscopy (LIOAS) circumvents this problem by measuring the heat stored by intermediates longer–lived than the heat integration time of the experiment. A piezoelectric detector (a ceramic lead zirconium titanate transducer) senses the acoustic wave generated in the solution by the heat burst due to radiationless deactivation of the molecules excited by a 6–ns laser pulse.[2]

By changing the diameter of the laser beam the acoustic transit time, τ'_a (i.e., the heat integration time) was varied between ca. 150 and 600 ns. The intermediate K, formed in 5 ps from J[3] and living for ca. 2 µs, stores the fraction of energy not delivered to the solution within these heat integration times.

Since the LIOAS signal maximum, H, depends on the thermoelastic properties of the solution (those of the solvent for dilute solutions), on the geometrical conditions of the experiment, and on the sample absorbance, the signal amplitude was calibrated with calorimetric references, $CoCl_2$ or $CuCl_2$ (depending on the wavelength), which convert all the absorbed energy into heat within τ'_a. Thus, the fraction of heat dissipated promptly by the sample, α, for the particular solvent, absorbance, and geometrical arrangement results from the ratio of the LIOAS amplitudes for the BR solution, S, and for the reference, R, i.e., $\alpha = H_S/H_R$.[2] By measuring the heat stored $(1 - \alpha)$ as a function of τ'_a a lifetime of K, $\tau_K = (1.8 \pm 0.3)$ µs, was determined, which is well within the previously reported value (2 µs).[4]

The dependence of α on the excitation energy fluence was modeled mathematically for each of the six excitation wavelengths (532–630 nm). Due to the establishment of a photochromic equilibrium between K and BR within the 6 ns of the pulse, the heat arises from excited BR (BR*) as well as from excited K (K*). Consequently, four differential equations describing the kinetic behaviour of all four species, BR, K, BR*, and K*, present in the system within the pulse duration were considered. Included was the absorbance ratio of K over BR which varied between 0.7 and 2.7 for the various wavelengths.[4] A fifth equation described the heat evolution within τ_a,[5] in terms of the radiationless deactivation of BR^* and K^* by internal conversion (eqs. 1 and 3) and by reaction (eqs. 2 and 4), and of the decay of K to the next intermediate L (eq. 5):

$$BR^* \rightarrow BR \qquad k_1 \qquad\qquad (1)$$
$$BR^* \rightarrow K \qquad\; k_2 \qquad\qquad (2)$$
$$K^* \rightarrow K \qquad\quad\;\; k_3 \qquad\qquad (3)$$
$$K^* \rightarrow BR \qquad\;\; k_4 \qquad\qquad (4)$$
$$K \;\; \rightarrow L \qquad\quad\;\; k_5. \qquad\qquad (5)$$

The term for the decay of K to L was neglected in the energy balance equation since the energy difference between K and L should be much smaller than the energy differences involved specially in reactions (1), (3), and (4).

Since the LIOAS data afford the energy stored by K as the product of its formation quantum yield times its energy content ($\Phi_{BR \rightarrow K} \cdot E_K$), the 0–0 band of BR (190 kJ/mol), as derived from the crossing of the absorption and fluorescence spectra,[3] was taken as a maximum value for the energy content of K, E_K, assuming the reaction sequence BR→J→K to be exothermic. In this manner a lower limit for $\Phi_{BR \rightarrow K}$ was derived.

The best fit of the experimental points for α as a function of the laser energy fluence for each of the excitation wavelengths was obtained for a quantum yield of formation of K, $k_2/(k_1 + k_2) = \Phi_{BR \rightarrow K} = 0.6$, a quantum yield of formation of BR from K, $k_4/(k_3 + k_4) = \Phi_{K \rightarrow BR} = 0.5$, and a lifetime of K^*, $\tau_{K^*} = 40$ ps. The radiative decay constants did not need to be considered. The relation $\tau_{K^*} > \tau_{BR^*}$ (0.7 ps)[3] is consistent with the higher fluorescence and the smaller Stokes shift reported for this intermediate, when compared to BR.[6]

We are indebted to Professor Kurt Schaffner for his encouragement and constant support. M.R. is recipient of a fellowship from the Krupp von Bohlen und Halbach–Stiftung.

REFERENCES

1. A.K. Dioumaev, V.V. Savransky, N.V. Tkachenko, and V.I. Chukharev, J. Photochem. Photobiol. B: Biology, 1989, 3, 385.
2. S.E. Braslavsky and K. Heihoff, "Handbook of Organic Photochemistry" (Ed. Scaiano, J.C.) CRC Press, Boca Raton, Florida, 1989, Vol. I, Chapter 14.
3. H.J. Polland, M.A. Franz, W. Zinth, W. Kaiser, E. Kölling, and D. Oesterhelt, Bioph. J., 1986, 49, 651.
4. G.H. Atkinson, T.L. Brack, D. Blanchard, and G. Rumbles, Chem. Phys., 1989, 131, 1.
5. G.M. Bilmes, J.O. Tocho, and S.E. Braslavsky, J. Phys. Chem., 1989, 93, 6696.
6. G.H. Atkinson, D. Blanchard, H. Lemaire, T.L. Brack, and H. Hayashi, Biophys. J., 1989, 55, 263.

HIGH RESOLUTION FT-IR SPECTROSCOPY STUDY OF RETINAL RHODOPSIN

D. Garcia-Quintana*, P. Garriga and J. Manyosa

Unitat de Biofísica, Departament de Bioquímica i de Biologia Molecular, Edifici M, Universitat Autònoma de Barcelona, 08193-Bellaterra, Barcelona, Catalonia/Spain

Some works have been previously published yielding spectroscopical quantitative results about the structure of rhodopsin[1,2], suggesting α-helix to be its major secondary structure type. Nevertheless, not a detailed study has been reported, so that several structural models coexist at present[3].

High resolution FT-IR spectroscopy has been applied, taking extreme care regarding buffer influence, signal-to-noise ratio, water subtraction and vapor elimination. Fourier self-deconvolution analysis reveals 9 major component bands which can be assigned as amide I vibrational modes. Quantification has been achieved by means of a least-squares curve-fitting iterative program over the deconvoluted spectrum. In opposition to the fit of the original absorption contour, fit of the deconvoluted spectrum has a single, unambiguous solution; reconstruction of the absorption spectrum contour by summation of the mathematically transformed (k-factor multiplied, Gaussian-to-80% Lorentzian adjusted) obtained component bands, yields a contour which is virtually undistinguishable from the original one, so that a non-significant distortion of relative areas can be assumed in the Fourier-deconvolved spectrum.

The amount of helical segments is calculated to be 60%, which is a value in accord with the aforementioned existing previous works reporting a helix content between 47-60%. Nevertheless, the FTIR spectral analysis suggests that not all helical segments in rhodopsin belong to α_I type, but that at least one fifth of the total helix can be assigned to the more extended 3_{10} type. This assignation and its structural implications (retinal pocket)[4] will be further discussed, as well as the presence of α_{II} type helix and loops. In addition, our results suggest that at least 11% of the amino acid residues are involved in extended, β-type, structures, as well as 14% in turns and bends and 11% in irregular ("random") segments. A minor area is calculated to be attributable to the spectral band arising from the Schiff-base vibration which appears inside the amide I region.

In order to calculate the proportion of residues in each kind of secondary structure which is exposed to the solvent, the spectral shifts induced by deuteration have been studied. The quantification of the amide I' component bands from the FTIR spectrum of leaky discs extensively incubated in ^2H_2O buffer suggests that although all secondary structure types are at least partially exposed to the solvent, most N-H amide groups involved in α_I helixes, and at least half of those presumably involved in 3_{10} helix show no deuteration shift. Regarding the total protein, 41% of the N-H amide groups is calculated to have remained unexchanged. Such value will be compared with the ones obtained from different approaches[5], and a discussion will be held in relation to the estimated[6,7] intramembranous protein moiety.

REFERENCES

1. G. W. Stubbs, H. G. Smith Jr. and B. J. Litman, Biochim. Biophys. Acta, 1976, 425, 46.
2. N. W. Downer, T. J. Bruchman and J. H. Hazzard, J. Biol. Chem., 1986, 261, 3640.
3. W. J. De Grip, Photochem. Photobiol., 1988, 48, 799.
4. J. B. C. Findlay and D. J. C. Pappin, Biochem. J., 1986, 238, 625.
5. P. I. Haris, M. Coke and D. Chapman, Biochim. Biophys. Acta, 1989, 995, 160.
6. E. M. Kosower, Proc. Natl. Acad. Sci. USA, 1988, 85, 1076.
7. Y. A. Ovchinnikov, FEBS Lett., 1982, 148, 147.

NMR AND VIBRATIONAL STUDIES OF A STABLE CARBINOL AMINIUM CHLORIDE, $(CH_3)_2-CH-CH(OH)-NH_2^+-CH(CH_3)_2,Cl^-$, IN THE SOLID STATE. INFLUENCE OF SOLVENTS ON CARBINOLAMINIUM/IMINIUM EQUILIBRIUM

M.H. Baron*, H. Zine and A. Goypiron

LASIR, CNRS, 2 rue Henri Dunant, 94320 Thiais (France)

The influence of water on the protonation and deprotonation of the imine function of retinyliden Schiff base (SB), chromophore of bacteriorhodopsins and rhodopsins, has been suggested.[1] Water molecules are assumed to stabilize the ion pair structure between SBH^+ and a negative counterion, H_2O being intercalated between the iminium and the anion. In another study,[2] the mechanism of SB formation and hydrolysis has been investigated in aqueous detergent micelle systems. A tetrahedral carbinolamine intermediate of N-retinylidene-n-butylamine has been observed directly by continuous flow, PH-jump resonance Raman spectroscopy.

Now we evidence that the carbinolaminium function resulting from an H_2O addition on the $-CH=NH^+-$ group of N-methyl-propylidene-propane-2-iminium chloride, $(CH_3)_2-C^2H-C^1H=NH^+-C^2H(CH_3)_2Cl^-$ (NPPH$^+$Cl$^-$), is stable in the solid state at room temperature. The pure substance has a 1:1 ratio of H_2O to NPPH$^+$Cl$^-$. Its NMR and vibrational spectra reveal all the specific signal expected for such $-C^1H(OH)-NH_2^+-(1'$-alcohol-1'-aminium) function.
(IR frequencies (cm^{-1}) : $\nu OH(3217)$, $\delta OH(1423)$, $\nu CO(1206)$, $\gamma OH(662)$, $\delta CCO(622)$, $\nu NH_2^+(2800)$, $\nu CN^+(1160)$, r,tw,wNH$_2^+$ (1403,978,564), $\delta CCN^+,\delta CN^+C(448,420)$, tCO(174)).
The $-CH(OH)(NH_2^+-$ function is assumed to be stabilized by electrostatic interactions between Cl^- and both the NH_2^+ and OH contiguous groups. The two isopropyle substituants frame an apolar environment large enough to increase the strength of the complex. For example, such stable entity cannot be isolated if it is the cetiminium $(CH_3)_2-C=NH^+-CH_3,Cl^-$ which is hydrated. Also, the importance of the activity of the anion Cl^- is evidenced by the fact that the carbinol aminium specy cannot be obtained from NPPH$^+$BF$_4^-$ and H_2O. Very stable in the solid state, the unstability of NP(OH)PH$_2^+$Cl$^-$ in organic solvents is proofed by the IR analysis of NP(OH)PH$_2^+$Cl$^-$ in various solvents. Assuming the following equilibrium :

$$-CHO, -NH_3^+ \rightleftarrows -CH(OH)-NH_2^+ \rightleftarrows -CH=NH^+- \rightleftarrows -CH=N-$$

$$\begin{pmatrix} Cl^-, H_2O \\ solvent \end{pmatrix} \qquad Cl^- \qquad Cl^- \qquad \begin{pmatrix} H_3O^+, Cl^- \\ solvent \end{pmatrix}$$

$$\qquad\qquad\quad \text{"solid"} \qquad (H_2O, solvent)$$

$$\mathrm{I} \qquad\qquad\quad \mathrm{II} \qquad\qquad \mathrm{III} \qquad\qquad\qquad \mathrm{IV}$$

compound II is no more detected in solution. In a proton donor solvent (HCCl3) the equilibrium is displaced toward I. On contrary, a weak proton acceptor solvent (CH3CN) essentially favours the III species while a strong proton acceptor solvent (DMSO) leads essentially to a mixture of III and IV. The water molecules which are no more involved in NPP after II dissolution interact with the solvents. These results point out that H2O addition does not necesserily lead to hydrolysis of aldimines and show how precisely the molecular environment may select one or another of the I, II, III or IV states.

In the recent studies about the cyclic photochemical transformation of the SBH$^+$ in bacteriorhodopsin, the occurance of a carbinolaminium intermediary is generally not considered. One can wonder if the hypothesis of a carbinolaminium intermediate would not be right to take in account the kinetic effects of pH,[4] H2O,[1,2] chloride salts[5] and organic solvants.[6] Such -CH(OH)-NH2$^+$- function could evidently favour the occurance of syn $\rightarrow$ anti isomerism. Moreover, in the particular case of conjugated imines, a δ^+ charge, resulting from the OH$^-$ departure between step II and III, may potentially shift along the polyenic chain and then favours isomerization around one C=C bond.

REFERENCES

1. C.N. Rafferty and H. Shichi, Photochem. and Photobiol., 1981, 33, 229 ; P. Hildebrandt and M. Stockburger, Biochemistry, 1984, 23, 5539 ; S. Badilescu, L.S. Lussier, C. Sandorfy, H. Le Thanh and D. Vocelle, Chem. Phys. Letters, 1987, 133, 63 ; C. Rhofir and D. Vocelle, Research. on Chemical Intermediates, 1989, 12, 131.

2. A. Cooper, S.F. Dixon, M.A. Nutley and J.L. Robb, J. Amer. Chem. Soc., 1987, 109, 7254.

3. J.A. Shudek, R. Foster and D. Young, J. Chem. Soc. Perkin Trans II, 1985, 1285.

4. M. Nakagawa, T. Ogura, A. Maeda and T. Kitagawa, Biochemistry, 1989, 28, 1347 ; G. Varo and J. Lanyi, Biochemistry, 1990, 29, 2241 ; 1990, 29, 6858 ; J.B. Ames and R. Mathies, Biochemistry, 1990, 29, 7181 ; T. Kouyama, A. Nasuda-Kouyama, A. Ikegami, M. Mathew and W. Steckenius, Biochemistry, 1988, 27, 5855 ; S. Fodor, J. Ames, R. Gebhard, E. Vand der Berg and W. Stockenius, Biochemistry, 1988, 27, 7097.

5. S.O. Smith, J. Courtin, E. Van der Berg, C. Winkel, J. Lugtenburg and J. Herzfeld, Biochemistry, 1989, 28, 237 ; L. Zimanyi, L. Keszthelyi and J. Lanyi, Biochemistry, 1989, 28, 5165.

6. P. Pande, R. Callender, R. Henderson and A. Pande, Biochemistry, 1989, 28, 5971.

V

NMR

ENZYME CHARACTERIZATION

Using NMR to study the Structure and Dynamics of Proteins

G. C. K. Roberts

Department of Biochemistry and Biological NMR Centre,
University of Leicester,
Leicester LE1 7RH, U.K.

The revolution in nmr techniques over the last few years, notably in two-, three- and now four-dimensional nmr experiments and the use of stable isotope labelling, has enormously increased the power of this method in the determination of three-dimensional structures of small proteins in solution, in studies of the specific interactions of proteins with other molecules and in the study of the dynamics of these structures and interactions.[1-4] Examples of the use of nmr in these areas will be drawn from our recent work on dihydrofolate reductase (DHFR), phospholipase A_2 (PLA$_2$) and chloramphenicol acetyl-transferase (CAT).

Structure determination by nmr. The structural information provided by nmr is essentially in the form of constraints. From the point of view of structure determination, the most important nmr parameter is the nuclear Overhauser effect (NOE), whose magnitude depends on inter-nuclear distances. This provides constraints in the form of a distance range (*e.g.*, 2.2 - 3.2 Å) for each pair of nuclei for which NOEs are observed. Further constraints are available, at least for small proteins, in the form of estimates of dihedral angles from scalar coupling constants. Provided that a sufficient number of such relatively weak constraints are available, typically several hundred for a protein of 100 amino-acid residues, they are sufficient to generate a model of the three-dimensional structure.[1-4] The limitations of the technique, for example in the size of protein that can be tackled and in the quantitative structural interpretation of the data, are gradually becoming less restrictive, with improvements in relaxation and NOE analysis, the continuing development of higher-field spectrometers, and the use of ^{13}C and ^{15}N labelling.[3,4] With the aid of stable isotopes, it is now possible to determine the solution structures of proteins of 200 amino-acid residues by nmr.

Nmr studies of protein-ligand interactions. The same kind of NOE-based structure analysis can be applied to the determination of the geometry of ligand-protein complexes.[5] Analysis of the detailed energetics underlying protein specificity depends upon making changes in the covalent structure of the protein (by mutagenesis) or the small molecule (by synthesis). The functional effects of these changes can only be understood in the context of studies of their conformational effects, and nmr has proved to be a convenient and sensitive way to study these (e.g. R5-11). There is considerable variation in the conformational consequences of chemical change, presumably arising from the differing degrees of flexibility of different parts of the structure. For example, nmr studies of the Asp26 -> Asn mutant of *L. casei* DHFR[9] show that any structural changes in the enzyme-methotrexate complex are very small and local, while in the case of the Thr63 -> Gln substitution, nmr analysis shows that the conformational effects are transmitted as much as 25Å through the protein.[10]

Just as isotope-labelling has provided a means of extending structure determination by nmr to larger proteins, so it has increased the range of systems in which nmr can be used to study the details of ligand binding. For example, studies of the binding of a ^{13}C-labelled substrate to CAT by ^{13}C-edited ^{1}H-^{1}H NOEs in combination with site-directed mutagenesis allowed us to define the mode of binding even in this large protein (M_r 75,000).[11]

Nmr studies of structural fluctuations. Motions in proteins and their complexes occur at rates spanning a wide timescale, from picoseconds to seconds, and nmr can provide detailed information over substantial parts of this range[12], with the major advantage that it allows one to identify which residue(s) are involved in the rate processes being studied. These range from local fluctuations, such as the 'flipping' of aromatic rings[13,14] whose rate is determined by the packing of amino-acid side-chains around the ring, to much slower and more extensive conformational equilibria. The rapid fluctuations may well play a role in the kinetics of substrate binding, while in DHFR and PLA$_2$ several conformational equilibria have been shown to play a role in the catalytic process.

Bovine and porcine PLA$_2$ differ at 20 positions in the sequence; the crystal structures of the two enzymes indicate that these sequence differences result in generally small differences in three-dimensional structure, with the exception of the loop formed by residues 59-70. There is only a single residue difference in this loop (Val63 -> Phe) and yet conformations it adopts in the two crystal structures are such that the position of Tyr69 differs by 5Å. This difference has been implicated in the different micelle binding properties displayed by the two enzymes. However, such a structural difference between the two enzymes in solution is not apparent from nmr.[15] Indeed, analysis of the observed NOEs suggests that the *time-average* solution state conformations of the two enzymes are very similar in this region. The crystallographic and nmr observations can most easily be reconciled by suggesting that there is a degree of flexibility in this loop, and that it adopts different conformations in the two crystal structures not because of any intrinsic preference but due to the effects of crystal packing forces. Analysis of the changes in NOEs between residues in the active site on ligand binding shows that substrate analogue binding is accompanied by a movement of Tyr69 so that its side-chain 'caps' the entrance to the active site[5], perhaps excluding water; the conformational equilibrium of the flexible loop is thus shifted towards one defined conformation on ligand binding.

The DHFR inhibitor methotrexate is a close structural analogue of the substrate folate, the important difference being a 4-amino rather than a 4-oxo substituent on the pteridine ring of methotrexate. In spite of this close similarity, methotrexate binds some 10^5 times more tightly than folate. Furthermore, the stereochemistry of reduction implies a difference of 180^0 in the orientation of the pteridine ring between substrate and inhibitor[8]. Nmr experiments have recently shown that both the enzyme-folate and the enzyme-folate-NADP$^+$ complexes exist in solution as a mixture of slowly interconverting conformations whose proportions are pH-dependent. NOE and mutagenesis experiments have shown that the crucial difference between these conformations is the orientation of the pteridine ring of the substrate, only one corresponding to a catalytically productive mode of binding for the substrate[8], and that the pH-dependence of the equilibrium is determined by the ionization of Asp26.[9]

In common with most dehydrogenases, the affinity of the enzyme for substrate analogues is markedly affected by the binding of coenzyme. In the case of DHFR, the substrate analogue methotrexate binds 600 times more tightly in the presence of the coenzyme than in its absence[16], and nmr is allowing us to begin to define the route of transmission, which appears to involve an axial movement of helix C, residues 42-49.[17] Interestingly, in a mutant, Thr63->Gln, having altered cooperativity there is again clear nmr evidence for the involvement of helix C.[10]

We have undertaken detailed studies of the kinetic mechanism of *L. casei* DHFR[18]. The most striking feature of this mechanism is that, in the kinetically preferred pathway for product release, tetrahydrofolate (FH_4) dissociates only *after* NADPH has bound. This originates from the marked negative cooperativity in binding between FH_4 and NADPH; FH_4 dissociates 300 times faster from the E.NADPH.FH_4 complex than from E.FH_4. Very similar negative cooperativity between NADPH and the stable product analogue 5-formyl-FH_4 was earlier demonstrated by nmr and binding constant measurements[19], and nmr studies show that these changes are qualitatively distinct from those associated with the positive cooperativity between NADPH and methotrexate.[17,19] Thus, notwithstanding the small size and apparent simplicity of dihydrofolate reductase, both large positive and large negative free-energy coupling between coenzyme and substrate analogue binding has been observed, each associated with a distinct conformational change in the protein.

Acknowledgements. I am most grateful to the many colleagues involved in this work whose names appear in the references, particularly to Lu-Yun Lian, Jim Feeney, and Berry Birdsall, and to SERC, MRC, the Wellcome Trust, NEDO (Japan), Du Pont and Fisons for support of the work described here.

1. K. Wuthrich, 'NMR of Proteins and Nucleic Acids', Wiley, New York, 1986.
2. A. Bax, *Ann. Rev. Biochem.*, 1989, *58*, 223.
3. N. J. Oppenheimer and T. L. James (eds.) 'Methods in Enzymology', Academic Press, New York, 1989, vols. 176 and 177.
4. G. Wagner, *Prog. NMR Spectrosc.*, 1990, *22*, 101..
5. W.U. Primrose, R.L. Magolda and G.C.K. Roberts, in preparation, 1991.
6. S.J. Hammond, B. Birdsall, J. Feeney, M.S. Searle, G.C.K. Roberts and H.T.A. Cheung, *Biochemistry*, 1987, *26*, 8585.
7. B. Birdsall, J. Andrews, G. Ostler, S.J.B. Tendler, J. Feeney, G.C.K. Roberts, R.W. Davies and H.T.A. Cheung *Biochemistry*, 1989, *28*, 1353.
8. B. Birdsall, J. Feeney, S.J.B. Tendler, S.J. Hammond and G.C.K. Roberts *Biochemistry*, 1989, *28*, 2297.
9. M.A. Jimenez, J.R.P. Arnold, J. Andrews, J.A. Thomas, G.C.K. Roberts, B. Birdsall and J.Feeney, *Protein Engineering*, 1989, *2*, 627.
10. Thomas, J.A. Ph.D. thesis, University of Leicester, 1991; J.A. Thomas, J.R.P. Arnold, J. Andrews, J. Feeney and G.C.K. Roberts, in preparation, 1991.
11. J. Derrick, L.-Y. Lian, G.C.K. Roberts and W.V. Shaw, in preparation, 1991.
12. O. Jardetzky and G.C.K. Roberts, 'NMR in Molecular Biology', Academic Press, New York, 1981.
13. G.M. Clore, A. Gronenborn, B. Birdsall, J. Feeney, and G.C.K. Roberts, *Biochem. J.*, 1984, *217*, 659.
14. M.S. Searle, M.J. Forster, B. Birdsall, G.C.K. Roberts, J. Feeney, H.T.A. Cheung, I. Kompis and A.J. Geddes, *Proc. Natl. Acad. Sci. US*, 1988, *85*, 3787.
15. J. Fisher, W.U. Primrose, G.C.K. Roberts, N. Dekker, R. Boelens, R. Kaptein and A.J. Slotboom, *Biochemistry*, 1989, *28*, 5939.
16. B. Birdsall, A.S.V. Burgen and G.C.K. Roberts, *Biochemistry*, 1980, *19*, 3723.
17. S.J. Hammond, B. Birdsall, M.S. Searle, G.C.K. Roberts and J. Feeney, *J. Mol. Biol.*, 1986, *188*, 81.
18. J. Andrews, C.A. Fierke, B. Birdsall, G. Ostler, J. Feeney, G.C.K. Roberts and S.J. Benkovic, *Biochemistry*, 1989, *28*, 5743.
19. B. Birdsall, E.I. Hyde, A.S.V. Burgen, G.C.K. Roberts and J. Feeney, *Biochemistry*, 1981, *20*, 7186.

GENETIC AND SPECTROSCOPIC CHARACTERIZATION OF STRUCTURAL CHANGES IN DEHYDROGENASE ENZYMES AND THEIR SUBSTRATES DURING ENZYME CATALYSIS.

A.R. Clarke, A. Cortes[§], D.J. Halsall, K.W. Hart and J. J. Holbrook.

University of Bristol Molecular Recognition Centre and Department of Biochemistry, School of Medical Sciences, Bristol BS8 1TD, U.K.

[§]Department of Biochemistry and Physiology, University of Barcelona, Martí y Franqués 1, 08028 Barcelona, Spain

INTRODUCTION

Analysis of the mechanism and structure of the NAD-dependent L-lactate dehydrogenases (LDH) from B. stearothermophilus, from authentic pig and cloned from human skeletal muscle reveals that slow movements of the protein framework enable discrimination between closely related small substrates. Only small and correctly charged substrates allow the protein to engulf the substrate in an internal vacuole that is isolated from solvent protons, in which water is frozen and hydride transfer is rapid. The vacuole provides discrimination between small substrates which have too few functional groups to be distinguished at a protein surface. This model successfully predicted the design and synthesis of new enzymes such as L-hydroxyisocaproate dehydrogenase and fully active malate dehydrogenase.[1] The disadvantage of a vacuole mechanism is that solvent friction limits the rate of forming the vacuole and thus the maximum rate of catalysis.[2]

The unambiguous detection of movements in specific regions of a large enzyme protein during catalysis requires a spectroscopic signal which (i) can be monitored during the lifetime of enzyme-substrate complexes (1-50 ms) and (ii) can be assigned to events at a defined position in the three dimensional structure of the large molecule. Site-specific mutagenesis enables the construction of proteins so that spectral changes can be unambiguously assigned and thus enable description in molecular detail the sequence of protein and substrate structure changes associated with enzyme catalysis.

Spectroscopic assignment of residues which distort substrate towards the transition state.

In collaboration with R.H. Callender and J. Burgner mutagenesis has enabled the roles of Arg-109 and of His195-Asp168 couple in polarizing the carbonyl bond in the abortive ternary E-NAD-pyruvate adduct to be evaluated. When the His195-Asp168 couple was changed to His195-Ala168 the Raman $>C=O$ was shifted only 20cm^{-1} towards -CH-O$^-$ i.e. 20% single bond character,[3] confirming the role of Asp168 in maintaining a protonated His195 (see fig.1 for residue positions). In contrast removing the full positive charge of Arg-109 (->Gln-109) had no effect on the carbonyl stretch, confirming this was only effective in the transition state. Mutagenesis unambiguously assigns the protein contributions in stabilizing the alcoholate-like transition state (H-$\rangle$C-O$^-$).

Fluorescence to assign which coenzyme form is bound to enzyme in the rate-limiting complex in catalysis.

To understand how proteins function as catalysts it is important to discover which event(s) in a reaction sequence are limiting. The thermodynamically favourable reaction catalyzed by LDH is pyruvate + NADH -> lactate + NAD$^+$. When this reaction is run at saturating concentrations of NADH and pyruvate the (cloned human) enzyme turns over maximally at 350s^{-1}. This rate is the same whether the bond being broken is C-H in NADH or C-D in NADD ($k_H/k_D = 1.1 \pm 0.2$). Thus the chemical reaction is much faster than the

223

rate-limiting reaction. Neither is the first catalytic turnover faster than the subsequent ones and thus the slowest step is neither product dissociation nor the chemical (bond-breaking) step and must be a slow rearrangement of the protein after the enzyme is saturated with both substrates and prior to hydride transfer. Thus the mechanism must include a pathway such as (E = an enzyme subunit, P = pyruvate, L = lactate, R = NADH, O = NAD$^+$):

$k_{1,2}/k_{2,1} = 10^6$ M^{-1}; $k_{2,3}/k_{3,2} = 10^4$ M^{-1}; $k_{conf+} = 650$ s^{-1}; $k_{conf-} = 180$ s^{-1}; $k_{chem+} = 10^5$ s^{-1}; $k_{chem-} = 3 \times 10^4$ s^{-1}

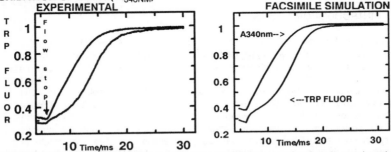

EXPERIMENTAL DETECTION OF THE STEADY STATE COMPLEX OF LDH BY MEASURING TRYPTOPHAN FLUORESCENCE DURING THE STEADY STATE OXIDATION OF NADH (-A$_{340NM}$)

SCHEME 1. Mechanism, rate constants and protein fluorescence signals used to simulate the oxidation of 100μM NADH by 1mM pyruvate and 20μM cloned human M$_4$ LDH. During the first 3 turnovers the enzyme has low fluorescence i.e. is in NADH-containing complexes. The simulated results and the stopped flow experimental results agree well.

Fluorescence spectroscopy dissected this mechanism. Most of the tryptophan fluorescence of the human and pig M$_4$ LDH arises from tryptophan-248 which lies very close to the nicotinamide ring of the cofactor (the relative orientation is similar to that shown for B. stearo. LDH in Fig. 1. All complexes of NADH have low indole fluorescence; those of free enzyme or NAD$^+$ have high fluorescence. During the first 3-4 turnovers when the enzyme is in a steady state the protein fluorescence is low and demonstrates the steady state compound contains enzyme-bound NADH (saturated with pyruvate) and rate is limited by protein shape change.

Structural description of the binary --> ternary rate-limiting conformation change.

What structural change takes place during the slow (3->5) kinetic step in the mechanism ? This change is only induced after the substrate collides with the E-NADH binary complex. There is no slow step involved in the combination of the apoenzyme with NADH. A high resolution crystallographic description of the binary -> ternary structural change is now available.[2] A mobile surface loop (residues 98-110) is much closer to the body of the protein (helices α-2G and α-3G) in the ternary than in the binary complex and the C-terminal surface helix (α-H) moves along its axis to provide better solvent exclusion in the ternary than in the binary. Thus entry of the negatively charged oxamate or pyruvate into the protein surface cleft induces a rearrangement by which the catalytic pathway[2] becomes completely solvated by protein. Does the structural change which takes place in the crystal also take place in solution ? To answer this question requires spectroscopic probes at known positions in space. Insertion can be chemical or genetic.

The fast shape change probed from the altered ionization of 3-NO$_2$-tyrosine-237.

Tetranitromethane specifically nitrates tyrosine-237 at the lip of the active site vacuole of the pig enzyme without much changing enzyme activity.[4] The tyrosine is at position W237 (Fig.1). The pK$_a$ of this yellow residue is sensitive to the approach of acidic amino acid side chains: 7.2 (apo), 7.4 (E-NADH binary) and 7.8 in E-NADH-oxamate/pyruvate ternaries. In crystals the apo -> ternary change shortens the distance from tyrosine-Oτ to Oϵ of glutamate-107 and to Oδ of aspartate-197 and either was a possible cause of the bleaching. By mutagenically removing Glu107 or Asp197 it was shown that approach of Asp197 (on βH) the Y-237 (on α-2G) causes the yellow colour change (Table 1). Temperature-jump experiments at pH 7.5 revealed the yellow colour was quenched in a step after initial formation of a collision complex of E-NADH with oxamate with an isomerization rate of 3000s^{-1}. The step can also be observed in a low temperature stopped flow mixing experiments (-15° in 30% DMSO) and via a linear Arrhenius plot, in agreement with the T-jump results, extrapolates to 3000s^{-1} at 25°C.[5] This was the first evidence that the first conformational event induced by substrate is approach of βG/βH carrying Asp197 to helix α-2G (Tyr237).

TABLE 1. Assignment of the acidic aminoacid responsible for bleaching the yellow colour 3-nitrotyrosine-237-LDH during conformation changes induced to 37μM enzyme-30μM NADH-0.5mM fructose-1,6-bisphosphate in triethanolamine-HCl,pH7.6 by 1mM oxamate.

ENZYME	$\delta\epsilon_{428nm}$ pH 7.5 / mol^{-1} cm^{-1}			
	E -->^{FBP}E	^{FBP}E-->^{FBP}E$_{NADH}$	^{FBP}E$_{NADH}$->^{FBP}E$_{NADH}^{OXAM}$	SUM
NATIVE	277	160	225	662
NATIVE (NAD/oxalate-protected nitration)	64	0	0	64
E107Q	192	288	96	576
D197N	148	45	0	193

At this stage in the analysis the chemical probes had revealed that entry of the substrate or its competitive inhibitor into the active site resulted in the fast approach of β-G/β-H (carrying Asp-197) to helix α-2G (carrying Tyr-237) followed by a slower rearrangement process, at the same rate as the maximum velocity of the enzyme detected by cysteine-S^{13}CN-165 on α-1F. The need was now to   define other contact positions when the mobile loop approached helix α-2G. Chemistry is not specific enough for such a challenge.

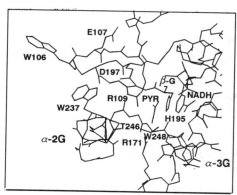

Figure 1. The sites at which single tryptophan (W) residues were genetically inserted into the tryptophanless B. stearothermophilus LDH framework to provide specific site signals for the substrate-induced formation of the catalytic vacuole. The first event is a fast (3000s^{-1})approach of Asp197 on βH to helix α-2G (Y-237 or W237). This is triggered by H-bond cross-bridges from Thr246 (α-3G) to the pyruvate -COO$^-$ and >C=O to His-195 (βG/βH) and onwards to α-H (see reference[2]). The slowest structural change (at Vmax 250s^{-1}) is LOOP (W106) closing onto helix α-2G (W237).

<u>A general method for inserting spectral probes at defined positions in proteins. The genetically inserted single tryptophan technique.</u>

The protocol uses site-directed mutagenesis to first replace all the natural tryptophan residues (80,150,203) of the protein by tyrosine(usually) and then checking the substitutions did not alter the catalytic mechanism or the regulatory properties of the enzyme. In the case of <u>B. stearo.</u> LDH they did not.[6] The next step is to reinsert a single tryptophan residue at a position in the structure where changed motion is suspected in response to a biological event - such as catalysis or regulation. The technique has also revealed the appearance of helices on the pathway by which proteins fold.[7] Both equilibrium and time-resolved indole fluorescence intensity and anisotropy are measured.

A genetically engineered single-tryptophan reporter identifies the movement of a peptide domain of LDH as the event which limits the maximum enzyme velocity.[8]

At which point in the mechanism does the mobile surface loop (98-110) closedown over helix α-2G ? To probe this event Gly-106 at the tip of the loop (Fig. 1) was replaced by tryptophan. The maximum rate of NADH-oxidation in this mutant was 160s^{-1} and the maximum rate at which oxamate could perturb the fluorescence of this loop-closure probe was also 160s^{-1}. This specific probe showed the tip of the mobile loop was induced to change state at the 'slow' (Vmax) conformation change rate. To determine the surfaces contacting in this shape change another single tryptophan enzyme was constructed with a tryptophan at position 237 (the same as a previously nitrated Tyr). The information from this probe was that when ternary complex formation was induced by adding oxalate to E-NAD$^+$ the fluorescence of the probe decreased in two phases at 2°C: the one at 90s^{-1} and the second at 25s^{-1}. When warmed to 25°C the slowest rate in the oxalate-induced shape change increased to 222s^{-1}, while the fast step became too rapid for stopped flow. Thus tryptophan 237 detects both the fast conformation change initially induced by substrate binding and the slower step which is also seen by a tryptophan probe at the tip of the mobile surface loop.

To conclude: together with two new x-ray structures,[2] the kinetics of the site specific spectroscopic probes suggest the first shape change when the substrate binds to E-NADH is the very rapid (3000s^{-1}) approach of ß-H to α-2G. Movement of ß-H is induced by the substrate forming strong hydrogen bonds which rapidly cross-link His-195 (ß-H) to Thr-246 (α-3G)[9] which then draws ß-H towards both α-2G and α-3G. The slower shape change is on the contact of Trp-106 and α-2G, the relation of Cys-165 to α-1F and the environment of a tryptophan at position 248 (under the coenzyme site) . The final slow clench of the recognition loop onto α2G and of α-H onto ßH is associated with water immobilization and is considerably retarded in a viscous medium (30% w/v glycerol).[2]

A.C. thanks the Spanish Dirección General de Investigación Científica y Técnica for a grant (BE 90-326) for sabbatical leave. Supported by SERC and (travel) NATO and British Council.

REFERENCES

1. A.R. Clarke, T. Atkinson and J.J. Holbrook in "Proteins: form and function" edited by R.A. Bradshaw and M. Purton, Elsevier Trends Journals, 1990, pp. 31-45.

2. C.R. Dunn, H.M. Wilks, D.J. Halsall, T. Atkinson, A.R. Clarke, H. Muirhead and J.J. Holbrook, Phil. Trans Roy. Soc. B. (London), 1991 accepted, issue July 26th

3. H. Deng, J. Zheng, J. Burgner, A.R. Clarke, J.J. Holbrook and R.H. Callender. Biophysical J., 1990, 57, 41.

4. D.M. Parker, D. Jeckel and J.J. Holbrook, Biochem. J., 1982, 201, 465.

5. A.R. Clarke, A.D.B. Waldman, K. Hart, and J.J. Holbrook, Biochim. et Biophys. Acta, 1985, 829, 397.

6. A.D.B. Waldman, A.R. Clarke, D.B. Wigley, K.W. Hart, W.N. Chia, D.A. Barstow, T. Atkinson, I. Munro and J.J. Holbrook, Biochim. et Biophys. Acta, 1987, 913, 66.

7. C. J. Smith, A. R. Clarke, W.N. Chia, L. Irons, T. Atkinson and J.J. Holbrook, Biochemistry, 1991, 30, 1028.

8. A.D.B. Waldman, K.W. Hart, A.R. Clarke, D.B. Wigley, D.A. Barstow, T. Atkinson, W.N. Chia and J.J. Holbrook, Biochem. Biophys. Research Commun., 1988, 150, 752.

9. D. Bur, A.R. Clarke, J.D. Friesen, M. Gold, K.W. Hart, J.J. Holbrook, J.B.Jones, M.A. Luyten and H.M. Wilks. Biochem. Biophys. Res. Commun., 1989, 161, 59.

RAMAN SPECTROSCOPIC STUDIES OF LIGANDS BOUND TO PROTEIN ACTIVE SITES: ENZYMES AND G-PROTEINS

Robert Callender
Physics Department
City College of City University of New York
New York, New York 10031

INTRODUCTION

The general question of how molecules or molecular fragments interact within proteins is of great importance. Generally, the chemistry that makes up life is tightly controlled and highly organized. Most of the chemical pathways depend on specific interactions between reactants, as for example between a protein and a bound ligand. In general, vibrational spectroscopy is a very powerful tool to investigate the molecular properties of small molecules.

A serious problem confronting the use of vibrational spectroscopy of small molecules embedded in large proteins is one of the 'needle and the haystack'. The bands observed in vibrational spectroscopy arise from all parts of the protein•ligand complex with many, generally overlapping, bands. It is thus difficult to unravel those from the coordinate of interest from all others. Resonance Raman spectroscopy has provided the needed selectivity in studies of bound chromophores. Unfortunately, this method is limited to residues and molecular groups whose absorption bands are relatively isolated from the absorption bands of other groups. For these and other reasons, we have extended classical (or non-resonance) Raman **difference** spectroscopy to look for very small signals embedded within large signals and have had substantial success in obtaining the Raman spectrum of ligands bound to proteins[1-4]. In these experiments, the Raman spectrum of the protein•ligand complex is obtained as is that of the protein itself. The spectrum of the bound ligand is very weak compared with that of the protein, the peak intensity of ligand signals being about 0.1 to 3.0% of the protein bands. Nevertheless, a high signal to noise difference spectrum can be calculated which yields the spectrum of the bound ligand. We are able to calculate reliable differences as small as 0.1% with our current difference spectrometer. With this technology in hand, the 'needle' can be found within the 'haystack' for most any problem.

227

We have applied the technique to a number of protein-ligand systems. These include measurements of substrate or inhibitor and coenzyme binding to the NAD dehydrogenases such as liver alcohol[5-6] (LADH), lactate[2,4,8,9] (LDH), and malate (MDH) dehydrogenases and dihydrofolate reductase (DHFR). We have also examined the binding properties of GDP and GTP to the EF-Tu elongation factor and the H-*ras* protein, p21. These proteins are strongly related to a large class of signalling proteins, the so-call G-proteins, which are involved in a diverse set of biochemical pathways from vision to growth. We summarize here a small sample of recent results with the intention of presenting more detailed poster papers at the meeting.

ENZYME-SUBSTRATE INTERACTIONS

The NAD-linked dehydrogenases form a broad class oxidation-reduction enzymes which are needed for a variety of metabolic functions. We have studied several enzymes of this class. The reactions catalyzed by alcohol and lactate dehydrogenases between substrate and the co-enzymes NADH and NAD+ are shown:

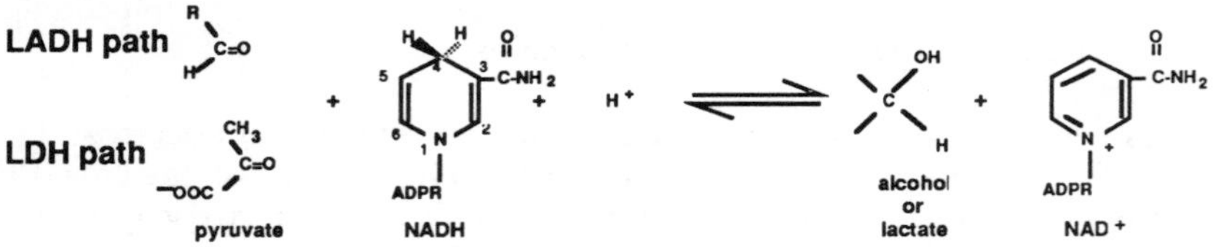

In both these enzymes, the reversible reaction involves the direct stereochemical transfer of a hydride, H-, ion from the R face of NADH to the carbonyl carbon of the substrate with a high degree of fidelity. For example, the hydride transfer occurs from NADH's pro-R face,"A side", with an error rate of only one part in 10^8 in LDH[10].

While we have obtained a wealth of information on the bound molecules, we will focus here on data that concerns the carbonyl, C=O, moiety of the substrates and the carboxamide moiety of the coenzymes. These groups have vibrational motions that are quite localized and are characterized by well defined Raman bands. We have extensive results on an aromatic aldehyde, p-(dimethylamino)benzaldehyde (DABA), bound to LADH and pyruvate bound to LDH. We observe a substantial change in the C=O stretching vibration as these substrates bind to their respective enzyme. We have also observed the C=O stretch and the -NH$_2$ rocking motion of NADH's (or NAD+) carboxamide moiety and have also observedchanges between their solution and bound positions to these two enzymes and a number of others. Table I summarizes this data. It is feasible to calculate the strength of the hydrogen bond, ΔE, between substrate and the carbonyl moiety based on these spectroscopic data, since Δv and ΔE are functionally related (see e.g. ref. 11).

There are a number of notable conclusions from the data in the Table. One is the existence of a very sizable interaction between the substrate's carbonyl moiety and the enzyme in both LADH and LDH. The transition state of the reaction coordinate involves a polar $^+$C-O$^-$

Table 1. The observed frequencies (cm^{-1}) of chosen groups.

Group	Δv (in situ - solution)	ΔE(kcal/mol)
C=O stretch of bound substrates		
DABA in LADH	-94	13.9
Pyruvate adduct in LDH	-34	21.5
C=O stretch of NAD's carboxamide's group		
LDH	-10	7.6
LADH	-10	7.6
sMDH	-9	7.3
mMDH	-9	7.3
DHFR	--	
-NH$_2$ rock of NAD's carboxamide's group		
LDH	+30	~2-4
LADH	+30	~2-4
sMDH	+35	~2-4
mMDH	-4	~0
DHFR	+35	~2-4

species, and this will certainly be stabilized by strong hydrogen bonding, resulting in a sizable C=O bond polarization[4]. Also, we have suggested previously that the hydrogen bonding patterns observed for the C=O and -NH$_2$ groups of NAD's carboxamide moiety are sufficiently strong to be a major factor in determining the observed high fidelity of stereospecific hydride transfer[9]. Recently, it has been estimated that the 'energy difference' between the transfer of LDH's A proton compared to the B proton is greater than 10.4 kcal/mol in LDH, based on the application of an Arrhenius analysis of measured error rates (V. Anderson, private communication). A molecular mechanics analysis performed by these authors also suggests that the cavity containing the nicotinamide head group has sufficient space and flexibility so that the syn and anti conformations are essentially accessible equally well sterically. In general, it is found that the pro-R hydrogen is transferred when NADH binds syn and pro-S for anti. The data in Table 1 rationalizes the large required energy difference between the syn and anti conformations in terms of the hydrogen bonding pattern of NADH's carboxamide group.

THE RAMAN SPECTRA OF GDP IN G-PROTEINS

G-proteins are generally very unstable without bound nucleotide. We therefore have obtained the spectra of GDP to EF-Tu and the H-*ras* proteins by incorporating an isotopically labelled nucleotide into these proteins. By substracting labelled protein•GDP complex from unlabeled, we obtain difference bands (both positive and negative) associated with motions that the label affects. Our first attempt has been to specifically deuterate the nucleotide ligand at the 8 position of the purine ring. A number of purine ring modes are shifted, and we have obtained the difference spectra of those bands with a signal to noise of over 30 to 1. We shall discuss the conformational and hydrogen bonding pattern changes resulting from the binding of GDP to these proteins that are suggested by this data.

ACKNOWLEDGMENTS

The summary reported here is the work of many individuals: Drs. J. Burgner, J. Holbrook, H. Deng, J. Zheng, D. Mannor, F. Jurnak, V. Balogh-Nair, S. Cosloy, D. Sloan, T. K. Yue, D. Chen, C. Martin, J. Lugtenburg, K. Rhee, and R. VanderSteen and students G. Weng, K. Delaria, Z. Chen, and C. Chen.

REFERENCES

1. K.T. Yue, J.P. Yang, C.L. Martin, S.K. Lee, D. Sloan & R. Callender. *Biochem. Biophys. Res. Comm.* 122, 225-229 (1984).
2. H. Deng, J. Zheng, J. Burgner, D. Sloan & R. Callender. *Biochem.* 28, 1525-1533 (1989).
3. K.T. Yue, H. Deng & R. Callender. *J. Raman Spec.* 20, 541-546 (1989).
4. H. Deng, J. Zheng, J. Burgner & R. Callender. *Proc. Natl. Acad. Sci. (USA)* 86, 4484-4488 (1989).
5. K.T. Yue, J.P. Yang, M. Charlotte, S.K. Lee, D. Sloan & R. Callender. *Biochem.* 23, 6480-6483 (1984).
6. R. Callender, D. Chen, J. Lugtenburg, C. Martin, K.W. Ree, D. Sloan, R. VanderSteen & Y.K. W. *Biochem.* 27, 3672-3681 (1988).
7. D. Chen, K.T. Yue, C. Martin, K.W. Rhee, D. Sloan & R. Callender. *Biochem.* 26, 4776-4784 (1987).
8. R. Callender, H. Deng, D. Sloan, J. Burgner & T.K. Yue. in *Raman Difference Spectroscopy and the Energetics of Enzymatic Catalysis* 1, 154-160 (International Society for Optical Engineering, Los Angeles, 1989).
9. H. Deng, J. Zheng, J. Burgner & R. Callender. *J. Phys. Chem.* 93, 4710-4713 (1989).
10. R.D. LaReau, W. Wan & V.E. Anderson. *Biochem.* 28, 3619-3624 (1989).
11. Z. Latajka & S. Scheiner. *Chem. Phys. Letts.* 174, 179-184 (1990).

NEW OPPORTUNITIES OFFERED BY FTIR AND RESONANCE RAMAN
STUDIES OF PROTEIN COMPLEXES

P. J. Tonge and P. R. Carey

Institute for Biological Sciences
National Research Council of Canada
Ottawa, Ontario K1A OR6 Canada

Introduction

Ground state strain in the reaction mechanism of enzymes.

Enzymes catalyse reactions, in part, by reducing the
free energy difference (ΔG) between the ground and
transition states on the reaction coordinate. This
reduction in ΔG can be accomplished either by
stabilisation of the transition state or by
destabilisation of the ground state. Whilst transition
state stabilisation is accepted as an integral part of
enzyme catalysis, the degree and even the existence of
ground state destabilisation remains a central question
in mechanistic enzymology.

Using resonance Raman spectroscopy we have obtained
the vibrational spectrum of the carbonyl group ($\nu_{C=O}$)
for a series of acyl-chymotrypsins and acyl-subtilisins
which encompass a 500-fold range of deacylation rate
constants. The acyl-enzymes were formed from $\alpha-\beta$
unsaturated arylacryloyl substrates (Figure 1) which
absorb in the 300-400 nm region. Laser excitation at 324
or 337.5 nm into the substrates' $\pi-\pi^*$ electronic
transition provides the RR spectrum of the bound
substrate.

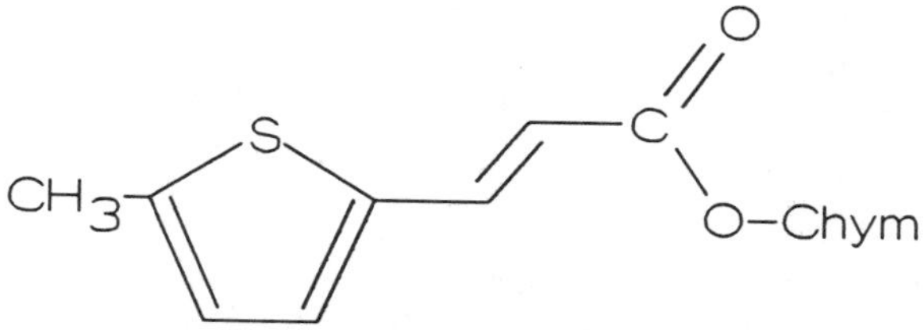

Figure 1 A typical $\alpha-\beta$ unsaturated acyl-enzyme;
5-methylthienylacryloyl-chymotrypsin (5-MeTA-Chym).

For the series of acyl-enzymes an increase in the deacylation rate has been correlated with a decrease in $\nu_{C=O}$ for the acyl-enzyme indicating an increase in polarisation of the reactive carbonyl group in the active site. Using protein engineering it has been shown that the increase in carbonyl polarisation results from stronger hydrogen bonding of the carbonyl group in the enzymes' oxyanion hole. Finally, using an empirical correlation between $\nu_{C=O}$ and carbonyl bond length, the 500-fold increase in rate observed through the acyl-enzyme series has been shown to correlate with a 0.015 Å increase in the carbonyl bond length. This represents ca. 7% of the expected increase in bond length on going from the ground state to the transition state in the deacylation reaction. There is thus now a quantitative measure of the degree of ground state destabilisation in the reaction mechanism of serine proteases (1).

FTIR spectroscopy of acyl-enzymes

The above RR studies are limited to chromophoric acyl-enzymes. Potentially, however, the vibrational spectrum of the substrate can also be obtained by FTIR spectroscopy, permitting the investigation of non-chromophoric (e.g. peptide-based) acyl-enzymes.

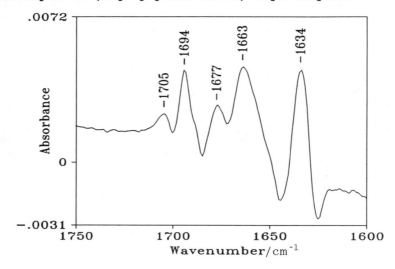

Figure 2 FTIR spectrum of free enzyme subtracted from the FTIR spectrum of phenylmethylsulfonyl-chymotrypsin. Both samples 1.4 mM in H_2O (2).

Subtraction of the FTIR spectrum of free enzyme from the spectrum of the acyl-enzyme, results in a spectrum dominated by enzyme-associated bands which effectively mask the spectrum of the bound substrate. This is illustrated in Figure 2 where the FTIR difference spectrum of phenylmethylsulfonyl-chymotrypsin is shown. Phenylmethylsulfonyl-fluoride, the 'substrate' used to generate the complex, shows no bands in this region of the spectrum and thus the difference spectrum in Figure 2 results from enzyme bands. In all probability substrate binding results in a minor structural change in the enzyme which is reflected by the conformationally sensitive amide I modes. The 'novel' features seen in Figure 2 may provide access to the subtle conformational changes associated with substrate binding which are a necessary prelude to catalysis.

Resonance Raman studies of acyl-enzymes: Photoisomerisation.

Figure 3 compares the FTIR carbonyl spectrum of 5-MeTA-Chym free from enzyme-associated bands with the RR spectrum. Whilst the carbonyl profiles in the 1700 cm^{-1} region are comparable, the carbonyl band at 1727 cm^{-1} in the RR spectrum is absent in the FTIR spectrum.

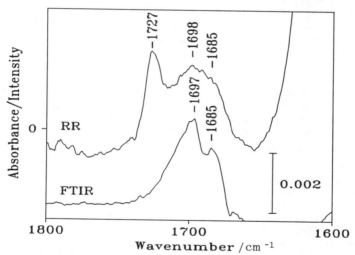

Figure 3 Top. RR spectrum of 5-MeTA-Chym in ^2H_2O at p^2H 3.4. Bottom. FTIR spectrum of 5-MeTA-Chym at p^2H 3.4 obtained by subtracting the spectrum of $^{13}C=O$-labelled acyl-enzyme (1.4 mM) from the spectrum of unlabelled acylenzyme (1.4 mM) (2).

However, a band near 1720 cm^{-1} can be induced in the FTIR spectrum by actinic light in the near UV region (Figure 4). Thus the band observed at 1727 cm^{-1} in the RR spectrum is assigned to a laser-induced population of acyl groups. This does not affect the bond length-rate correlation discussed in the Introduction as the 1727 cm^{-1} band was previously assigned to an inactive population of acyl groups.

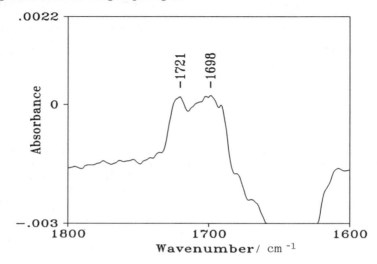

Figure 4 FTIR spectrum of 5-MeTA-Chym at p^{2}H 3.4 after 1 min irradiation with a mercury arc lamp generated by subtracting the spectrum of irradiated ^{13}C=O-labelled acyl-enzyme (1.4 mM) from the spectrum of irradiated unlabelled acylenzyme (1.4 mM) (2).

It is hypothesised that a light-induced trans to cis isomerization of the acyl group is occurring. As trans-cis isomerization involves a major structural change in the acyl-group, it requires a significant transient alteration in the conformationally restricted active site. Any fluctuation of the active site will be intimately related to the dynamics of the protein molecule. Thus modulation of active site dynamics via alteration in, for example, sample temperature or viscosity can potentially be monitored via substrate photoisomerism.

References
1. P.J. Tonge and P.R. Carey, Biochemistry, 1990, 29, 10723.
2. P.J. Tonge, M. Pusztai, A.J. White, C.W. Wharton and P.R. Carey, Biochemistry, 1991, in press.

THE "LIGAND SHUTTLE" REACTIONS OF CYTOCHROME OXIDASE: SPECTROSCOPIC EVIDENCE, DYNAMICS, AND FUNCTIONAL SIGNIFICANCE

William H. Woodruff*[1a], R. Brian Dyer[1b], Ólöf Einarsdóttir[2], K. A. Peterson[1b], P. O. Stoutland[1a], K. A. Bagley[1a], G. Palmer[3], J. R. Schoonover[3], D. S. Kliger[2], R. A. Goldbeck[2], T. D. Dawes[2], J.-L. Martin[4], J.-C. Lambry[4], S. J. Atherton[5], and S. M. Hubig[5]

[1] Los Alamos National Laboratory, Los Alamos, NM 87545
(a. INC-4, Mail Stop C345; b. CLS-4, Mail Stop J567)
[2] Department of Chemistry, University of California, Santa Cruz
[3] Department of Biochemistry, Rice University, Houston, TX
[4] Laboratorie D'Optique Apliquée, École Polytechnique, Palaiseau
[5] Center for Fast Kinetics Research, University of Texas, Austin

* Author to whom correspondence should be addressed

INTRODUCTION

Time-resolved electronic absorption, infrared, resonance Raman, and magnetic circular dichroism spectroscopies are applied to characterization of the intermediate which is formed within 20 ps after photodissociation of CO from cytochrome a3 of reduced cytochrome oxidase. This intermediate decays with the same halflife (ca. 1 μs) as the post-photodissociation Cu_B^+-CO species previously observed by time-resolved infrared (1). The transient UV-Vis spectra, kinetics, infrared, and Raman evidence suggest that an endogenous ligand is transferred from Cu_B to Fe_{a3} when CO binds to Cu_B, forming a cytochrome a3 species with axial ligation which differs from the reduced unliganded enzyme. The time-resolved magnetic circular dichroism results suggest that this transient is high spin and therefore five coordinate. Thus we infer that the ligand from Cu_B binds on the distal side of cytochrome a3 and displaces the proximal histidine imidazole. This remarkable mechanistic feature is an additional aspect of the previously proposed (1) "ligand shuttle" activity of the Cu_B/Fe_{a3} pair. We suggest that the ligand shuttle may play a functional role in redox-linked proton translocation by the enzyme. More detail on this work is presented elsewhere (1-5).

RESULTS AND DISCUSSION

Transient Absorbance. Using UV-Vis measurements between 100 fs and 100 ms, we have studied the photodissociation/recombination reactions of CcO-CO (3,4). The timescale of the initial photoevent (CO dissociation) is close to the Fe-C vibrational period, ca. 64 fs. The dependence of k_{obs} for the rebinding CO to the heme upon [CO] yields the rate constant k_2 for transfer of CO from Cu_B^+ to Fe_{a3}^{2+}, 700 s^{-1} and the pre-equilibrium constant K_1, 130 M^{-1}(3). On shorter timescales the α-band shows two transients. The first, an increase in absorbance, has a halflife of 20 ps (4). By comparison, the risetime of the Cu_B-CO absorbance measured by TRIR is 1 ps or less (5). The second is a decrease with the same rate constant observed by TRIR for loss of CO by Cu_B^+-CO (1). We infer that a ligand L is transferred within 20 ps from Cu_B^+ to Fe_{a3}^{2+}, triggered by the 1 ps binding of photodissociated CO to CuB^+. Conversely, we suggest that L returns from Fe_{a3} to CuB in ca. 1 μs, triggered by loss of CO from CuB^+-CO into solution. We have also observed that the recombination of CO with cytochrome a3 is accelerated by increasing the probe light intensity of the transient spectrophotometer. This is easily understood in the mechanism suggested above. If the Fe-L bond is photolabile, light will accelerate both k_2 and k_{obs}.

<u>Time-Resolved Resonance Raman</u>: The TR^3 spectra were obtained at pulse-probe time delay of 200 ns; both high probe laser pulse energy (1 mJ) and low energy (ca. 10 μJ) were employed. In the low-frequency region the high-power spectrum clearly shows the Fe-N(Im) stretch of 5-coordinate high-spin cytochrome a3 at 223 cm^{-1}. However, this peak is missing in the low-power trace. The low-power TR^3 spectrum could be due to a 6-coordinate low-spin species with both L and imidazole as axial ligands, wherein the Fe-N stretch is missing. Alternatively, it may be due to a 5-coordinate high-spin cytochrome a3-L complex wherein L is not imidazole and induces cleavage of the Fe_{a3}-N(proximal histidine) bond. The disappearance of the 223 cm^{-1} peak then occurs simply because the axial ligand L is not imidazole.

<u>Time-Resolved Magnetic Circular Dichroism</u>: The MCD spectrum of reduced unliganded CcO has a large contribution from high-spin cytochrome a3. This is characterized by an intense positive Soret MCD feature at 455 nm. In the TRMCD a signal which is similar to that of unliganded CcO develops within the excitation laser pulsewidth ($\leq$ 10 ns), and remain the same until replaced by that of CcO-CO on the millisecond timescale. The TRMCD results are strong evidence that the transient observed in the UV-Vis and TR^3 experiments is high spin.

<u>Conclusions</u>: The evidence presented above favors the scheme in Figure 1 as a minimal mechanism needed to account for the kinetics and the spectroscopic facts (ignoring the hypothetical protonation/deprotonation reactions). The ligand L is exchanged between CuB^+ and Fe_{a3}^{2+} as CO binds to and dissociates from CuB. The UV-Vis kinetics (3) set the timescales for the binding of L to the heme ($\leq$ 20 ps) and its return to CuB (~ 1 μs) as CO binds to CuB (<1 ps) and dissociates into solution (~ 1 μs) (1,5). The TRMCD suggests that the transient is high spin, therefore 5-coordinate. Thus the binding of L to Fe_{a3} causes the release of the proximal histidine. The five-coordinate Fe_{a3}^{2+}-L complex is suggested to be responsible for the short timescale α-band transients, the TRMCD spectra, and the low-power TR^3 spectrum.

We expect that the phenomena observed for CO are representative of other π-acid ligands, including NO, isocyanides, and most importantly O_2. If this is the case the "ligand shuttle" activities of CuB^+ (i) in acting as an intermediate binding site for ligands entering and leaving the site of O_2 reduction, (ii) in synchronously transferring L between CuB and cytochrome a3, and (iii) in engendering the release and rebinding of the proximal histidine imidazole, undoubtedly have functional significance. One such function is suggested (schematically, presuming L to be a group with a labile proton) in Figure 1. Directed protonation/deprotonation reactions of L and the proximal histidine, accompanying the transfer of L from one metal to another on one side of the heme plane and the release and rebinding of the proximal histidine on the other, may act as microscopic "gates" for active proton translocation. The proton gating mechanism based upon the ligand shuttle embodies the novel feature of being linked to the coordination chemistry of the Cu_B/Fe_{a3} pair, rather than exclusively to the redox reactivity.

We emphasize that the driving force for proton translocation must arise from the energy of the electron transfer reactions associated with the reduction of O_2 (6). However, the ligand shuttle as a proton gate may be coupled to electron transfer by changes in heme and copper redox potentials and electron transfer rates as a consequence of changing the identity or protonation state of the ligands. In this regard we note the recent result of Brzezinski and Malmström (7), who reported the electron transfer kinetics of mixed-valence CcO following photodissociation of CO from Fe_{a3}^{2+}. The rate constant for electron transfer from (Fe_{a3}^{2+}, Cu_B^+) to Cu_A^{2+} in the "E_1 state" was 600 s^{-1} ($\Delta H^{\ddagger}$ = 7.7 kcal/mol, $\Delta S^{\ddagger}$ = -18 eu). By comparison we have measured the rate of dissociation of the Fe_{a3}^{2+}-L bond (k_2, or the conversion of intermediate 2 to 3 in Figure 1) to be 700 s^{-1} ($\Delta H^{\ddagger}$ = 10 kcal/mol, $\Delta S^{\ddagger}$ = -12 eu) (3). The similarity of these kinetics parameters suggests that the rate determining step for the electron transfer reaction may be the cleavage of the Fe_{a3}^{2+}-L bond, as suggested in Figure 2. Considering this, it may be that a major microscopic difference between the E_1 and E_2 proton translocating states postulated by Brzezinski and Malmström (6b, 7) is whether L is bound to Fe_{a3} or Cu_B.

Acknowledgement: This research was supported by NIH Grant DK 36263 (WHW), GM 38549 (DSK), GM 21337 (GP), and RR 00886 (CFKR). Work at Los Alamos National Laboratory was performed under the auspices of the U. S. Department of Energy.

REFERENCES

1. Dyer, R. B., Einarsdóttir, ´O., Killough, P. M., López-Garriga, J. J., and Woodruff, W. H. (1989) J. Am. Chem. Soc. 111, 7657-7659.
2. Woodruff, W. H., Einarsdóttir, ´O., Dyer, R. B., Bagley, K. A., Palmer, G., Atherton, S. J., Goldbeck, R. A., Dawes, T. D., and Kliger, D. S. (1991) Proc. Nat. Acad. Sci. U. S. A., in press.
3. Einarsdóttir ´O., Killough, P. M., Dyer, R. B., López-Garriga, J. J., Hubig, S. M., Atherton, S. J., Palmer, G., and Woodruff, W. H. (1990) submitted for publication in Biochemistry.
4. Stoutland, P. O., Lambry, J.-C., Martin, J.-L., and Woodruff, W. H. (1990) manuscript in preparation.
5. Dyer, R. B., Peterson, K. A., Stoutland, P. O., and Woodruff, W. H. (1990) submitted for publication in J. Am. Chem. Soc.
6. (a) Wikström, M., Krab, K., and Saraste, M. (1981) "Cytochrome Oxidase - A Synthesis," Academic Press, New York; (b) Malmström, B. G. (1990) Arch. Biochim.Biophys., 280 233-241; (c) Chan, S. I., and Li, P. M. (1990) Biochem. 29, 1-12.
7. Brzezinski, P., and Malmström, B. G. (1987) Biochim.Biophys. Acta, 894 29-38.

237

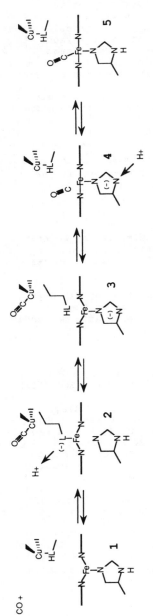

Figure 1. Pictorial representation of the mechanism of CO binding to and dissociation from cytochrome oxidase. Only CuB and Fea3 are shown. The species 1, 2, 3, 4, and 5 represent the thermal mechanism. All of the reactions are first order, except for the bimolecular reaction with CO in solution; the halflives are either known or reliably estimated from experiment, or else they are omitted. The protonation/deprotonation reactions shown are hypothetical, not inferred from the evidence presented here (see text).

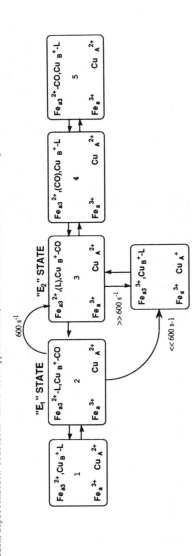

Figure 2. Suggested mechanism for the electron transfer reaction from Fea3 to CuA following photodissociation of the mixed-valence CO complex of cytochrome oxidase. See "conclusions" section of the text and Reference 7. The species 1,2,3,4, and 5 are the mixed-valence analogues of the corresponding species in Figure 1.

HYDROGEN BOND INDUCED GROUND STATE ELECTRONIC STRAIN AND CATALYSIS OF DEACYLATION IN ACYL-α-CHYMOTRYPSINS. INVESTIGATION BY FOURIER TRANSFORM INFRARED SPECTROSCOPY.

A.J.White, K.Drabble, S.Ward, and C.W.Wharton*

School of Biochemistry
University of Birmingham
Birmingham   B15 2TT
United Kingdom

In acyl-chymotrypsins, the ester acyl C=O is H-bonded to two backbone amides giving the catalytically important "Oxyanion Hole" (OAH). OAH strength and significance was assessed using FTIR to examine acyl C=O stretch vibration in a series of non specific acyl-chymotrypsins[1].

Acyl C=O spectra (c.1700cm^{-1}) were obtained by subtraction of free enzyme from acyl enzyme spectra to eliminate the large protein band c.1750-1600cm^{-1} and so disclose the small acyl C=O band unique to the acyl enzyme. The conformational difference between free and acyl enzymes leads to perturbation of the protein band such that clean subtraction is not possible and artefact bands remain. Substitution of ^{13}C for ^{12}C at the acyl C=O allows eradication of perturbation artefacts[2] via subtraction of {^{13}C=O}acyl enzyme from {^{12}C=O}acyl enzyme spectra. The isotopic substitution causes a shift for the acyl C=O band of 38cm^{-1} (model studies and theoretical calculations). Both of these methods have provided useful data[1,2], and two or three acyl C=O bands have been observed for the acyl-chymotrypsins so far examined. These bands have been assigned to a number of different conformations with respect to the acyl group, contrary to crystallographic evidence[3], but in accord with FTIR studies of enzyme-substrate complexes of triosephosphate isomerase[4], aldolase[5], and citrate synthase[6]. The conformers have been assigned[1] as productive,P (OAH), non-productive,NP (acyl C=O in H-bond bridge to His57 via H$_2$O - as per Henderson[3]), and non-bonded,NB (no acyl C=O H-bonds).

Ground state electronic strain (GSES) is present in acyl-chymotrypsins through the OAH H-bonds which enhance the acyl C=O dipole, thereby weakening the bond and raising its electrophilicity for H$_2$O attack during deacylation. The acyl C=O frequencies for P and NP, and NB in one case, for the acyl-chymotrypsins studied have been used to attempt to demonstrate the importance of GSES for the deacylation catalysis. The comparison NB-P represents non H-bonded versus full OAH state, ie, the full OAH interaction, while comparisons NP-P represent weaker H-bonded Henderson[3] versus full OAH states, ie, only a portion of the full OAH interaction. An important assumption underlying this analysis is that all of the OAH interaction, ie, the OAH H-bonds and their strain effect on the acyl C=O, is actually utilised as GSES directed towards transition state formation and thus catalytic action. Two approaches were adopted[1], one based on model systems, and the other a theoretical treatment.

Lady and Whetsel[7] studied H-bonding thermodynamics in complexes of aniline and ethylacetate using N-H stretch (c.3480cm^{-1}). This has been adapted so as to use C=O stretch, and frequency shift,$\Delta V_{C=O}$ correlated with ΔH_{H-bond} values originally obtained using ΔV_{N_TH}. On this basis[1], dehydrocinnamoyl(dc) NP-P $\Delta V_{C=O}$=13cm^{-1} gives ΔH_{H-bond}=13kJ/mol. Therefore, *trans*-cinnamoyl(tc) NP-P $\Delta V_{C=O}$=10cm^{-1} represents a weaker and hydrocinnamoyl(hc) NP-P $\Delta V_{C=O}$=18cm^{-1} a stronger H-bond, and hc NB-P $\Delta V_{C=O}$=39cm^{-1} represents a very strong H-bond. These H-bonds seem to be of significant strength on a biological scale (c.5-40kJ/mol). $\Delta V_{C=O}$ and thus ΔH_{H-bond} are observed[1] to be directly linearly related to kinetic specificity, showing the catalytic importance of these H-bonds and thus GSES. This suggests that specific substrates will give high $\Delta V_{C=O}$ and ΔH_{H-bond}. We have initiated experiments using a rapid scanning (c.7 scans/s at 2cm^{-1} resolution) instrument and stopped and continuous flow fast delivery/mixing apparatus in order to study more specific acyl-chymotrypsins.

Using simple harmonic motion as an approximation, it can be directly shown[1] from hc NB-P $\Delta V_{C=O}$=39cm^{-1} that GSES provides a rate enhancement factor of 53000 in this case. This is highly significant and demonstrates the importance of GSES in catalysis.

In conclusion, it is thus proposed that GSES plays a significant role in generating catalysis of deacylation in acyl-chymotrypsins through powerful H-bonding to the acyl C=O. It is interesting to speculate why the NB and NP states are present to any extent at all if they are so much less stable than P. Clearly, there must be other factors stabilising NB and NP and/or destabilising P. Moreover, the existence of such states is likely to be of some as yet unclear biochemical significance, because they otherwise would seem to only decrease the efficiency of the enzyme. This is especially so since they appear to be a common occurrence[1,4,5,6].

1. A.J.White and C.W.Wharton, *Biochem.J.*, 1990, *270*, 627.
2. A.J.White *et al.*, *Biochem.Soc.Trans.*, 1991, *19/2*, In Press.
3. R.Henderson, *J.Mol.Biol.*, 1970, *54*, 341.
4. J.G.Belasco and J.R.Knowles, *Biochemistry*, 1980, *19*, 472.
5. J.G.Belasco and J.R.Knowles, *Biochemistry*, 1983, *22*, 122.
6. L.C.Kurz and G.R.Drysdale, *Biochemistry*, 1987, *26*, 2623.
7. J.H.Lady and K.B.Whetsel, *J.Phys.Chem.*, 1967, *71*, 1421.

FOURIER TRANSFORM INFRARED (FTIR) SPECTROSCOPIC DETECTION OF SMALL CONFORMATIONAL CHANGES IN THE CATALYTIC CYCLE OF SARCOPLASMIC CALCIUM ATPASE

A. Barth, W. Kreutz and W. Mäntele

Institut für Biophysik und Strahlenbiologie
Albertstr. 23
D–7800 Freiburg

INTRODUCTION

Ca^{2+}–transport from the cytoplasm of muscle cells into *Sarcoplasmic Reticulum* (SR), necessary for muscle relaxation, is performed by the Ca^{2+}–ATPase, an intrinsic membrane protein of about 110 kDa molecular mass. The energy required for this active transport process is provided by hydrolysis of ATP. Although the Ca^{2+}–ATPase is one of the best examined transport proteins, the transport mechanism, especially the coupling between ATP hydrolysis and active calcium transport, is still unknown. The investigation of molecular changes connected with the catalytic function should clarify this point.

Infrared (IR) spectroscopy is a powerful tool to reveal even very small structural changes in a catalytically active protein, provided that it is possible to start the protein reaction in the IR cuvette. This was demonstrated in the last years for photoreactions and electron transfer reactions [1,2]. However, the required sample thickness of 10 μm makes it extremely difficult to start a protein reaction by simply mixing the substrate to the IR sample. In order to circumvent this problem, we have used the photolytic release of ATP from caged ATP to start the reaction cycle of the ATPase [3]. Caged ATP is an inactive, photolabile ATP–derivative that releases ATP upon ultraviolet illumination.

RESULTS AND DISCUSSION

FTIR spectroscopy was used to investigate structural changes in the Ca^{2+} ATPase of SR during the catalytic cycle. The ATPase reaction was started in the IR cuvette by release of ATP from caged ATP. Absorption spectroscopy in the visible spectral region using the Ca^{2+}–sensitive dye Antipyrylazo III ensured that the IR samples were able to transport Ca^{2+} in spite of their low water content required for mid–IR measurements (1800–950 cm^{-1}). Small, but characteristic and higly reproducible IR absorbance changes were observed upon ATP release (Fig. 1). According to their kinetic properties 3 signal types can be distinguished (mainly in the specified wavenumber regions):

(a) PERMANENT CHANGES (1524, 1342, 1270–950 cm^{-1}). Comparison to a sample prepared without ATPase, but with caged ATP, showed that they are related to photolysis of caged ATP [3,4].

(b) SLOW CHANGES (1300–1000 cm^{-1}), rising on the timescale from several seconds to minutes depending on temperature and sample composition. They are due to hydrolysis of ATP as seen by the difference between the absorption spectra of ATP and [ADP+P$_i$] [3,4].

241

(c) TRANSIENT CHANGES (1750–1520 cm⁻¹) due to a protein absorbance change. This assignment is based on control samples which contained inactivated ATPase or caged ADP instead of caged ATP and which did not show transient signals comparable to the "normal" ATPase sample [4].

The absorbance changes of the ATPase were mainly observed in the region of Amide I and Amide II protein absorbance and presumably reflect the molecular processes upon phosphoenzyme formation. Since the absorbance changes were small compared to the overall ATPase absorbance, no major rearrangement of ATPase conformation as the result of catalysis could be detected.

Several differences are evident between the absorbance changes of the standard sample and a sample containing approx. 20% DMSO. They might be caused by the structural change from the Ca_2E_1P to the E_2P state. The absorbance changes of a standard sample in D_2O are difficult to interpret in terms of isotopic band shifts, since the ATPase might accumulate in another step of the catalytic cycles.

Time–resolved FTIR measurements of ATPase phosphorylation are in progress. In addition, other steps of the reaction cycle will be investigated.

ACKNOWLEDGEMENTS

We would like to thank Prof. W. Hasselbach and Dr. R. S. Goody for generous gifts of Ca^{2+}–ATPase and caged ADP.

REFERENCES

1.   F. Siebert, W. Mäntele and W. Kreutz, <u>FEBS Lett.</u>, 1982, <u>141</u>, 82.
2.   D. Moss, E. Nabedryk, J. Breton, and W. Mäntele, <u>Biophys. J.</u>, 1989, <u>55</u>, 394a.
3.   A. Barth, W. Kreutz, and W. Mäntele, <u>FEBS Lett.</u>, 1990, <u>277</u>, 147–150
4.   A. Barth, W. Mäntele, and W. Kreutz, <u>Biochim. Biophys. Acta</u>, 1991, in press

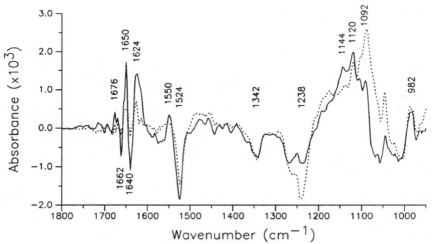

Fig. 1. Change of IR absorbance due to ATP release in an ATPase sample at 0°C. Spectra recorded 4 s (full line) or 2 min (dotted line) after ATP release.

FOURIER-TRANSFORM INFRARED DIFFERENCE SPECTROSCOPY OF ELECTROCHEMICALLY TRIGGERED REDOX TRANSITIONS IN PROTEINS

W. Mäntele[*], M. Bauscher, M. Leonhard, F. Baymann, and D.A. Moss

Institut für Biophysik und Strahlenbiologie der Universität Freiburg, Albertstraße 23, D-7800 Freiburg FRG

Infrared (IR) difference spectroscopy has become an established tool for the investigation of those proteins in which a transition could be triggered without disturbing the sample. This has been particularly successful in case of photobiological systems, because it is possible to trigger a transition in the protein by illumination[1,2]. Since only few enzymes undergo a light-induced reaction, while many change their redox state during their catalytic cycle, we have developed electrochemical techniques to trigger enzyme redox reactions. To study electrochemically generated redox states of proteins in aqueous solutions, IR and Optically Transparent Thin-Layer Electrochemical (IOTTLE) cells were developed[3]. For a fast reaction rate and sufficient transmittance, the optical path length has been decreased, so that the entire solution in the measuring beam constitutes a diffusion layer (10 μm). A three electrode system with a transparent gold grid working electrode, a platinum foil counter electrode and a silver/silver chloride reference electrode enabled us to perform controlled potential electrolysis as well as diagnostic electrochemical experiments, such as cyclic voltammetry. A Fourier transform (FT) IR spectrophotometer has been used to yield a high signal/noise ratio in a short scan time. As a control, VIS/NIR spectra can be recorded simultaneously with an IR scan. Data acquisition is performed by a computer program which controls electrochemistry, IR and VIS/NIR spectroscopy.

Three factors can severly limit the rate of electron transfer between electrodes and proteins: Slow diffusion rates, inaccessibility of redox centers, and the tendency of most proteins to stick to metal surfaces. These problems can be overcome by the use of surface-modified electrodes[4], mediators[5], or both.

Quantitative and reversible redox reactions have been obtained for a number of soluble, membrane-bound or detergent-solubilized redox enzymes from photosynthetic or respiratory electron transport chains[6-8]. IR difference spectra calculated from spectra of two different redox states show highly-structured band features in the mid-infrared, as demonstrated by reduced-minus-oxidized difference spectra for a number of c-type[5] and c_2-type[6] cytochromes or for the electron donor[7] and acceptor[8] in bacterial photosynthetic reaction centers. The difference spectra exhibit a level of sensitivity which corresponds to individual bonds in the redox enzymes, and are thus as rich in information on intramolecular protein processes as are light-induced IR difference spectra. Furthermore, redox-induced IR difference spectroscopy has the advantage, that absorption changes can be redox titrated, leading to unequivocal identification of the redox component from which the IR signals arise, even in complex multicomponent systems.

REFERENCES

1   J.O. Alben, P.P. Moh, F.G. Fiamingo and R.A. Altschuld, Proc. Natl. Acad. Sci. USA, 1981, 78, 234.

2   W. Mäntele, E. Nabedryk, B.A. Tavitian, W. Kreutz and J. Breton, FEBS Lett., 1985, 187, 227.

3   D. Moss, E. Nabedryk, J. Breton and W. Mäntele, Eur. J. Biochem., 1990, 187, 565.

4   M.J. Eddowes and H.A.O. Hill, J. Chem. Soc. Chem. Commun., 1977, 721.

5   W.R. Heineman, Anal. Chem., 1978, 50, 390A.

6   F. Baymann, D.A. Moss and W. Mäntele, this volume.

7   M. Leonhard, D.A. Moss, M. Bauscher, E. Nabedryk, J. Breton and W. Mäntele, this volume.

8   M. Bauscher, D.A. Moss, M. Leonhard, E. Nabedryk, J. Breton and W. Mäntele, this volume.

RESONANCE RAMAN STUDIES OF THE TERNARY INHIBITOR COMPLEX OF E.COLI THYMIDYLATE SYNTHASE WITH 5-FLUORO-2'-DEOXYURIDYLATE

J. C. Austin, T. G. Spiro and J. E. Villafranca*

Department of Chemistry, Princeton University, Princeton, NJ 08544, U.S.A.,
*Aguron Pharmaceuticals, 11025 N. Torrey Pines Rd., La Jolla, CA 92037, U.S.A.

Thymidylate synthase (TSase) catalyses the reductive methylation of 2'-deoxyuridylate (dUMP) to thymidylate (dTMP), using the cofactor 5,10-methylenetetrahydrofolate (CH_2-H_4folate) as the methylating agent:-

$$dUMP + CH_2\text{-}H_4folate \longrightarrow dTMP + H_2folate$$
$$(7,8\text{-dihydrofolate})$$

The unique role of TSase in the synthesis of dTMP, and its high activity in tumor cells have made TSase a target for cancer chemotherapy. Extensive investigations have contributed valuable insights to the mechanism of TSase action [1, 2], and have also aided the understanding of the activity of a number of anticancer drugs[2].

Nucleotide analogs have been found to effectively inhibit TSase action; in particular 5-fluoro-2'-deoxyuridlate (FdUMP) has been shown to form a covalent ternary inhibitor complex with TSase and CH_2-H_4folate ($K_i = 10^{-11}$ M^{-1}).

The proposed structure[3] of the complex is shown in Figure 1 below:-

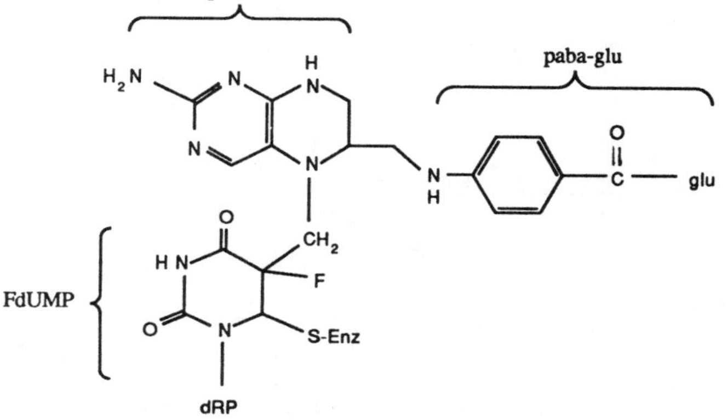

Figure 1: Proposed Stucture of the TSase Ternary Complex with FdUMP

A previous RR study of the ternary inhibitor complex of FdUMP with TSase from L.Casei purported to show evidence for alternative equilibrium structures of the complex in solution[4]. The alternative structure was assigned on the basis of a RR spectrum obtained in resonance with the longest wavelength absorption band of the complex, centered around 375 nm ($\varepsilon = 6500$ M^{-1}cm^{-1}). The present data show that a photoproduct species gives a RR spectrum at longer excitation wavelengths (356 or 363 nm); the spectrum of the photoproduct was the basis for the alternate structure assignment. Both the absorption and RR spectra of the complex change after prolongèd laser irradiation.

To understand the complex vibrational spectrum of the TSase ternary complex, and to aid the identificaation of the photoproduct species, RR spectra have been obtained with isotopically labelled CH$_2$-H$_4$folate. ^{13}C and D labels on the p-aminobenzoylglutamate (paba-glu) moiety of the cofactor produce large changes in the RR spectra obtained with 337 nm excitation (see Figure 2) whereas the RR spectrum of the photoproduct species (observed with 363 or 356 nm excitation) appears to be largely unaffected. This evidence leads to the conclusion that the photoproduct species does not arise from a photochemical modification of the paba-glu moiety, but is possibly a species with an oxidised pterin ring.

Figure 2: 337 nm-excited RR spectra of E.Coli TSase ternary complex with FdUMP and (a) unlabelled cofactor, (b) D-labelled cofactor - at 3,5-positions on paba ring, (c) ^{13}C-labelled cofactor - at paba-benzoyl position.

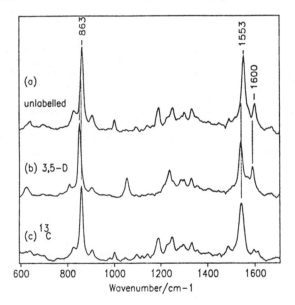

References
[1] D.V. Santi and P.V. Danenberg in "Folates in Pyrimidine Nucleotide Biosynthesis", R. L. Blakeley and S. J. Benkovic, Eds., pp.350-398 Wiley, New York (1984).
[2] R. J. Cisernos, L. A. Silks and R. B. Dunlap Drugs of the Future 13 859-881 (1988).
[3] T. L. James, A. L. Pogolotti, Jr. , K. M. Ivanetich, Y. Wataya, S. S. Lam and D. V. Santi Biochem. Biophys. Res. Commun. 72 404 (1976).
[4] A. L. Fitzhugh, S. Fodor, S. Kaufman and T. G. Spiro J. Am. Chem. Soc. 108 7422 (1986).

KETOSTEROID ISOMERASE: NEW INSIGHTS FROM ULTRAVIOLET RESONANCE RAMAN SPECTROSCOPY

J. C. Austin, A. Kuliopulos#, A. S. Mildvan# and T. G. Spiro

Department of Chemistry, Princeton University, Princeton NJ 08544, U.S.A.
Department of Biological Chemistry, The Johns Hopkins School of Medicine, Baltimore MD 21025, U.S.A.

Ketosteroid Isomerase (KSI) catalyses the isomerization of a Δ^5-ketosteroid to a Δ^4-ketosteroid. The reaction is thought to proceed via a dienol intermediate, by the concerted action of two enzyme functional groups[1]. A schematic representation of the reaction mechanism is shown in Figure 1.

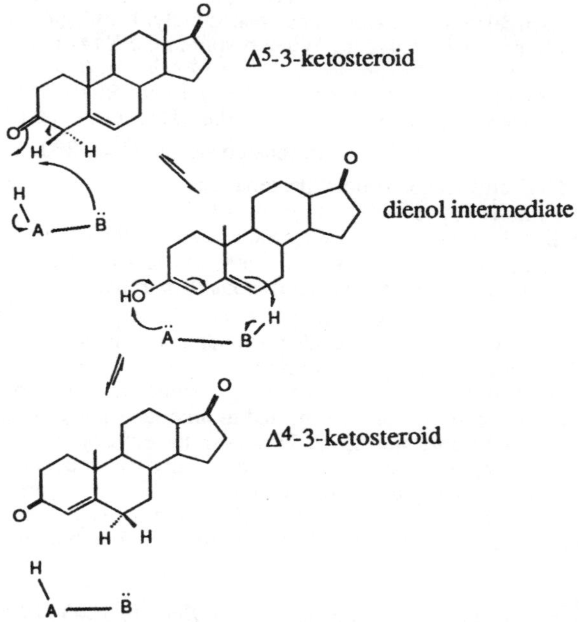

Figure 1: Representation of the KSI reaction mechanism.

From spectroscopic and kinetic studies of specific KSI mutants, it is believed that tyrosine-14 (Tyr-14) acts as a general acid, protonating the steroid carbonyl, whilst aspartate-38 acts as a base, deprotonating the steroid C_4 position[2]. Ultraviolet resonance Raman (UVRR) spectroscopy has been used to investigate the binding of steroid inhibitors to two mutants of KSI. The two

247

mutants are the Y14$_o$ mutant, in which Tyr-55 and Tyr-88 are both mutated to phenylalanine (leaving the active site Tyr-14 as the only tyrosine), and the Y14F mutant in which Tyr-14 is mutated to phenylalanine.

When the competitive inhibitor 19-nortestosterone (19-nor) or its 4-fluoro-derivative is bound to Y14$_o$ or wild type KSI, the steroids' UV absorption is 10 nm red-shifted - from 248 nm to 258 nm in the case of 19-nortestosterone. This red shift is mimicked by placing 19-nortestosterone in strong acid, but is not observed in KSI if Tyr-14 is mutated to phenylalanine. The latter observations have led to the suggestion that 19-nortestosterone may be protonated in the enzyme active site, leaving Tyr-14 deprotonated[2]. UVRR spectra of Y14$_o$ KSI with and without bound steroid have been obtained at 230 nm and 254 nm to determine the state of the steroid and Tyr-14 in the resting and complexed enzyme. Excitation at 230 nm selectively probes the tyrosine residue(s) with minimal interference from the phenylalanine residues, whereas 254 nm enhances the RR spectrum of the bound steroid.

The UVRR data obtained with 230 nm excitation show no evidence for the formation of tyrosinate when 19-nortestosterone (or 4-fluoro-19-nor) is bound to Y14$_o$ KSI. Thus, if 19-nortestosterone is protonated in the active site of KSI, then Tyr-14 is being rapidly reprotonated. There is evidence for a change in the hydrogen bonding state of Tyr-14 as the steroid binds, but there is no evidence for a strong steroid carbonyl - tyrosine hydrogen bond.

The UVRR spectrum of 19-nortestosterone in aqueous solution contains two strong RR bands at 1614 and 1638 cm^{-1}, arising from the (predominantly) vinyl and carbonyl stretching vibrations, respectively. Both bands shift to higher wavenumber in non-polar solvents. When bound to the Y14$_o$ mutant of KSI, the two bands downshift to 1596 and ca. 1615 cm^{-1} (a similar downshift is observed for enzyme-bound 4-fluoro-19-nor). Interestingly, the RR spectrum of 19-nortestosterone in acid does not resemble the RR spectrum of 19-nortestosterone bound to KSI. The protonated steroid (λ_{max} 260 nm) has a single RR band at 1605 cm^{-1}. Thus simple protonation cannot explain the large effects observed in both the RR and absorption spectra.

The dienolate species formed by the reaction of the natural substrate with base has an absorption maximum at 256 nm[3]. The UVRR spectrum of this short-lived species has been obtained in a continuous flow apparatus. The RR spectrum of the dienolate shows a strong band at 1628 cm^{-1}, and shows no similarity to the bound 19-nortestosteone spectrum. We conclude that a dienolate species is not formed as 19-nortestosterone binds to KSI.

Both hydrogen bonding to the carbonyl oxygen and conformational changes in the enone group can be expected to produce changes in the absorption and RR spectra of 19-nortestosterone. However, the environment provided by the enzyme for 19-nortestosterone is not easily mimicked by model environments - varying the solvent polarity and hydrogen bonding ability. The present RR data suggest that there is a hydrogen bond from Tyr-14 to the steroid carbonyl, but that there may also be a conformational change in the steroid on binding to the enzyme.

References
[1] J. M. Schwab and B. S. Henderson *Chem. Rev.* 90 1203 (1990).
[2] A. Kuliopulos, A. S. Mildvan, D. Shortle and P. Talalay *Biochem.* 28 149 (1989).
[3] T. C. M. Eames, D. C. Hawkinson and R. M. Pollack *J. Am. Chem. Soc.* 112 1996 (1990).

The Effect of Substrate Addition on the Vibrational Modes of the Fe_4S_4 Cluster of Aconitase: A Resonance Raman Study

LaTonya Kilpatrick, Roman Czernuszewicz, Thomas G. Spiro
Dept. of Chem., Princeton University, Princeton, New Jersey
08544

Mary Claire Kennedy and Helmut Beinert
Dept. of Biochemistry, Medical College of Wisconsin, Milwaukee,
WI 53226

The enzyme, aconitase, catalyzes the isomerization of citrate to isocitrate through a cis-aconitate intermediate. (Fig. 1) The active protein contains a diamagnetic $[4Fe-4S]^{2+}$ cluster which binds substrate as an equilibrium mixture of citrate (88%), isocitrate (8%), and cis-aconitate (4%) during the enzymatic reaction. Mössbauer, EPR, and ENDOR results have shown that substrate coordinates to a specific iron of the cluster through the C-2 carboxyl and/or -OH group of citrate.[1,2,3] This iron is released under aerobic conditions with concomitant formation of a $[3Fe-4S]^+$ cluster. The crystal structure of pig heart aconitase further reveals that this unique iron is not bound to a terminal cysteine residue.[4] Instead, the fourth ligand is a hydroxyl group from solvent.[5]

We have exploited the resonance Raman (RR) effect to examine the novel structure of this Fe_4S_4 cluster. We have also used this vibrational technique to monitor the structural perturbations that occur following substrate addition.

Our RR data show that the symmetry of aconitase's Fe_4S_4 cluster is lower than D_{2d} symmetry due to the asymmetric ligation of the terminal ligands. Upon addition of substrate, however, the symmetry about the cluster site increases. In the RR spectrum, the FeS^b (FeS bridging) 269/275 cm^{-1} and the

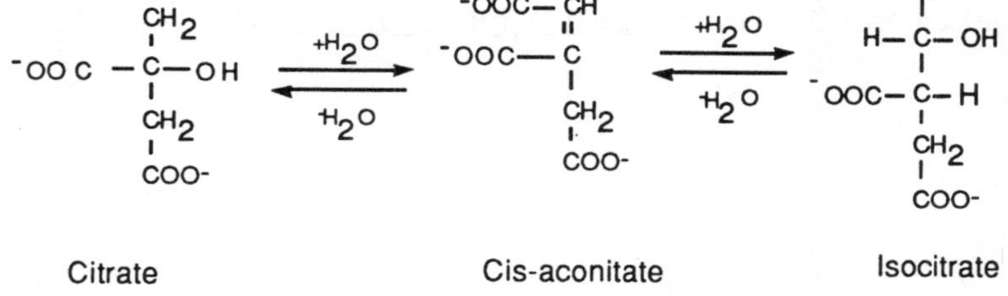

Figure 1 Isomerization of citrate to isocitrate

FeS^t (FeS terminal stretching) 252/260 cm^{-1} vibrations collapse to form two very intense modes at 272 and 356 cm^{-1}. These new modes are ^{34}S sensitive. However, no frequency shifts are observed when the protein is equilibrated with $H_2{}^{18}O$ or D_2O,

thus, eliminating the possibility that these modes are υ(Fe-OH$_x$) stretching interactions.

In addition, we have been able to detect hydrogen bonding interactions between the cluster sulfur atoms and NH groups of the protein. Several FeS^t and FeS^b stretching vibrations experience 1 to 2 cm^{-1} downshifts upon labelling the amide hydrogens with deuterium. Moreover, we have observed a deuterium sensitive vibrational mode at 466 cm^{-1}. This mode, however, is not ^{34}S sensitive thus ruling out a high frequency υ(FeS) stretching vibration. We have assigned this band to an amide skeletal bending vibration, based on a normal coordinate analysis reported for N-methylacetamide.[6]

Although several attempts were made to detect a substrate derived υ(Fe-OH) stretching vibration, we did not observe any bands that could be assigned to this interaction. Excitation lines from the UV (354.6 nm) to the red (676.4 nm) were used in an unsuccessful attempt to find the OH-->Fe charge transfer transition. While we did observe a very weak mode at 565 cm^{-1} with 413.1 nm laser excitation, this band exhibited no $H_2{}^{18}O$

sensitivity and thus could not be assigned to a υ(Fe-OH) stretching vibration.

References

1. T. A. Kent, M. H. Emptage, H. Merkle, M. C. Kennedy, H. Beinert, and E. Münck, J. Biol. Chem., 1985, 260, 6371.
2. M. H. Emptage, T. A. Kent, M. C. Kennedy, H. Beinert, and E. Münck, Proc. Nat'l Acad. Sci. U.S.A., 1983, 80, 4674.
3. J. Telser, M. H. Emptage, H. Merkle, M. C. Kennedy, H. Beinert, and B. M. Hoffman, J. Am. Chem. Soc., 1989, 110, 1935.
4. A. H. Robbins and C. D. Stout, Proc. Natl. Acad. Sci. U.S.A., 1989, 86, 3639.
5. M. Werst, M. C. Kennedy, A. L. Houseman, H. Beinert, and B. M. Hoffman, Biochemistry, 1990, 29, 10533.
6. J. Jakes and S. Krimm, Spectrochimica Acta , 1971, 27A, 19.

STRUCTURAL PROPERTIES OF STREPTOKINASE

H. Welfle, R. Misselwitz, H. Fabian, K. Welfle, W. Pfeil,
W. Damerau, R. Kraft, S. Kostka and D. Gerlach*

Central Institute of Molecular Biology, D-1115 Berlin, FRG
*Central Institute of Microbiology and Experimental
Therapy, DO-6900 Jena, FRG

INTRODUCTION

Streptokinase (SK), a catabolic by-product of certain
strains of ß-hemolytic streptococci, can activate human
plasminogen and merits interest because of its clinical use
as a thrombolytic agent. Activation of plasminogen to the
fibrinolytic enzyme, plasmin, proceeds via the formation of
a stoichiometric complex of SK and plasminogen[1] by an up to
now unknown mechanism which is probably connected with conformational changes of both proteins. SK is composed of 414
amino acids and has a molecular mass of 47 000 Da.[2] Its
primary sequence was determined both by protein sequencing[2]
and from elucidation of the nucleotide sequence of the
cloned SK gene.[3] Conformational properties of SK obtained
by spectroscopic and calorimetric techniques and by secondary structure prediction methods have been reported and
have suggested the classification of SK as an $\alpha + \beta$ protein.[4]
     Being interested in the details of the plasminogen ac-
tivation mechanism we have started an analysis of the
structural properties of the components of this system and
present here results of biochemical and biophysical studies
on the solution structure of SK.

RESULTS

Limited proteolysis of SK by trypsin and thermolysin was
performed under different incubation conditions and analyzed by polyacrylamide gel electrophoresis. At low ionic
strengths the C-terminal region of SK is much more susceptible to proteolytic cleavage than the relatively stable
region comprising amino acids Ser_{60} to Lys_{294}. The N-terminal sequences of fragments Tr27, Tr18, Th26 and Th16 and
the C-terminal sequences of Tr27 and Th26 were identified
by partial sequencing. In this way Tr27 and Th26 were localized within the Sk sequence comprising amino acids Ser_{60}
to Lys_{294} and Phe_{63} to His_{292}, respectively.
The N-terminal amino acids of Tr18 and Th16 were found to
be Glu_{148} and Ile_{161}, respectively.

A solvent accessibility of about 70% of the 22 Tyr residues was found by UV perturbation spectroscopy. Fluorescence measurements indicated also the surface localization of the single Trp_6 residue. Circular dichroic, infrared and Raman spectra were analyzed in order to estimate the content of secondary structure elements of SK. Values in the range of 14-23% α-helices, 38-46% ß-structures, 10-30% turns and 12-23% residual structures were found. Effects of temperature, pH and detergents were studied by CD spectroscopy. Structural effects were induced at temperatures above 40 °C, pH values below 3.0 and urea concentrations above 2 M. At temperatures above 70 °C, at pH 2.1, and at urea and Gu·HCl concentrations of 7 M and 5 M, respectively, no further structural changes are revealed in the spectra. At temperatures around 50 °C, at pH 3.0, and detergent concentrations of about 1 M Gu·HCl and 2 M urea, CD effects were observed in the near-UV region indicating an increase in the asymmetry for aromatic amino acids in comparison to the structure of SK in low ionic strength buffers at neutral pH, 20 °C and in the absence of detergents. These effects were most pronounced for the temperature dependence of the CD spectra. ESR spectroscopy on spin-labeled SK has shown that loosening of the protein surrounding of the spin label begins already at 1 M urea and that the mobility of the spin label points to a structural change in SK at 46 °C.

The melting pattern of SK at neutral pH and low ionic strength is characterized by two distinct heat absorption peaks and by a significant increase of heat capacity. Thermal unfolding of SK is almost reversible as judged by reheating of the same sample. Quantitative analysis of the calorimetric recordings gives two separate two-state transitions with average transition temperatures and enthalpy changes of T_{trs1} = 45.9 + 0.4 °C, Δ H1 = 431 + 18 kJ/mol, T_{trs2} = 60.1 + 1.3 °C, and ΔH2 = 306 + 16 kJ/mol. The partial specific heat capacity of native SK was determined to be Cp = 1.42 + 0.17 J/g/K, and the denaturational heat capacity change associated with the two transitions, Δ Cp_1 = 0.21 J/g/K and Δ Cp_2 = 0.38 J/g/K, respectively. The overall melting pattern of SK remains almost unchanged at a variety of tested solvent compositions, except at pH values below 4 and above 10 and in the presence of denaturants. The two domains show different susceptibility to urea. Microcalorimetric studies on isolated fragments points to a localization of the less stable energetic folding unit of SK in the N-terminal region of the protein.

REFERENCES

1. K.W. Jackson and J. Tang, Biochemistry, 1982, 21, 6620
2. H. Malke, B. Roe and J.J. Ferretti, Gene (Amst.) 1985, 34, 357
3. F.J. Castellino, Bioscience, 1983, 33, 647
4. J.T. Radek and F.J. Castellino, J. Biol. Chem., 1989, 264, 9915

FLUORESCENCE INVESTIGATION OF THIAMINE KINASE PROPERTIES DURING INTERACTIONS WITH SUBSTRATES, METAL IONS AND ALLOSTERIC EFFECTORS

A.A.Maskevich[*], I.P. Chernikevich[1] , G.A. Gachko,
L.N. Kivach, I.L. Korotaeva

State University of Grodno
Grodno 220023, USSR
[1]Institute of Biochemistry
Byelorussian Academy of Sciences
Grodno 220009, USSR

Tryptophane fluorescence of thiamine kinase from brewer's yeast and its complex with the probe 2-toluidinyl-naphthalene-6-sulfonate (TNS) was used to study the structure and dynamics of the protein at its binding to substrates (thiamine and ATP), effectors (pyruvate and its structural analogs) and divalent metal ions (Mg^{2+}, Mn^{2+}, Ca^{2+}). On the basis of the analysis of the decay curves the fluorescent tryptophanyls can be classified as three groups of chromophores with different lifetimes ($T_1 = 6.10 \pm 0.40$, $T_2 = 2.76 \pm 0.25$ and $T_3 = 1.10 \pm 0.12$ ns) and position of individual spectra (λ_1 -330, λ_2 - 325, λ_3 -320 nm).

The fluorescence decay of TNS in a complex with thiamine kinase can be represented by a sum of two components with $T_1 = 11.20 \pm 0.41$, $T_2 = 2.3 \pm 0.12$ ns which are due to the existence of two different sites for binding of the protein to the probe. TNS molecules having a shorter lifetime and a higher mobility are located at the site of enzyme allosteric regulation. On the contrary chromofores which greater lifetime values are located at the protein active site.

The binding of the enzyme to ATP is accompanied by quenching and a short-wave shift (≈ 1 nm) in the tryptophanyl fluorescence maximum. In this situation the fluorescence intensity of the probe is considerably increased (≈ 1.7- fold) and the fluorescence spectrum is shifted from 442 to 432 nm.

The interaction with thiamine brings about a considerable quenching of intensity and a long-wave shift in the protein tryptophan fluorescence spectrum. The fluorescence of the probe undergoes appreciable quenching (≈ 1.8-fold) and a 10-nm short-wave spectral shift. The obtained results show that thiamine and ATP are bound by pyrimidine and adenine cycles at the hydrophobic site of the enzyme.

The addition of $5 \cdot 10^{-3}$ M pyruvate leads to local conformational changes in the protein structure accompanied by more efficient quenching of fluorescence of tryptophanyls located at the most hydrophobic sites in the macromolecule. This effector also induces considerable, mainly static, quenching of probe molecules, located at the both binding sites. The obtained results confirm the assumption made from the biochemical studies on the presence of two sites for binding of pyruvate to thiamine kinase: the first one is relatively hydrophobic inhibitory, whereas the second one is hydrophilic activating. The latter is in close proximity to thiamine and provides the interaction between the pyruvate carbonyl group and the substrate.

The divalent metal ions raise the substrate and effector affinity for the protein, concurrently contributing to the increase of the hydrophobicity of its binding site. The obtained results suggest the following arrangement of substrate and effector molecules within the enzyme active site at the onset of the enzyme process (Fig. 1):

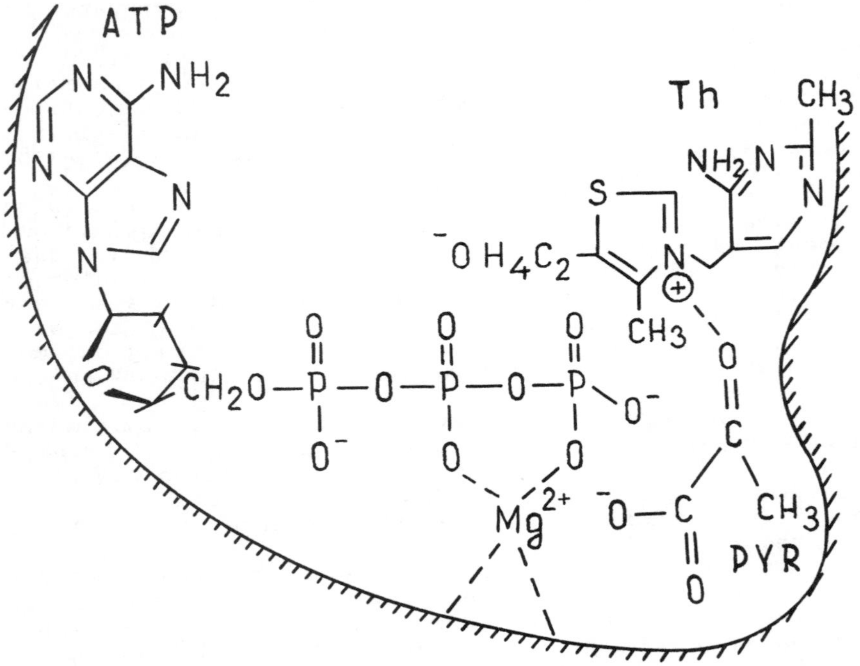

Fig. 1. The arrangement of substrate and effector molecules within the enzyme active site at the onset of the enzyme process. Th - thiamine, PYR - pyruvate ($CH_3COCOOH$)

THE STUDY OF STRUCTURE AND DYNAMICS OF ACTIVE CENTER OF PYRUVATE DECARBOXILASE FROM BREWER'S YEAST

I.P. Chernikevich[1], S.K. Basharin[*], L.N. Kivach,
S.A. Maskevich, N.D. Strekal, V.E. Voronich

State University of Grodno
Grodno 220023, USSR
[1]Institute of Biochemistry
Byelorussian Academy of Sciences
Grodno 220009, USSR

This paper presents the continuation of the study of active center structure of Pyruvate Decarboxilase (PDC) by Surface Enhanced Raman Scattering (SERS) in model systems which was begun in [1] and also Raman and fluorescence spectra of native enzyme.

It was noted that pyruvate forms complexes with thiaminepyrophosphate (ThPP), thiamine (Th), and 4'-hidroxythiamine during the adsorbtion process on Ag-electrode. The structure of these complexes are different Although all Th-derivatives are binding with pyruvate by thiazole component, effeciency of this interaction and their interaction with electrode surface significantly increase if atom N is protonated in site 1' of the pyrimidine cycle. This confirms the earlier [2] assumption about participation of atom N1' in organization of PDC active site structure.

The band of Amid-I (1647 cm^{-1}) in the Raman spectra of native holo-PDC (HPDC) and apo-PDC (APDC) is very intensive, since the band of Amid-III is very weak. This means the enzyme has mainly α-helical structure.

The results of the fluorometrical investigations of HPDC and APDC are shown in table 1. Decay curves analisys of tryptophan fluorescence of enzyme shows that they are approximated by sum of 3 components. There appeared two types of luminescence centers with different fluorescence decay times for the TNS probe which was bound to the enzyme. Values of T_1 and T_2, and also preexponential coefficients (they are shown within brackets in table 1) are different for HPDC and APDC and are sensitive to interaction between enzyme, ThPP and pyruvate. There are two types of binding sites of TNS and PDC apparent in probe fluorescence spectra. From table 1 it follows that Mg^{2+} ions change the structure of these sites very strong and its hydrophobicity is raised due to this. Based on the results obtained one can think that TNS has one more site of binding to PDC besides the active ones. Probably it is located close

to the pyruvate. The obtained values of the polarization degree (P) indicate higher mobilities of micro-
surroundings of this site as compared with the active one.

Experimental data analisys was making using
IBM PC/AT by software which was developed at State
University of Grodno.

Table 1   Fluorescence parameters of HPDC, APDC and
           its comlexes with fluorescence probe TNS,
           pyruvate PYR), Mg^{2+} and ThPP.

Complex	λ_{max}, nm	Int.	P	T, ns
HPDC [1]	331.0	—	0.23	5.7; 3.2; 1.3
HPDC+TNS [2]	428.0	40	0.30[4]	12.2; 5.2 (1:1)
			0.32[5]	9.3; 3.3 (1:1)
APDC [1]	332.5	—	0.22	6.9; 3.3; 1.6
APDC+Mg^{2+} [1]	332.0	—	0.23	—
APDC+TNS [2]	431.5	77	0.30	11.0; 4.8 (1:1)
APDC+Mg^{2+}+TNS [2]	424.5	100	0.31	12.1; 6.0 (1:2)
APDC+Mg^{2+}+TNS [3]	423.5	45	—	—
APDC+Mg^{2+}+ThPP+TNS [2]	422.0	80	0.29	12.6;5.7 (1:2.5)
APDC+Mg^{2+}+ThPP+TNS [3]	419.0	41	—	—
APDC+Mg^{2+}+PYR+TNS [2]	426.0	65	0.32	13.8; 6.0 (1:3)
APDC+Mg^{2+}+PYR+TNS [3]	420.0	36	—	—

λ_{ex}: [1]-295; [2]-350; [3]-390 nm. λ_{em}:[4]- 440; [5]-410 nm

Concentrations: HPDC, APDC - $1,5 \cdot 10^{-6}$; TNS - $2 \cdot 10^{-5}$;
                ThPP-$1.5 \cdot 10^{-5}$; Mg^{2+}-$1 \cdot 10^{-3}$; PYR-$1 \cdot 10^{-3}$ M

REFERENCES

1. S.A. Maskevich, K.V. Sokolov, L.N. Kivach et al.,
   'Surface-enhanced Raman spectroscopy of cocarboxilase
   and pyruvate complexes'. Proceedings of the Twelfth
   International Conference on Raman Spectroscopy.
   13-17 August 1990, Columbia, South Carolina.

2. Schelenberger A., Angew.Chem., 1967,23,1050-1061.

RAMAN SPECTROSCOPIC STUDIES ON HEN EGG WHITE LYSOZYME N-ACETYLGLUCOSAMINE INTERACTIONS

A. Bertoluzza[1], S. Bonora[1], G. Fini[1] and M.A. Morelli[2]

Centro di Studio Interfacolta'
  sulla Spettroscopia Raman
[1] Sezione di Chimica e Propedeutica
  Biochimica, Dipartimento di
  Biochimica, via Selmi 2, Bologna, Italy
[2] Dipartimento di Chimica "G. Ciamician"
  via Selmi 2, Bologna, Italy

The effect of the binding of specific inhibitors to an enzyme results in some modifications on the secondary structure as well as on the microenvironment of the representing groups in the active site of the enzyme and also in side groups that are relatively remote from the active site, usually as a result of conformational changes associated with the binding. It has been shown by many different chemico-physical techniques that N acetylglucosamine (NAG) binds to the same sites which are part of the active site cleft in the lysozyme (LYS) molecule.

Addition of NAG (10%w/w) to a saline water solution of LYS (10%w/w) increases the percentage of alpha helices (about 40%), calculated by the method of Lippert[1] and also stabilizes the folded conformation against thermal denaturation. The denaturation temperature measured by DSC increases from 72 to 80°C in the presence of NAG (NAG/LYS molar ratio 30).

Analogous experiments were carried out with other mono- and disaccharides (as D-mannose, D-glucose and saccharose), but with these chemicals the spectral features of LYS do not change even in the presence of a great amount of sugar; the results indicate the poor affinity of these substrates with regards to the enzyme.

Raman features regarding the lateral side chains, can offer interesting conclusions particularly on the Trp microenvironment, because three of the six Trp residues of LYS are part of the binding sites of the enzyme. By examination of the figure some interesting conclusions can be drawn. The frequency of a Trp band increases from 876 to 879 cm^{-1} indicating a weakening of the hydrogen bond given by N1H groups. Moreover, the addition of NAG induces a decrease of the $I(1360$ cm$^{-1})/I(1340$ cm$^{-1})$ intensity ratio, indicating that the hydrophobicity of Trp environment decreases[2] by inhibitor interaction.

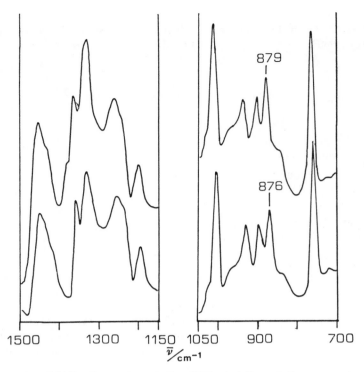

Raman spectra of LYS saline water solution (10%w/w) (lower); Raman spectra of LYS saline water solution in the presence of NAG (10% w/w) (upper).

The experimental results lead us to conclude that the interaction LYS-NAG involves mainly the active sites of LYS containing Trp residues as a result of the competitive inhibition of NAG; the interaction takes place mainly via weak hydrogen bonds.

The orientation of the indole ring of Trp does not change as a consequence of the interaction as it is clearly demonstrated by the steadiness of the frequency of the W3 band of Trp[3] located at 1563 cm[-1].

ACKNOWLEDGEMENT
This work was supported by C.N.R. (Italy) progetto finalizzato Chimica Fine II (Grant n. 89.00799.72).

REFERENCES

1. J.L. Lippert, D. Tyminski and P.J. Desmeules, J. Am. Chem. Soc., 1976, 98, 1075.
2. T. Miura, H. Takeuchi and H. Harada, Biochemistry, 1988, 27, 88.
3. P. Miura, H. Takeuchi and H. Harada, J. Raman Spectrosc., 1989, 20, 667.

PHOTOPHYSICAL PROCESSES IN THE NADH-ALCOHOL DEHYDRO-GENASE COMPLEX

N. L. Vekshin

Institute of Biophysics of
 Cell, Acad.Sci. of the
USSR, 142292, Moscow Region
Pushchino

Fluorescent analysis showed a high sensitivity of the NADH-alcohol dehydrogenase complex to UV-irradiation. The complex reacts to UV-illumination with a "photoresponse": rapid decrease in NADH fluorescence intensity to some stationary level. No such decrease was found in the NADH solution without protein. The maximum rate of photoresponse was observed at pH=8.0, i.e. in the biochemical optimum of the enzyme. It is suggested that the main reason of the photoresponse is desorption of NADH from protein due to vibrational and conformational relaxation after photoexcitation. The data obtained point to the possibility of controlling "dark" enzymatic processes with light.

Materials and Methods.

Horse liver alcohol dehydrogenase (ADH) used in the experiments.

Photoexcitation of the system in cuvette was carried out by a xenon 450-W lamp. The filter transmitting the UV-light over the range 250 to 380nm (UVL-1) or 320 to 380nm (UVL-6) was placed between the lamp and the cuvette. The fluorescence induced by the same light flow was recorded on a "SLM-4800" through monochromator with 4nm slits and the light filter SZS-19 or SZS-22 in order to cut off the exciting light in the fluorescence channel. The fluorescence wavelength was 450nm.

Irradiation was turned on by opening the shutter between the lamp and the cuvette. The monochromator shutter was already opened. In the course of experiment the dark and light alternated when it was necessary.

For recording the fluorescence in the absence of intensive irradiation ("dark fluorescence") we have used such position of the shutter between the lamp and the cuvette when the intensity of the exciting light was 100-200 times lower. In this case the sensitivity of amplifier or the voltage on photomultiplier were increased.

Results and Discussion.

As described in [1,2], photoexcitation of tryptophanyl or the probes sorbed on the protein induced protein structural dynamics in nanosecond time range. The
induced mobility of protein structure (photoconformati-
onal relaxation) could probably lead to such processes
as activation of enzymes or desorption of coenzymes
or substrates.
Under UV-irradiation the effective energy transfer
from trp-314 to NADH in NADH-ADH complex takes place[3].
Table 1 presents some data on the variations in
the intensity of NADH fluorescence under different experimental conditions on exposure to illumination.

Table 1. Change in fluorescence of NADH in the presence
of enzyme and without it on exposure to intensive UV-
light.

Time, min.	Illumination	Change in the intensity of NADH fluorescence in system:			
		NADH	NADH+ADH	NAD^++ADH $+C_2H_5OH$	NADH+ BSA
1	No	0.71	0.93	0.80	0.85
2	No	0.71	0.93	0.80	0.85
3	Light	0.69	0.82	0.63	0.79
4	Light	0.69	0.72	0.55	0.79
5	No	0.69	0.81	0.59	0.80
6	No	0.69	0.90	0.64	0.81
7	Light	0.67	0.77	0.53	0.77
8	Light	0.67	-	0.48	-

Footnote: Illumination is through the filter UVL-1.
Concentrations: NADH - 26 μM, alcohol dehydrogenase - 3 μM, NAD^+- 75 μM, ethanol
- 30 mM, bovine serum albumin (BSA)-3 μM.
In the system NAD^++ADH+ethanol the equilibrium ratio was reached; NADH-ADH complex
at a concentration of about 3 μM. Solutions were thermostated at 20°C.

REFERENCES

1. N.L. Vekshin, 'Photonics of Biological Structures',
   Acad.Sci., Pushchino, 1988 (in Russian).
2. N.L. Vekshin, Studia Biophys., 1987, 118, 173.
3. N.L. Vekshin, Eur.J.Biochem., 1984, 143, 69.

FLUORESCENCE SPECTROSCOPY OF MEMBRANE-BOUND NA,K-ATPase

SELECTIVELY LABELLED WITH A FLUOROPHORE

E. Grell, E. Lewitzki, R. Warmuth and H. Ruf

Max-Planck-Institute for Biophysics, D-6000 Frankfurt,

K. Brand and F. W. Schneider

Institute for Physical Chemistry, D-8700 Würzburg

Stationary and time-resolved fluorescence studies are carried out on an active transport enzyme, membrane-bound Na,K-ATPase (purified from pig kidney). In order to characterize its conformational properties, the protein is covalently labelled with fluoresceine isothiocyanate (FITC) (1) or studied in the presence of eosin Y, which reversibly interacts with the enzyme (2). N^{ϵ}-FITC,N^{α}-acetyl-L-lysine methylamide acts as a suitable model compound.

The stationary fluorescence emission intensity decreases markedly with decreasing pH in case of both the enzyme and the model compound. The corresponding pK value is 6.3 for the model compound and ranges between 6.2 and 6.6 for the enzyme, being only slightly dependent on the ionic composition. The emission intensity of the FITC-enzyme, eg. in imidazole buffer containing 100 mM choline chloride is about two times lower than that of the model compound, which may reflect intramolecular quenching effects due to protein residues. According to the results, it is assumed that the enzyme-bound fluorophore is located in a water exposed part of the protein.

The emission intensity of FITC-Na,K-ATPase exhibits marked changes upon varying the ionic composition. The highest intensity (F1) is found in the presence of eg. 100 mM NaCl or most other 1:1 and several 2:2 electrolytes (Figure). However, addition of millimolar concentrations of K^+, Rb^+ Cs^+, Tl^+ or NH_4^+ leads to a large intensity decrease (up to about 50%; F_2 in Figure) under conditions where the excitation maximum remains essentially unchanged. A similarly low intensity level (F_3) is found for the FITC-enzyme in histidine buffer. No comparable intensity changes are detectable for the model compound in the same media. In accordance with earlier proposals (1,3), the level F1 is attributed to the main conformational state E_1 and F_2 to the state E_2. As concluded from the

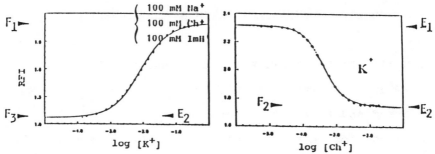

Figure Titrations of FITC-Na,K-ATPase in 10 mM histidine-HCl pH 7.5 with choline (Ch$^+$) chloride (similar results obtained with Na$^+$ or protonated imidazole ImH$^+$) [left] and in 25 mM imidazole-HCl pH 7.5 with KCl [right] : Plot of rel. fluor. intensity (495/518) versus logarithm of cation concentration.

kinetic studies, the level F_3 is also assigned to E_2. The spectral properties allow to perform titrations providing the thermodynamic parameters of the various cation complexes. These data are discussed on the basis of an electrostatic model concerning binding of most cations and of a model exhibiting cooperativity for the binding of the larger alkali ions.

In addition, fluorescence decay measurements (single photon counting) are carried out. The FITC-enzyme is characterized by two decay times (1.6 nsec and 4.1 nsec with amplitudes of similar magnitude) and the model compound by a single one of 3.8 nsec. These decay times are insensitive to the presence of Na$^+$ or K$^+$ and thus do not depend on the conformational state of the FITC-enzyme. We propose that the conformational change of the protein leads to a local absorbance change of the bound fluorophore which affects the measurable fluorescence intensity but not the decay times. In case of the eosin Y-Na,K-ATPase system, life time measurements allow to distinguish between free and enzyme-bound fluorophore and to differentiate between conformational states, which is discussed.

REFERENCES

1. S.J.D. Karlish, _J. Bioenerg. and Biomembr._,1980, _12_, 111; C. Hegyvary and P.L. Jorgensen, _J. Biol. Chem._, 1981, _256_, 6296.
2. J.C. Skou and M. Esmann, _Biochim. Biophys. Acta_, 1981, _647_, 232.
3. M. Mezele, E. Lewitzki, H. Ruf and E. Grell, _Ber. Bunsenges. Phys. Chem._, 1988, _92_, 998.

DEOXYGENATION IN THE BIOSYNTHESIS OF AROMATIC POLYKETIDES

K. Ichinose,[*] M. Sugimori, A. Itai, Y. Ebizuka, and U. Sankawa*

Faculty of Pharmaceutical Sciences,
The University of Tokyo,
Hongo, Bunkyo-ku,
Tokyo 113,
Japan.

Deoxygenation reaction ("Post-aromatic process") involved in the biosynthesis of fungal melanin was investigated by using synthesis, theoretical chemistry and enzymology.

Biosynthetic Pathway of Melanin in Phytopathogenic Fungi

[1] Reinvesitigation of a biomimetic synthesis of scytalone.

A biomimetic synthesis of scytalone (1),[1] was re-investigated[2] and it was revealed that scytalone (1) is formed from 1,3,6,8-THN (2) by sodium borohydride only in the presence of sodium methoxide. The reactive species of this reaction was clarified as a non-symmetrical keto-tautomer, THN4H , from the NMR spectroscopical data of a CD_3ONa-CD_3OD solution of 1,3,6,8-THN (2). A complete NMR assignment of the species is shown in the next table.

NMR Spectral Assignments of 1,3,6,8-THN in CD$_3$ONa-CD$_3$OD.

Position	^{13}C-NMR (ppm)[a]	^{1}H-NMR[b] (ppm)
C-1	192.6	-----
C-2	102.1	5.16
C-3	192.2	-----
C-4	41.0	c)
C-4a	143.4	-----
C-5	112.2	6.02
C-6	174.5	-----
C-7	105.3	5.88
C-8	165.0	-----
C-8a	107.7	-----

a) Relative to TMS, b) measured in CD$_3$OD-CH$_3$OH (1:1) solution of sodium methoxide, c) overlapped with methanol signal.

[2] Theoretical investigation of a biomimetic synthesis of scytalone.

The structure of the most stable ionic species of 1,3,6,8-THN (2) in CH$_3$ONa-CH$_3$OH was studied[3] by the semi-empirical (AM1) and ab initio (Gaussian 86: RHF; 3-21+G//3-21G) molecular orbital calculations. The most stable species was a non-symmetrical keto-tautomeric trianion, and the naphthalene type isomers were less stable by ca. 10 kcal/mol. The result is in good agreement with the observations by NMR spectroscopy.

Furthermore, transition structures and stabilities were simulated (AM1) for the hydride attack to the carbonyl carbons at C-1 and C-3 of THN4H, suggesting[4] that C-3 carbon was more favourable for the reduction, which coincided the experimental fact.

[3] Purification and characterization of hydroxy-naphthalene Reductase.

The enzymatic activities which reduce 1,3,6,8-THN and 1,3,8-THN to scytalone (1) and vermelone (3) respectively were detected in a cell-free extract from shaken cultures of Phialophora lagerbergii. Both reductive activities were observed in the same enzyme fractions in the partial purification using ammonium sulfate precipitation, DEAE cellulose column chromatography, and gel filtration. Further purification and characterization of the enzyme are now in progress.

References:
1) B.W. Bycroft, M.H. Cashap, and T.K. Leung, J. Chem. Soc., Chem. Commun., 1974, 443.
2) K. Ichinose, Y. Ebizuka, and U. Sankawa, Chem. Pharm. Bull., 1989, 37, 2873.
3) K. Ichinose, M. Sugimori, A. Itai, Y. Ebizuka, and U. Sankawa, Tetrahedron Lett., 1990, 31, 5905.
4) K. Ichinose, M. Sugimori, A. Itai, Y. Ebizuka, and U. Sankawa, Chemistry Lett, 1991, 219.

PROBING PROTEIN DYNAMICS BETWEEN 4 AND 300 K

Paul R. CAREY* and Munsok KIM

Institute for Biological Sciences
National Research Council of Canada
Ottawa, Ontario, Canada K1A 0R6

INTRODUCTION

Protein dynamics has become an important field of biophysical study in the past decade. A recurring theme in experimental studies is to follow a property of the protein over a wide range of temperature. A number of groups using different techniques have observed a transition in the monitored property near 200 K (1,2). This has been ascribed to the protein becoming glass-like below 200 K. However, controversy exists as to whether this is a natural transition of the protein itself or it is caused by the accompanying water solvent perturbing the protein. In this work we report resonance Raman measurements on enzyme-substrate complexes in ice matrices which allow us to probe simultaneously a thermal transition in the enzyme's active site and the properties of the bulk solvent.

As discussed recently (3-5),it is possible to obtain high quality RR data from enzyme-substrate complexes of the kind $RC(=O)NHCH_2C(=S)S$-papain which are trapped in ice matrices. The RR spectrum for $PhC(=O)NHCH_2C(=S)S$-papain seen in Fig. 1 shows an intense RR feature near 1130 cm^{-1}, due to a coupled $v_{C=S} + v_{C-S} + v_{C-C}$ motion. The peak was used to probe the active site properties as a function of temperature. At the same time the CH_3CN peak near 920 cm^{-1} was used to probe the bulk solvent properties (20% CH_3CN is present since this solvent was used to carry the substrate into the original reaction mixture).

Figure 2 plots the full widths at half height of the 1126-cm^{-1} enzyme-substrate RR feature and the 920-cm^{-1} CH_3CN solvent band. Interestingly, they both show a similar temperature dependence suggesting that the protein transition seen in the active site is strongly coupled to the properties of the bulk solvent.

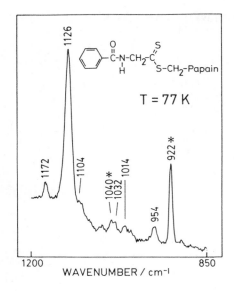

Fig.1 The 324-nm excited RR spectrum of the dithioacyl enzyme in an ice matrix (20% CH$_3$CN) at 77K. * solvent peak.

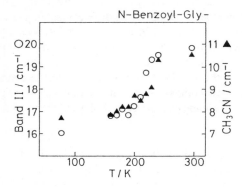

Fig. 2 Plots of the full widths of half height of the intense dithioacyl enzyme RR feature at 1126 cm^{-1} (open circles) and the CH$_3$CN solvent peak at 922 cm^{-1} (solid triangles) as a function of temperature.

REFERENCES

1. H. Frauenfelder, F. Parak and R. D. Young, Ann. Rev. Biophys. Biophys. Chem., 1988, 17, 451.

2. W. Doster, S. Cusack and W. Petry, Nature, 1989, 337, 754.

3. L. R. Sans Cartier, P. J. Tonge and P. R. Carey, Ind. J. Phys., 1989, 63, 5170.

4. P. J. Tonge, H. Lee, L. R. Sans Cartier, B. P. Ruzsicska, and P. R. Carey, J. Am. Chem. Soc., 1989, 111, 2496.

5. M. Kim and P. R. Carey, J. Mol. Structure, 1991, 242, 421.

TIME-RESOLVED INFRARED STUDIES OF THE DYNAMICS OF LIGAND BINDING TO THE CYTOCHROME a_3-Cu_B SITE OF CYTOCHROME c OXIDASE.

R. Brian Dyer and Kristen A. Peterson
Photophysics and Photochemistry Group
CLS-4, Mail Stop J567

Page O. Stoutland and William H. Woodruff
Isotope and Structural Chemistry Group
INC-4, Mail Stop C345

Los Alamos National Laboratory
Los Alamos, NM 87545 USA

Time-resolved infrared spectroscopy (TRIRS) has been employed to study the ligation dynamics of the cytochrome a_3-Cu_B site of cytochrome c oxidase (CcO). The reactions of small molecules such as CO and CN with these metal centers exemplify the mechanisms available to O_2, thus revealing the characteristics which are essential to the function of the enzyme in activating O_2, in effecting its reduction to water and perhaps in conserving the energy of the redox reaction.[1] All phases of these reactions have been investigated, from ultrafast phenomena (hundreds of femtoseconds) to relatively slow processes (hundreds of milliseconds). Time-resolved infrared techniques are uniquely suited as probes of ligation dynamics and structures, particularly for Cu_B^+ which is generally unobservable with other spectroscopies. We recently reported real-time ($>10^{-7}$ s) TRIRS and kinetic measurements on the fully reduced CO derivative of CcO which demonstrate that coordination to Cu_B^+ is an obligatory mechanistic step for CO departing the protein after photodissociation and for the subsequent rebinding process.[2] These measurements also reveal that the protein erects a long lived (milliseconds) barrier to CO recombination with Fe_{a3}^{2+}. We have attributed the formation of this barrier to the binding of an endogenous ligand L, triggered by the transfer of CO to Cu_B^+. We propose that the rate of recombination of photodissociated CO is determined by the rate of thermal breaking of the Fe_{a3}^{2+}-L bond. Further studies with other ligands have been undertaken to test the generality of these mechanisms. We have observed the photodissociation of CN^- from Fe_{a3}^{2+} in a real-time TRIR experiment. A transient IR bleach is observed by pumping at 532 nm and monitoring within the 2059 cm^{-1} band (ν_{CN} of Fe_{a3}^{2+}-CN). This is the first report of the dynamics of photodissociation of CN^- from a heme. The rise of the bleach is instrument limited (30 ns) while the recovery is fit well by a single exponential having $\tau = 430$ s^{-1}. This rate is comparable to the saturation value of 700 s^{-1} for the recombination of photodissociated CO, suggesting that the rate determining step for recombination is similar. We propose that photodissociated CN^- binds to Cu_B^+ and displaces L, which is subsequently transferred to Fe_{a3}^{2+}. Consequently, the rate of recombination of the CN^- is also determined by the thermal rate of the Fe_{a3}^{2+}-L bond dissociation.

Ultrafast ligation dynamics of photodissociated CcO-CO have also been studied using TRIR techniques with subpicosecond time resolution. The 5 ps TRIR spectrum is shown in the Figure. The bleach of the Fe_{a3}-CO band at 1963 cm^{-1} occurs in less than 1 ps. The appearance of the transient Cu_B^+-CO absorption at 2061 cm^{-1} also occurs in less than 1 ps, indicating that the photo-initiated transfer of CO is remarkably fast. The results suggest that an unhindered pathway or channel is required to expedite ligand transfer from one metal center to the other. The heme pocket must be constructed in such a manner that it restricts the motion of bound CO (narrow CO linewidths) but facilitates rapid transfer of the photodissociated CO between metal centers. This feature of the protein is significant to the role of Cu_B as a ligand shuttle to Fe_{a3} in the functional dynamics of the protein.

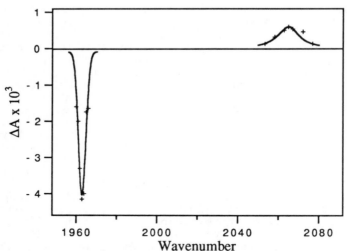

Figure. Five picosecond TRIR spectrum of photo-dissociated carbonmonoxy CcO.

[1]Woodruff, W. H.; Einarsdóttir, Ó.; Dyer, R. B.; Bagley, K. A.; Palmer, G.; Atherton, S. J.; Goldbeck, R. A.; Dawes, T. D.; Kliger, D. S. *Proc. Natl. Acad. Sci.* **1991**, in press.
[2]Dyer, R. B.; Einarsdóttir, Ó.; Killough, P. M.; López-Garriga, J. J.; Woodruff, W. H. *J. Am. Chem. Soc.* **1989**, *111*, 7657-7659.

FOURIER TRANSFORM INFRARED STUDY OF CYANIDE BINDING TO IRON AND COPPER SITES OF BOVINE HEART CYTOCHROME C OXIDASE

Motonari Tsubaki* and Shinya Yoshikawa

Department of Life Science, Faculty of Science, Himeji Institute of Technology, Himeji, Hyogo 671-22, Japan

INTRODUCTION Cytochrome c oxidase (CcO) is a complex protein with an approximate molecular weight of 200,000 spanning the mitochondrial inner membrane and works as a terminal oxidase of the mitochondrial electron transport system. By utilizing the electron transfer reaction this enzyme can generate a proton gradient across the inner mitochondrial membrane partly by the consumption of H^+ in the matrix side for the reduction of dioxygen to produce water and partly by the vectorial proton transport across the membrane coupled to the electron transfer reaction.   To catalyze the electron transfer reaction, CcO contains two hemes A, two copper atoms as prosthetic groups, in addition to   zinc and   magnesium ions and possibly an additional copper ion with unknown function. In the present study we have investigated the structural change of the Fe_{a3}-Cu_B reaction center by Fourier transform infrared spectroscopy using cyanide, a very strong and specific inhibitor for cytochrome c oxidase, comparable to carbon monoxide, as an infrared probe. Although a thorough study on the cyanide binding to cytochrome c oxidase conducted by dispersive infrared spectroscopy has been reported recently by Yoshikawa and Caughey[1], their conclusions were contradictory to the conventional view of the cyanide binding to the Fe_{a3}-Cu_B site. Present work was done to solve this problem on the basis of newly-collected data obtained by Fourier transform infrared spectroscopy.   Due to much higher quality of the collected data in the present study compared to those obtained by dispersive infrared spectroscopy, some of the previous assignments were corrected and a new structural model at the Fe_{a3}-Cu_B center of CcO is proposed here accordingly.

RESULTS AND DISCUSSION In the fully oxidized resting state, cyanide ion forms a bridging structure between heme a_3^{3+} and Cu_B^{2+} centers.   This state could be characterized by a sharp C-N absorption at 2152 cm^{-1}.   This bridging structure was very rigid and the bridging cyanide was hardly exchangeable with exogenous ligand added afterward.   Upon partial reduction of redox center(s) other than heme a_3 with sodium dithionite, this bridging structure was destroyed possibly due to the conformational change around the binuclear center, however, the cyanide ion remained bound to the ferric heme a_3 showing a transient C-N absorption at 2132 cm^{-1}. At this transient

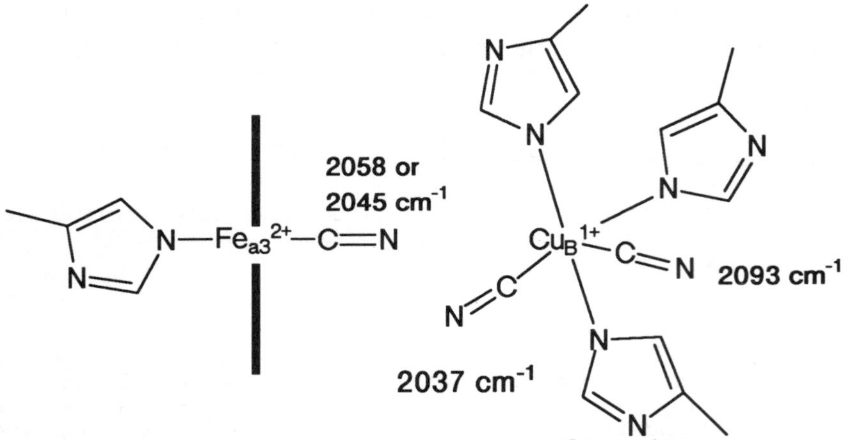

Figure 1 Cyanide bindings to the heme $a_3{}^{2+}$-$Cu_B{}^{1+}$ binuclear site of CcO

state, the heme a_3-bound cyanide could be easily exchanged with cyanide isotope in the medium added afterward. As the reduction of the heme a_3 proceeded further two new infrared bands appeared at 2058 and 2045 cm^{-1} concomitantly. At higher concentration of cyanide (more than 5 mM) additional two infrared bands appeared at 2093 and 2037 cm^{-1}. The former two bands are due to the ferrous heme a_3-bound cyanide, since both of these bands could be removed upon introduction of carbon monoxide; whereas the latter two bands are possibly due to the $Cu_B{}^{1+}$-bound cyanide since both of the latter bands were not affected by the presence of carbon monoxide at least at higher concentration of cyanide. At lower concentration of cyanide, however, the 2037 cm^{-1} band disappeared in the presence of carbon monoxide, indicating that the cyanide ion bound to the $Cu_B{}^{1+}$ that gives the 2037 cm^{-1} band has, at least, some competitive interaction with the ferrous heme a_3-bound carbon monoxide. These observations suggest that upon further reduction of the heme a_3 redox center (and the Cu_B redox center as well), the distance between the heme a_3 and the Cu_B becomes further larger allowing the simultaneous bindings of one cyanide to the heme $a_3{}^{2+}$ and two cyanide ions to the $Cu_B{}^{1+}$, one of the cyanide ions being oriented towards the heme a_3 (Figure 1). This tertiary structural change around the heme a_3-Cu_B moiety upon reduction may have some important roles for cytochrome c oxidase-catalyzed reactions, such as the reduction of oxygen to produce water and the proton pumping.

REFERENCE

1.   S. Yoshikawa and W. S. Caughey, J. Biol. Chem., 1990, 265, 7945.

EFFECTS OF SURFACTANTS ON THE STRUCTURE OF CYTOCHROME C OXIDASE AS STUDIED BY FT-IR.

J. Castresana*, J.M. Valpuesta, A. Muga, F.M. Goñi and J.L.R. Arrondo.

Department of Biochemistry, University of Basque Country, P.O. Box 644, 48080 Bilbao, Spain.

Introduction

Cytochrome c oxidase is an integral membrane protein constituting the oxidizing end of the mitochondrial electron transport chain. Numerous studies have been undertaken on the structure and role of this protein (for a review see [1]). These studies involve solubilization and reconstitution of the protein with endogenous or exogenous lipids[2]. The requirement of surfactants in both steps reinforce the importance of assessing their effects on membrane protein structure.

Fourier-transform infrared spectroscopy (FT-IR) has been used to study surfactant-protein interaction in both soluble and membrane proteins. Combined information from the AmideI absorption band and thermal stability provides a further insight into surfactant-induced changes of secondary and tertiary structures.

Materials and Methods

Cytochrome c oxidase was purified as described previously[3]. Surfactants, lauryl maltoside and N,N-dimethyldodecylamine-N-oxide (LDAO) were purchased from Calbiochem and Fluka, respectively. The protein was incubated with detergent at room temperature in a D_2O medium for 1 hour. The samples were placed onto CaF_2 windows and mounted in a thermostatted cell. 200 scans were obtained using a shuttle device in a Nicolet 520 spectrometer with a resolution of 2 cm^{-1}. Resolution enhancement techniques have been described previously[4].

Results and Discussion

Fig. 1 shows the deconvolved spectra of cytochrome c oxidase in the absence and in the presence of the surfactants lauryl maltoside and LDAO. Each of these surfactants has a characteristic effect on the protein. Wher-

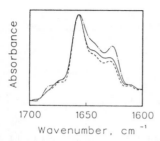

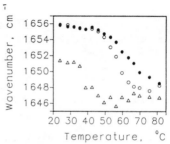

Figure 1 Deconvolved infrared spectra in the Amide I region of native cytochrome c oxidase (——), solubilized with 5% lauryl maltoside (---) and LDAO (-·-·-). Protein concentration was 20 mg/ml.

Figure 2 Amide I absorption maximum as a function of temperature of native cytochrome c oxidase (●), solubilized with lauryl maltoside (○) and LDAO (△).

eas LDAO is a strong solubilizing detergent, lauryl maltoside has been widely used as a mild solubilizing agent which does not alter enzyme activity[5]. This behaviour is also reflected in the infrared spectra: LDAO produces more profound alterations than lauryl maltoside in the spectral lineshape. The main modification in the presence of LDAO is an increase of a band around 1620 cm-1, which has been assigned to protein aggregation[4].

The thermal denaturation profiles of the protein (Fig. 2) show that while the presence of lauryl maltoside decreases slightly the protein denaturation temperature, LDAO reduces significantly its thermal stability.

Thus, it can be concluded that there is a close relationship between the solubilizing strength of the detergents used in this study and their effect on the protein structure, as monitored by FT-IR.

References

1. Saraste, M., Quarterly Reviews of Biophysics, 1990, 23, 331.

2. Tanford, C., "The Hydrofobic Effect", Wiley-Interscience, New York, 1980.

3. Valpuesta, J.M., Henderson, R. and Frey,T.G. J. Mol. Biol., 1990, 214, 237.

4. Surewicz, W.K. and Mantsch, H.H. Biochem. Biophys. Acta, 1988, 952, 115.

5. Li, Y., Naqui, A., Frey, T.G. and Chance, B. Biochem. J., 1987, 242, 417.

RESONANCE RAMAN CHARACTERIZATION OF THE ANIONIC SEMIQUINONE RADICAL OF HANSENULA ANOMALA LACTATE:CYTOCHROME c OXIDOREDUCTASE

M. Tegoni[1] and A. Desbois[2]*

[1] Laboratoire de Cristallographie et de Cristallisation des Macromolécules Biologiques, Université Aix-Marseille 2, Boulevard Pierre Dramard, 13326 Marseille Cedex, France
[2] Section de Biophysique des Protéines et des Membranes, Département de Biologie Cellulaire et Moléculaire, Centre d'Etudes de Saclay, 91191 Gif-sur-Yvette Cedex, France

Yeast mitochondrial L-lactate:cytochrome c oxidoreductase (EC 1.1.2.3) or flavocytochrome b2 (flavocyt b2) catalyzes the transfer of two electrons in two one-electron steps from its substrate (L(+)-lactate) to cytochrome c. In the reaction mechanism of this flavohemoprotein, a necessary intermediate is the semiquinone form of the four prosthetic flavins.[1] In the presence of the reaction product, pyruvate, this red anionic radical is thermodynamically stabilized and accumulated as a dead-end complex up to a level of 100 %. This inhibition, detected under steady-state conditions, provides a regulation of enzyme activity.[2]

Recent studies using resonance Raman (RR) spectroscopy excited at 441.6 nm have demonstrated the feasibility of monitoring the structure of both oxidized flavins and ferrihemes of the enzyme.[3] These Raman activities likely rely on quasi-resonance and pre-resonance conditions on π-π* electronic transitions of flavins (450 nm) and hemes (413 nm), respectively. Since the semiquinone forms of flavins and flavoproteins exhibit the strongest electronic absorptions in the 350-400 nm region, the use of near-UV excitations should be therefore particularly adapted to the observation of their RR modes.

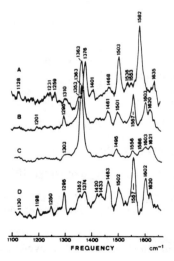

The 1100-1650 cm^{-1} region of the RR spectrum of oxidized flavocyt b2 excited at 363.8 nm is presented in Fig. 1(A). Its comparison to that of the cyt b2 core (not shown) indicates that modes of both flavins and hemes are active using this excitation. In particular, the 1376 and 1582 cm^{-1} bands can be used as markers of the oxidized state of heme and FMN, respectively. Using the experimental conditions described above,[2] the "flavin semiquinone form" of the enzyme have been generated. Its spectrum is displayed in Fig. 1(B). A comparison with

Figure 1 RR spectra (1100-1650 cm^{-1}) of flavocyt b2 excited at 363.8 nm. A) Oxidized form; B) "Semiquinone" form; C) Totally reduced form; D) Difference spectrum: D = 2(B - 0.9C); this spectrum was computed in order to subtract the contribution of reduced hemes in spectrum B (band at 1363 cm^{-1}).

273

Table 1 Comparative RR frequencies (cm^{-1}) for anionic flavin semiquinone ([a] Glucose oxidase from [4] ; [b] D-aminoacid oxidase from [5] ; [c] from [5])

Flavocyt b2	GO[a]	DAO[b]	Proposed assignment[c]
1198	1188		Ring III
	1222	1217	Ring III
1295	1290	1292	v(C4a-N5)
1352	1346	1361	
1374	1372		
1420		1422	
1433	1430		
1463	1454	1448	
1502	1500	1516	v(C4a-C10a)
1557		1555	
1602		1602	Ring I
1620	1620		

spectrum A indicates no significant spectral contribution from ferrihemes or oxidized FMN. However, considering the spectrum of the dithionite-reduced enzyme (C in Fig. 1), essentially ferrous hemes are present in the "semiquinone form" spectrum. Moreover, spectrum B reveals new intense bands at 1295, 1363, 1557, 1603 and 1620 cm^{-1} which are not observed in the spectra of the oxidized or totally reduced enzyme (A and C in Fig. 1) as well as in those of oxidized or reduced cyt b2 core (not shown). Therefore, the difference spectrum D can be attributed to the anionic form of FMN. The frequencies observed in spectrum D significantly differ from those previously reported for the anionic semiquinone of glucose oxidase and D-aminoacid oxidase using visible excitations (Table 1).[4,5] The observed variations very likely result from differences in hydrogen bonding on the radical. Further studies on the sensitivities of RR modes of the semiquinone of flavocyt b2 to deuteriation are in progress. Moreover, the RR spectra of lactate oxidase are under investigation. Indeed, this flavoprotein has a flavin site structurally homologous to flavocyt b2, uses L-lactate as substrate and oxidizes it to pyruvate, under anaerobic conditions. As in flavocyt b2, pyruvate is known to modulate the enzyme reactivity.[6]

In conclusion, the present short report outlines the usefulness of the near-UV RR spectroscopy in deducing structural information on the semiquinone forms of flavoproteins. This information will be particularly valuable in view of difficulties in obtaining crystals suitable for X-ray structural analyses of these relevant intermediates.

REFERENCES

1. C. Capeillère-Blandin, M.J. Barber and R.C. Bray, Biochem. J., 1986, 238, 745-756.
2. M. Tegoni, J.M. Janot and F. Labeyrie, Eur. J. Biochem., 1990, 190, 329-342.
3. A. Desbois, M. Tegoni, M. Gervais and M. Lutz, Biochemistry, 1989, 28, 8011-8022.
4. C.R. Lively, W.G. Gustafson and J.T. McFarland, "Flavins and Flavoproteins" (D.E. Edmondson & D.B. McCormick, Eds), Walter de Gruyter, Berlin, 1987, pp. 283-286.
5. Y. Nishina, H. Tojo and K. Shiga, J. Biochem., 1988, 104, 227-231.
6. S. Ghisla and V. Massey, "Chemistry and Biochemistry of Flavoenzymes" (F. Muller, Ed), CRC Press, Vol. 2, in press.

VI

METALLOPROTEINS

WHOLE CELLS

BIOMEDICAL APPLICATIONS

MAGNETO-OPTICAL STUDIES OF COPPER CENTRES IN PROTEINS

Andrew J. Thomson, Jacqueline A. Farrar and Stephen J. Bingham

Centre for Metalloprotein Spectroscopy and Biology,
School of Chemical Sciences,
University of East Anglia,
Norwich NR4 7TJ. U.K.

The enzyme nitrous oxide reductase (N_2OR) contains copper sites of unusual optical, magnetic and redox properties. Characterisation of these centres has been carried out using a combination of low-temperature electron paramagnetic-resonance (EPR) and magnetic circular dichroism (MCD) spectroscopy and a novel form of paramagnetic resonance by optical detection (PROD). A comparison has been made with the Cu_A site in bacterial and mammalian cytochrome c oxidase and also with copper substituted into one of the zinc sites in liver alcohol dehydrogenase in which it is coordinated by a pair of cysteine, one histidine and one water ligand.

Magnetic circular dichroism (MCD)

The MCD intensity of paramagnetic ions is inversely dependent on absolute temperature and directly dependent on the applied magnetic field. The advantages of using MCD spectroscopy to probe the electronic properties of transition-metal ions in proteins is that at low temperatures, ~1.5 K, the MCD signals are intense and only the paramagnetic centres are detected. This gives a selectivity based upon paramagnetic moment. The field and temperature dependence of the MCD signals are related to the ground state electronic spin, g-values and zero-field splittings. Hence a discrimination is also made on the basis of the electronic ground state. The wavelength range accessible to study centres in proteins is wide, 200-3000 nm since the optical bands of paramagnets can be detected below the protein absorption in the UV between 200-300 nm and in the near infrared between 1400-3000 nm where vibrational overtones normally obscure the optical absorption spectrum. Finally, since the MCD signal is a signed quantity a more detailed optical fingerprint is obtained (1).

Paramagnetic resonance by optical detection (PROD).

This technique uses the MCD signal intensity as an optical detector of ground state microwave resonance. The ground-state g-value is detected via the optical transitions.

The line-shape of the optically detected g-tensor depends upon the linear polarisation of the optical transition. This permits the polarisation of the optical transition to be determined relative to the g-tensor axes for a randomly oriented molecule in frozen aqueous solution. The experiment can also be used for the selective quenching of electronic transitions (SQUELTS) by measuring the MCD spectrum at the magnetic field, B_R, at which the microwaves are in resonance in the presence and absence of microwave power. The difference spectrum is the MCD of the paramagnet in microwave resonance (2).

Nitrous oxide reductase (*Pseudomonas stutzeri*)

The denitrifying bacterium, *Pseudomonas stutzeri*, reduces nitrous oxide reductase, N_2O, to di-nitrogen as part of a respiratory pathway of energy conservation which is coupled to ATP generation. The enzyme responsible for catalysing this two-electron reduction is nitrous oxide reductase (N_2OR), which is composed of two identical subunits (M_r ~70000) of known amino-acid sequence (3) and has a stoichiometry of about 4 copper ions per monomer (4). A variety of spectroscopic methods have been applied to the study of the copper centres but no clear structural model has yet been proposed for them. Two oxidation levels of the enzyme have been observed previously (5). The oxidised state, referred to as the 'purple' form, has EPR signals representing up to 30% of the total copper and absorption bands at 480 nm and 540 nm. On reduction with a variety of agents including dithionite, a second copper site is observed with an absorption band at 650 nm, a broad EPR signal with g-values close to 2.0 and a resonance Raman spectrum similar to that of type 1 copper sites in proteins. This is known as the 'blue' form (6). The EPR spectrum of the purple form shows a distinctive 7-line hyperfine pattern which, it has been proposed, arises from a mixed-valence copper dimer in the oxidation state ($Cu^{+1.5}...Cu^{+1.5}$) with an unpaired electron delocalised between both copper nuclei ($I_N = {}^3/2$) (7). Each copper ion in the dimer must therefore have an identical chemical environment. Low-temperature MCD, optical and EPR spectra of the oxidised, semi-reduced and fully reduced forms of N_2OR show that the oxidised state contains both the mixed-valence dimer, centre $\underline{A}$ and a second dimeric copper structure containing Cu(II)/Cu(II) ions antiferromagnetically coupled, centre $\underline{Z}$. On treatment with ascorbate centre A is reduced to the Cu(I)/Cu(I) state whereas centre Z is not reduced. Addition of a mediator such as phenazine methosulphate PMS causes centre Z to accept an electron to generate an EPR active semi-reduced form. The MCD spectra suggest thiolate coordination for both of these centres. Figure 1 compares the spectra of the oxidised state of N_2OR with that of the reduced enzyme. PROD spectroscopy has allowed the polari- sation of all the CT bands to be measured. Centre Z is likely to be the site of N_2O binding. Addition of the inhibitor N_3^- causes this centre to be fully reduced to the Cu(I)/Cu(I) state in which it can presumably carry out the two electron reduction of substrate, N_2O.

Cu_A in cytochrome c oxidase

The centre called Cu_A is located in subunit II of mammalian and bacterial cytochrome c oxidase and transfers electrons from cytochrome c via cytochrome a and the haem-copper pair of the oxygen reducing centre. Sequence data suggests that the Cu_A site possesses at least two cysteine and two histidine residues as potential ligands for copper (8). A comparison between the low temperature MCD spectra of Cu_A and the dimeric copper centre A in N_2OR shows them to be identical. This finding raises questions about the nature of Cu_A in cytochrome c oxidase, whether it is monomeric or dimeric.

Cu site in liver alcohol dehydrogenase.

Copper(II) ion can, under rigorously anaerobic conditions, replace the catalytic zinc ion in LADH to generate a site of coordination by two cysteine, one histidine and a water ligand (9). The MCD spectra of this site matches well in many respects that of the reduced state of the copper Z centre of N_2OR.

Conclusions

N_2OR appears to contain a pair of dimeric copper centres which can cycle through different oxidation levels, see scheme. Both centres contain some cysteinate ligation of copper. We propose that centre A, which has a role in electron transfer, possesses two copper ions of identical ligation possibly bridged by thiolate. Centre Z, the site of substrate reduction, contains two copper ions, one of which is ligated by thiolate. This dimer in the semi-reduced state forms a trapped valence state Cu(I)/Cu(II). In the fully reduced state, Cu(I)/Cu(I), the linear substrate N-N-O may be able to bridge between these two copper sites so that a concerted two electron reduction takes place.

Acknowledgments. This work has been supported by the Molecular Recognition and TRAMPS Initiative of the SERC. N_2OR samples have been generously supplied by Professor W.G. Zumft (Karlsruhe), copper substituted LADH by Dr. D.M. Dooley (Amherst) and cytochrome c oxidase by Professor C. Greenwood (Norwich).

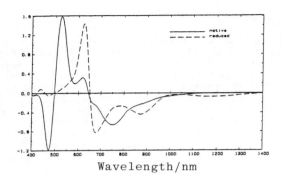

FIGURE 1.

The MCD spectra at 4.2 K and 5 Tesla of the oxidised and reduced states of N_2OR.

Wavelength/nm

SCHEME

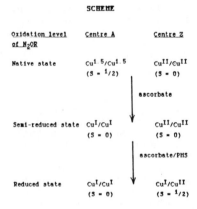

Oxidation level of N_2OR	Centre A	Centre Z
Native state	$Cu^{1.5}/Cu^{1.5}$ ($S = {}^1/2$)	Cu^{II}/Cu^{II} ($S = 0$)
	ascorbate	
Semi-reduced state	Cu^I/Cu^I ($S = 0$)	Cu^{II}/Cu^{II} ($S = 0$)
	ascorbate/PMS	
Reduced state	Cu^I/Cu^I ($S = 0$)	Cu^I/Cu^{II} ($S = {}^1/2$)

The one-electron reduced state, Cu^I/Cu^{II}, of centre Z is not enzymatically active. It is likely that in the presence of substrate centre Z accepts two electrons to generate the active state in $Cu^I/N_2O/Cu^I$.

REFERENCES

1. A.J. Thomson and M.K. Johnson, Biochem.J., 1980, 411.
2. C.P. Barrett, J. Peterson, C. Greenwood and A.J. Thomson, J.Amer.Chem.Soc., 1986, 108 3170.
3. A. Viebrook and W.G. Zumft, J.Bact. 1988, 170, 4658.
4. C.L. Coyle, W.G. Zumft, P.M.H. Kroneck, H. Körner and W. Jakob, Eur.J.Biochem., 1985, 153, 459.
5. J. Riester, W.G. Zumft and P.M.H. Kroneck, Eur.J.Biochem., 1989, 178, 751.
6. D.M. Dooley, R.S. Moog and W.G. Zumft, J.Am.Chem.Soc., 1987, 6730.
7. P.M.H. Kroneck, W.A. Antholine, J. Riester and W.G. Zumft, FEBS Lett., 1988, 24, 70.
8. C. Greenwood, B.C. Hill, D. Barber, D.G. Eglinton and A.J. Thomson, Biochem.J., 1983, 215, 303.
9. W. Maret, H. Dietrich, H.H. Ruf and M. Zeppezauer, J.Inorg.Biochem., 1980, 12, 241.

X–RAY ABSORPTION SPECTROSCOPY AND THE STRUCTURE OF TRANSITION METAL CENTRES IN PROTEINS

C. D. Garner

The Department of Chemistry
Manchester University
Manchester M13 9PL, UK

INTRODUCTION

The fact that many of the metals in biological systems are present only in trace amounts in no way diminishes their significance. Whenever nature has a difficult task to accomplish, a metal ion (or a cluster of ions) is invariably employed. This is especially true for metabolic processes involving small molecules. Thus, the chemical transformations which constitute the oxygen and nitrogen cycles involve catalyses at transition metal centres. Electron flow in natural systems is overwhelmingly from transition metal centre to transition metal centre. Displacement of charge in biological systems is also achieved by the displacement of alkali and alkaline earth cations, the mobility of which is carefully regulated by ionophores and used to signal, trigger, and control inter- and intracellular processes. Furthermore, Zn^{II}, Mn^{II}, and Fe^{II} cations are central to the control of both metabolic and genetic processes. Therefore, it is essential that we understand the biochemistry of these metals; especially how the chemistry of a metal is modulated by binding to an organic matrix to achieve a specific functionality. This challenge is clearly manifest by asking how symbiosis of transition metal and protein functions produces such elegant specificity.

Structural information alone will not describe protein function. However, even to ask intelligent questions, it is a prerequisite that we know the detailed structural arrangement of a biomolecule and the environment about the transition metal atom.

X-ray crystallography is rightly regarded as the most powerful structural technique to provide the architectural details of a protein molecule. However, sometimes the resolution of crystallographic data is insufficient to draw meaningful conclusions concerning the detailed nature of a metal centre within a protein. Also, some proteins refuse to crystallise or yield crystals suitable for a high resolution structure determination. Furthermore, it is important to establish structural details for proteins, and especially their catalytic centres, when maintained at conditions similar to their working environment -

typically, in aqueous media in the presence of substrate, inhibitor, or suitable redox partner.

In the mid 1970's, with the availability of intense X-ray synchrotron sources, a powerful new technique, X-ray absorption spectroscopy (XAS),emerged. This is a local structural probe, the information content of which derives from electron diffraction. For a metalloprotein, the electron source and detector is the metal atom which is probed, since selective excitation is achieved by scanning a range of X-ray wavelengths particularly appropriate to the element of central interest. The selectivity and the local nature of the diffraction process give the technique its major strength. For example, metal-ligand distances can be determined to an accuracy of *ca.* ±0.02Å. In addition, XAS does not require crystalline materials; thus, aqueous protein samples are readily probed under a variety of conditions.

Since 1975, when the first experiment using a synchrotron radiation source was accomplished,[1] XAS has become established as an important technique for probing the environment of transition metals in proteins. Numerous studies have been accomplished and many significant advances made. The majority of these have been identified and collated in various reviews.[2-6] This article will concentrate upon the application of XAS to determine aspects of the local structure of some iron-oxygen and iron-sulphur aggregates which occur within proteins.

IRON-OXYGEN AGGREGATES IN STORAGE PROTEINS

Iron plays an essential role in many important biological processes such as oxygen transport, electron transfer, DNA synthesis and cellular growth and development. However, large quantities of free iron cannot be stored under physiological conditions since iron(III) ions would rapidly precipitate as iron(III) oxide, from which iron would be poorly bioavailable. Therefore, nature has evolved two iron-storage proteins, ferritin and haemosiderin, in which iron can be sequestered and mobilised when required. The predominant iron-storage protein, ferritin, has been extensively studied.[7] The molecules consist of a hollow proteinaceous shell which surrounds a core of iron(III) oxyhydroxide with smaller amounts of associated phosphate, leading to a composition which is typically *ca.* $(FeOOH)_8(FeO_2PO_2H_2)$. Previous Extended X-ray Absorption Fine Structure (EXAFS) data for ferritin[8] have indicated that iron is ferritin is bound to approximately six oxygen atoms at *ca.* 1.9Å and, beyond this shell, iron atoms are present an average distance of *ca.* 3.1Å.

Clinical iron overload is due to either a recessively inherited abnormality in the intestinal uptake of iron, leading to enhanced absorption of dietary iron (primary haemochromatosis) or to multiple blood transfusions for congenital anaemias, e.g. beta thalassaemia (secondary haemochromatosis), leading to progressive accumulation of iron as there is no effective excretory mechanism for excess tissue iron. In both of these conditions the predominant form of stored iron is haemosiderin, and not ferritin as in normal individuals. Little is known about the biochemical and biophysical properties of haemosiderin but [57]Fe Mössbauer spectroscopy and electron diffraction[9,10] have shown that there are at least three different types of iron core, depending on the origin of the haemosiderin.

The EXAFS associated with the iron K-edge has been measured and interpreted for ferritin and haemosiderin extracted from horse spleen, and haemosiderin extracted from the livers of humans with treated primary haemochromatosis, and from the spleens of humans with treated secondary haemochromatosis. For the ferritin, the data are consistent with, on average, each iron atom being in an environment comprised of *ca.* six oxygen atoms at 1.93±0.02Å, *ca.* 1.5 iron atoms at 2.95±0.02Å and *ca.* 1.1 iron atoms at 3.39±0.02Å, with a further shell of oxygens at *ca.* 3.6Å. Iron in horse spleen haemosiderin is in an essentially identical local environment to that in horse spleen ferritin. In contrast, the EXAFS data for primary haemochromatosis haemosiderin indicate that the iron-oxide core is amorphous; only a single shell of *ca.* 1.94±0.02Å being apparent. Secondary haemochromatosis haemosiderin shows an ordered structure with *ca.* 1.4 iron atoms at both 2.97±0.02 and 3.34±0.02Å. This arrangement of iron atoms is similar to that in horse spleen haemosiderin but the first oxygen shell is split with *ca.* 2.9 atoms at 1.90±0.02Å and ca. 2.7 at 2.03±0.02Å, indicative of substantial structural differences between secondary haemochromatosis haemosiderin and horse spleen haemosiderin.

STUDIES OF FeMoco AND FeVaco

Nitrogen fixation occurs only in certain prokaryotes and is catalysed by the nitrogenases which have two protein components. These are the Fe-protein, which acts as a specific reductant of the larger MoFe-protein. Within the latter there are several metal-containing prosthetic groups, including two iron-molybdenum cofactors (FeMoco) which are the catalytic centres of the enzyme.

One of the earliest successful applications of EXAFS to probe a metalloenzyme was the study of the molybdenum site of nitrogenase. Studies of both the MoFe-proteins and on isolated FeMoco indicate that molybdenum is present as part of a polynuclear cluster containing sulphur and iron, with Mo–S and Mo–Fe distances of *ca.* 2.36 and 2.72Å, respectively and additional oxygen ligation at *ca.* 2.10Å.[11]

Genetic suppresion of the "normal", molybdenum-dependent nitrogenase of certain classes of *Azotobacter* allows expression of the vanadium-dependent enzyme. The vanadium and molybdenum nitrogenase systems show many similarities and, in particular, an iron-vanadium cofactor (FeVaco), analogous to FeMoco, has been isolated. Clear evidence of a strong similarity between active sites in the MoFe- and VFe-proteins has been provided by vanadium K-edge XAS studies. The EXAFS results[12,13] are consistent with vanadium in the VFe-protein of *A. chroococcum* being ligated by a *ca.* three oxygen (or nitrogen), sulphur, and iron atoms at *ca.* 2.13, 2.33, and 2.75Å, respectively. The vanadium would thus appear to substitute for molybdenum in FeMoco to form FeVaco.

The average environment of iron in FeMoco and FeVaco has been investigated by XAS. These data provide clear evidence for the two cofactors possessing the same molecular topology and give further insights into their internal organisation. Specifically, a longer range structural order is identified ca. 3.68Å which corresponds to an Fe---Fe separation.[14,15]

REFERENCES

1. B.M. Kincaid and P. Eisenberger, Phys. Rev. Lett., 1975, 34, 1361.
2. S.P. Cramer and K.O. Hodgson, Prog. Inorg. Chem., 1979, 25, 1.
3. L. Powers, Biochem. Biophys. Acta, 1982, 693, 1.
4. R.A. Scott, Meth. Enzymology, 1985, 117, 414.
5. S.S. Hasnain and C.D. Garner, Prog. Biophys. Mol. Biol., 1987, 50, 47.
6. C.D. Garner, in 'Applications of Synchrotron Radiation', Ed. C.R.A. Catlow and G.N. Greaves, Blackie, London, 1990, p. 268.
7. G.C. Ford, P.M. Harrison, D.W. Rice, J.M.A. Smith, A. Treffry, J.L. White, and J. Yariv, Phil. Trans. Roy. Soc. (London), 1984, 304B, 551.
8. J.S. Rohrer, T.I. Quazi, G.D. Watt, D.E. Sayers, and E.C. Theil, Biochemistry, 1990, 29, 259 and references therein.
9. S.H. Bell, M.P. Weir, D.P.E. Dickson, J.F. Gibson, G.A. Sharp, and T.J. Peters, Biochim. Biophys. Acta, 1984, 787, 227.
10. D.P.E. Dickson, N.M.K. Reid, S. Mann, V.J. Wade, R.J. Ward, and T.J. Peters, Biochim. Biophys. Acta, 1988, 957, 81.
11. S.D. Conradson, B.K. Burgess, W.E. Newton, L.E. Mortenson, and K.O. Hodgson, J. Amer. Chem. Soc., 1987, 109, 7507.
12. J.M. Arber, B.R. Dobson, R.R. Eady, P. Stevens, S.S. Hasnain, C.D. Garner, and B.E. Smith, Nature, 1987, 325, 372.
13. G.N. George, C.L. Coyle, B.J. Hales, and S.P. Cramer, J. Amer. Chem. Soc., 1988, 110, 4057.
14. J.M. Arber, A.C. Flood, C.D. Garner, C.A. Gormal, S.S. Hasnain, and B.E. Smith, Biochem. J., 1988, 252, 421.
15. I. Harvey, J.M. Arber, R.R. Eady, B.E. Smith, C.D. Garner, and S.S. Hasnain, Biochem. J., 1990, 266, 929.

OXO-BRIDGED DIMANGANESE ACTIVE SITES: A RESONANCE RAMAN STRUCTURAL INVESTIGATION

Roman S. Czernuszewicz*, Bakulkumar Dave, and J. Graham Rankin

Department of Chemistry
University of Houston
Houston, TX 77204, USA

Manganese is required in living systems to perform diverse redox transformations ranging from photosynthetic splitting of water to catalytic disproportionation of hydrogen peroxide and dismutation of superoxide radicals.[1] Our knowledge about the function of Mn sites in proteins and enzymes has been expanding considerably over the past few years, thanks to extensive studies by a variety of biochemical and physical methods, but much less is known about their structural motifs. It is believed that four Mn atoms per photosystem-II (PS-II) reaction center in green plants are essential for oxygen evolution.[2] The EXAFS data reveal two sets of Mn-Mn distances, 2.7 and 3.3Å, in the PS-II 4-Mn cluster, and indicate that only bridging oxide and terminal O- and/or N-donor ligands are coordinated to the Mn atoms.[3] Spectroscopic and biochemical studies[3,4] have given evidence that the H_2O photooxidation requires four steps in which the tetrameric Mn cluster transfers $4e^-$ to the PS-II reaction center, and that the bound substrate H_2O is freely exchangeable with H_2O in solution in all metastable S_n states $n = 0, 1, 2, 3$ prior to O_2 release (Scheme 1).

The PS-II 4-Mn complex can also disproportionate added H_2O_2 in the dark following formation of the S_2 state by a single flash.[5] The proposed mechanism for this reaction involves initial $2e^-$ reduction $S_2 \rightarrow S_0$ with O_2 formation followed by reoxidation to the S_2 state with H_2O release (Scheme 1). This same reaction is catalyzed in bacteria by the non-photosynthetic Mn catalases from *L. plantarum*, *T. thermophilus* and *T. album*.[6] The metal sites of these enzymes consist of two Mn ions, which are probably bridged by μ-oxo and μ-carboxylato linkages, with terminal imidazole (histidine) protein ligands completing the coordination sphere.

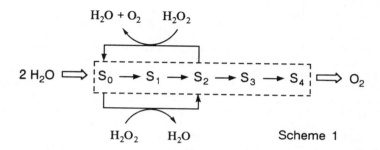

Scheme 1

285

The first examples of functional mimics of Mn catalysts were prepared, which shed some light on the most likely kinetic, electronic and structural requirements for catalysis.[7] In that study, the dimanganese(II) complexes, $Mn_2(L)Cl_3$ and $Mn_2(L)OHBr_2$, of a septadentate ligand tetrakis(2-methylenebenz-imidazolyl)-1,2-diaminopropan-2-ol were shown to be spontaneously and auto-catalytically oxidized by H_2O_2 from Mn(II)-Cl-Mn(II) to Mn(III)-O-Mn(III) + H_2O + Cl⁻ followed by re-reduction to the starting oxidation state + O_2 + H_2O. The formation of a stable µ-oxo dimeric Mn structure was critical for the H_2O_2 decomposition, as no catalysis was observed for mononuclear Mn. Although indirectly, this also suggests that an oxo or hydroxo bridge forms during the $S_0 \rightarrow S_2$ reaction in PS-II. The photosynthetic H_2O oxidation might, however, differ from H_2O_2 oxidation in that electron removal and H_2O binding/oxo bridge formation should be separable processes. Such a chemistry have recently been observed in solution upon oxidation of structurally characterized dimanganese(III) complexes of hydrotris(1-pyrazolylborate) (HBpz$_3$⁻)[8] and 1,4,7-triazacyclononane (TACN)[9] which contain the µ-oxo-di-µ-carboxylate bridge (Scheme 2). The µ-oxo bridges in these dimers and their oxidation Mn_2(III, IV) products may readily be exchanged with $H_2{}^{18}O$ via protonation followed by reversible ring opening. One-electron oxidation to the mixed valence states leads to spontaneous replacement of one of the carboxylate with oxo linkage.[8b,c]

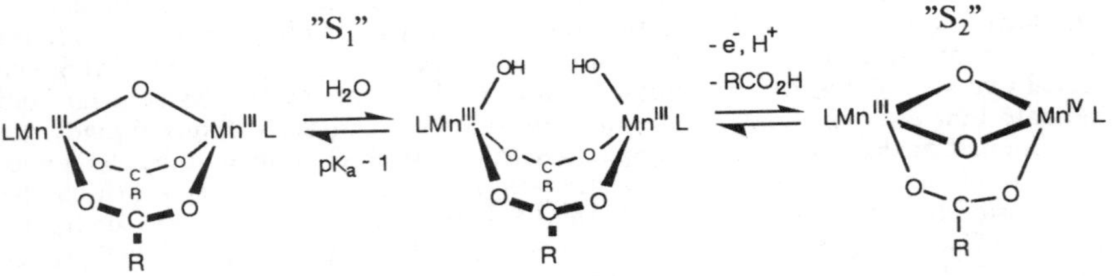

Scheme 2

As evidence implicating the presence of oxo bridges in multinuclear Mn centers in biology rapidly accumulates, there is an increased need for a direct probe of such structural features. This led us to initiate a systematic resonance Raman (RR) spectral investigation of various dimanganese clusters in different metal oxidation states which contain one or two µ-oxo bridges, with the aim of developing the RR spectroscopy as a structure probe of multinuclear Mn active sites.

a. *µ-Oxo-Bridged Dimanganese Centers.* The electronic absorption spectrum of Mn catalases closely matches those observed for the model complexes containing µ-oxo-di-µ-carboxlylato Mn(III) dinuclear core, $[Mn_2O(O_2CR)_2(HBpz_3)]$ (R=CH$_3$, C_2H_5, or H) and $[Mn_2O(O_2CCH_3)_2(TACN)_2]^{2+}$, suggesting that a similar $\{Mn(III,III)_2O(O_2CR)_2\}^{2+}$ structural unit presumably occurs in the enzymes. The RR spectra of HBpz$_3$ dimer and its µ-^{18}O isotopomer in the region 250-1500cm⁻¹, which are shown in Figure 1, are informative and display a strong enhancement of the Mn-O stretching modes of the µ-oxo Mn bridge. One of these occurs at 558 cm⁻¹ (541 cm⁻¹, ^{18}O) and is assigned to the Mn-O-Mn symmetric stretch (v_s), whereas the second one located at 717 cm⁻¹ (680 cm⁻¹, ^{18}O) is attributed to the corresponding asymmetric stretch (v_{as}).[8a] The assignment of v_{as} is further supported by a Raman band at 1432 cm⁻¹, which shifts to 1360 cm⁻¹ in the ^{18}O spectrum. The 1432 cm⁻¹ frequency is ca. 2v_{as} (717 cm⁻¹) and, because it is ^{18}O-sensitive, is identifiable as the first overtone of the asymmetric Mn-O-Mn stretch. This overtone is expected to be highly Raman active from symmetry considerations

(B_2 x B_2 = A_1 in C_{2v} point group). A similarly strong $2\nu_{as}$(Fe-O-Fe) was previously observed at 1504 cm^{-1} in the RR spectrum of the analogous diiron complex,[10] a model complex of O_2 binding hemerythrin.

b. _Di-μ-Oxo-Bridged Dimanganese Centers._   Fronko et al.[6d], who studied the EPR spectra of _L. plantarum_ catalase have recently established that this enzyme contains a mixed valence Mn_2 cluster, most likely a Mn_2(III, IV) dimer. A mixed valence Mn(III)(μ-O)$_2$Mn(IV) ring is formed upon oxidation of [$Mn_2O(O_2CR)_2$-(HBpz$_3$)$_2$] in air or by KMnO$_4$ in a H_2O/CH_2Cl_2 mixture[8a] (Scheme 2). This structural unit has been confirmed in the corresponding TACN analog by X-ray crystallography.[9] The EPR spectra of [$Mn_2O_2(O_2CR)$(HBpz$_3$)$_2$] exhibited a g = 2 16-line pattern signal[8a], which is essentially identical with those found for the valence-trapped Mn_2(III, IV) cations, [Mn_2O_2(bipy)$_4$]$^{3+}$ (bipy = 2,2'-bipyridine) and [Mn_2O_2(phen)$_4$]$^{3+}$ (phen = 1,10-phenantroline).[11] These spectra also closely match the EPR spectrum of the _L. plantarum_ Mn catalase,[6d] which raises the possibility that its Mn_2 center might be comprised of two μ-oxo bridges. The RR spectra in Figure 2 of solid samples of [$Mn_2O_2(O_2CR)$(HBpz$_3$)$_2$] show four strong peaks in the region characteristic of a Mn(III)(μ-O)$_2$Mn(IV) planar structure (705, 634, 603 and 515 cm^{-1}), as revealed by their isotopic shifts upon substitution with ^{18}O (not shown). Similar patterns are also seen in the spectra of TACN, bipy, and phen di-μ-oxo dimers. The observed RR frequencies (690, 654, 630 and 590 cm^{-1}) and ^{18}O isotope shifts for the latter two are readily reproduced by normal coordinate calculation based on a planar Mn_2O_2 four-body model and stretching force constants (k_d(Mn^{IV}-O) = 3.27, k_d(Mn^{III}-O) = 2.80 and k_{dd}(MnO/MnO) = 0.1 mdyne/Å) which scale well with bond distances around Mn(III) and Mn(IV) sites, consistent with a 0.1 Å difference in the Mn-O distances. The most distinguishing feature in the RR spectra of di-μ-oxo- versus mono-μ-oxo-bridged dinuclear Mn clusters is a very strong appearance of the μ-^{18}O sensitive band near 700 cm^{-1}, assigned to the breathing mode of the planar Mn_2O_2 ring. These results clearly demonstrate that resonance Raman spectroscopy can be a powerful structural probe for understanding dinuclear Mn sites in chemistry and biology.

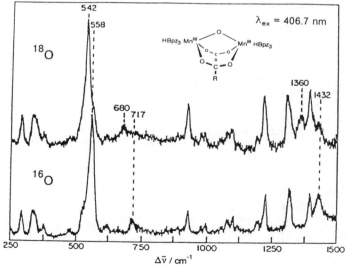

Figure 1.   RR spectra of [$Mn_2O(O_2CCH_3)_2$(HBpz$_3$)$_2$] and its μ-^{18}O isotopomer obtained via backscattering in KCl pellets at 77K with 406.7-nm Kr$^+$ laser excitation and 8-cm^{-1} slit widths.

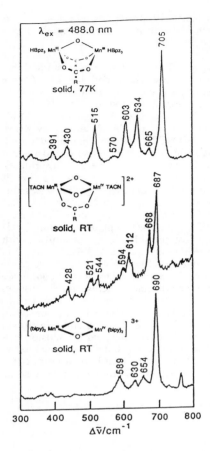

Figure 2. RR spectra of $[Mn_2O_2(O_2CCH_3)(HBpz_3)_2]$, $[Mn_2O_2(O_2CCH_3)(TACN)_2]^{2+}$, and $[Mn_2O_2(bipy)_4]^{3+}$ obtained via backscattering in KCl pellets with 488.0-nm Ar^+ laser excitation and 8-cm^{-1} slit widths.

REFERENCES

1. V.L. Scheram and F.C. Wedler, eds., 'Manganese in Metabolism and Enzyme Function', Academic Press, New York, 1986.
2. G.C. Dismukes, Photochem. Photobiol., 1986, 43, 99.
3. V.K. Yachandra, R.D. Guiles, A. McDermott, R.D. Britt, S.L. Dexheiner, K. Sauer, and M.P. Klein, Biochim. Biophys. Acta, 1986, 850, 324.
4. (a) H. Kretschmann, J.P. Dekker, O. Saygin, and H.T. Witt, Biochim. Biophys. Acta, 1988, 932, 358. (b) M. Sivaraja, J.S. Philo, J. Lary, and G.C. Dismukes, J. Amer. Chem. Soc., 1989, 111, 3221. (c) R. Radmer and O. Ollinger, FEBS Lett., 1986, 195, 285. (d) G.C. Dismukes and Y. Siderer, Proc. Nat. Acad. Sci., 1981, 78, 274.
5. (a) B. Velthuys and B. Kok, Biochim. Biophys Acta, 1978, 502, 211. (b) J. Mano, M-A. Takahashi, and K. Asada, Biochem., 1987, 26, 2495. (c) W.D. Frasch and R. Mei, Biochim. Biophys Acta, 1987, 891, 8.
6. (a) W.F. Beyer, Jr. and I. Fridovich, Biochem., 1985, 24, 6460. (b) V.V. Barynin and A.I. Grebenko, Dokl. Akad. Nauk, SSR, 1986, 286, 461. (c) G.S. Allgood and J.J. Perry, J. Bacteriol., 1986, 168, 563. (d) R.M. Fronko, J.E. Penner-Hahn, and C.J. Bender, J. Am. Chem. Soc., 1988, 110, 7554.
7. P. Mathur, M. Crowder, and G.C. Dismukes, J. Amer. Chem. Soc., 1987, 109, 5227.
8. (a) J.E. Sheats, R.S. Czernuszewicz, G.C. Dismukes, A.L. Rheingold, V. Petrouleas, J. Stubbe, W.H. Armstrong, R.H. Beer, and S.J. Lippard, J. Am. Chem. Soc., 1987, 109, 1435. (b) J.E. Sheats, B.C. Unni Nair, V. Petrouleas, S. Artandii, R.S. Czernuszewicz, and G.C. Dismukes, Progr. Photosynth. Res., 1987, 1, 721. (c) J.E. Sheats, G.C. Dismukes, R.S. Czernuszewicz, and S. Artandi, Proc. 24th ICC, Chimika Chronika, Y. Konstantatos, ed., 1986, 792.
9. K. Wieghardt, U. Bossek, B. Nuber, J. Weiss, J. Bonvoisin, M. Corbella, S.E. Vittols, and J.J. Girerd, J. Amer. Chem. Soc., 1988, 110, 7398.
10. R.S. Czernuszewicz, J.E. Sheats, and T.G. Spiro, Inorg. Chem., 1987, 26, 2063.
11. (a) P.M. Plaskin, R.C. Stoufer, M. Mathew, and G.J. Palenik, J. Am. Chem. Soc., 1972. (b) S.R. Cooper, G.C. Dismukes, M.P. Klein, and M. Calvin, J. Am. Chem. Soc., 1978, 100, 7248.

NEAR-INFRARED FT-RAMAN STUDIES OF PHOTOLABILE ORGANOCOBALT B₁₂ AND MODEL COMPOUNDS

Nai-Teng Yu[1,2], Shuming Nie[1] and Luigi G. Marzilli[3]

[1]School of Chemistry and Biochemistry, Georgia Institute of Technology, Atlanta, Georgia 30332.
[2]Department of Chemistry, The Hong Kong University of Science and Technology, Kowloon, HK.
[3]Department of Chemistry, Emory University, Atlanta, Georgia 30322.

INTRODUCTION

The recent advent of near-IR-excited FT-Raman spectroscopy has opened up many exciting opportunities in the spectroscopic investigation of biological molecules.[1-4] Laser excitation in the near-IR region (e.g., 1.064 µm) generally precludes electronic absorptions, thus leading to a nearly complete elimination of fluorescence interference. Additionally, interferometric detection offers the benefits of superior spectral resolution, frequency accuracy, and relatively high energy throughput relative to a dispersion-based Raman spectrometer.[5] In this report, we demonstrate that the near-IR FT-Raman technique is particularly well-suited for structural characterization of the photolabile B₁₂ coenzymes and model organocobalt compounds.

MATERIALS AND METHODS

The preparation and characterization of the B₁₂ model complexes have already been reported.[6] These materials are stable indefinitely when stored in the dark in a freezer. Sample purity was confirmed by ¹H NMR spectroscopy. Methylcobalamin and deuteriated methyl (CD₃) cobalamin were prepared according to literature procedures.[7]

FT-Raman and FT-SERS spectra were obtained by using a Bruker IFS 66 FT-spectrophotometer with the Bruker FRA 106 FT-Raman module attached. CW near-IR excitation at 1.064 µm was provided by a diode laser pumped Nd : YAG laser (Adlas, DPY 301).

RESULTS AND DISCUSSION

Biologically active cobalamins (viz., adenosylcobalamin, methylcobalamin) contain a stable organometallic Co-C bond. The homolytic cleavage of this bond is a key step in the most widely accepted mechanism of coenzyme B₁₂-dependent catalysis.[8] This organocobalt bond is highly photolabile, and previous Raman studies with UV-visible excitations provided no information about the Co-C bond.[9] The near-IR FT-Raman technique completely eliminates the Co-C bond photolysis problem, and the Co-C stretching mode in cobalamins and model compounds is now readily detected as a very intense and sharp line at ca. 500 cm⁻¹.[10-12] Figure 1

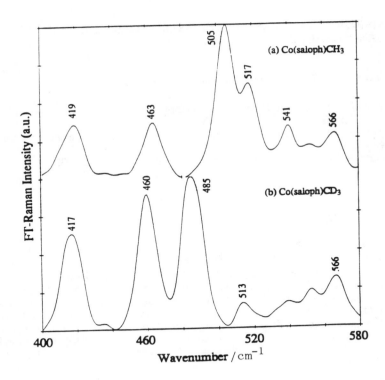

Fig.1. Solid state FT-Raman spectra of Co(saloph)CH₃ and its deuterated derivative, Co(saloph)CD₃, in the Co-C bond stretch spectral region. Excitation wavelength = 1.064 μm; laser power =0.1 W; coadded scans = 100; spectral resolution = 4.0 cm⁻¹.

depicts FT-Raman spectra of a model compound, Co(saloph)CH₃ (where saloph = dianion of disalicylidene-o-phenylenediamine,) and its deuterated derivative, Co(saloph)CD₃, obtained in the solid state. The cobalt-carbon stretching frequency in this five-coordinated B_{12} model is conclusively identified at 505 cm⁻¹, which exhibits a 20 cm⁻¹ downshift to 485 cm⁻¹ upon deuterium labelling of the axial methyl group. The saloph ligand has recently been shown to exhibit a large structural cis influence and a large kinetic cis effect compared to other classes of B_{12} models. It was also recently recognized that the trans and cis effects are synergistic. Structural studies of this model compound are of importance in their implications to the existence of stable five-coordinate cobalamin species in the B_{12}-dependent rearrangement process. The observed isotopic shift of the five-coordinate saloph compound, however, is substantially different from the 30 cm⁻¹ isotopic shift reported for model compounds of the type LCo(DH)₂CH₃ (DH = monoanion of dimethylglyoxime, L = donor ligand)[10,12] and for methyl cobalamins[11] in the solid state. In view of the theoretical isotopic shift of 36 cm⁻¹ calculated by approximating the methyl group as a point mass,[10] it appears that the Co-C bond stretch couples significantly with the equatorial saloph vibrational modes. For the six-coordinate Co(DH)₂ model compounds, such a vibrational effect is much smaller, resulting in a Co-C stretching mode that is essentially isolated.

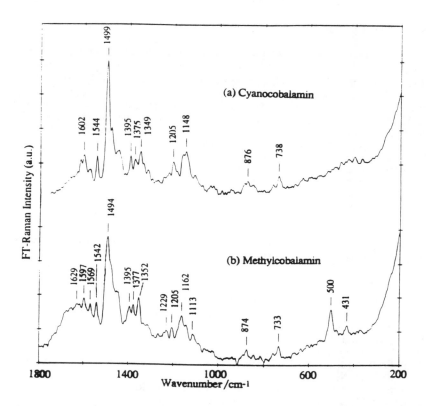

Fig.2. FT-Raman spectra of methylcobalamin and cyanocobalamin (vitamin B_{12}) obtained in aqueous solutions (~1 mM). Excitation wavelength = 1.064 μm; laser power = 0.3 W; coadded scans =500; spectral resolution = 4.0 cm⁻¹.

In Figure 2 are shown FT-Raman spectra of two cobalamins, methylcobalamin and cyanocobalamin, obtained in aqueous solution. The strong Raman line at 500 cm⁻¹ is isotopically sensitive, and its 30 cm⁻¹ downshift upon deuteration of the axial methyl is identical to the isotopic shift of the Co-CH₃ stretching mode observed in the model compounds, pyCo(DH)₂CH₃ and pyCo(DH)₂CD₃.[12] This line is thus unequivocally identified as the Raman-active Co-CH₃ stretching mode in biologically active methylcobalamin. The observed frequency of 500 cm⁻¹ is well within the range (522 to 480 cm⁻¹ depending on L) found in B₁₂ models of the type, LCo(DH)₂CH₃,[12] where L = neutral axial ligand. The similar values of the Co-C stretching frequencies for the cofactor and the models are consistent with the X-ray structural data, which show nearly identical Co-CH₃ bond lengths of ~2.0 Å.[13] In the cyanocobalamin spectrum, a moderately intense line at 2132 cm⁻¹ (not shown) can be assigned to the Raman-active cyano stretching mode, in view of the IR cyano stretching frequency[14] at 2132 cm⁻¹ and the absence of such a line in the FT-Raman spectra of the other cobalamins. A significant extension of the present work is to directly measure the Co-C bond strength in the coenzyme B₁₂-holoenzyme complexes.

CONCLUSIONS

Near-IR-excited Fourier transform (FT)-Raman spectroscopy is a newly developed technique that may have a major impact on the spectroscopic study of biological samples. Laser excitation in the near-IR region (1.064 μm) generally precludes electronic absorption, thus leading to a nearly complete elimination of photolytic sample decomposition and fluorescence interference. Moreover, the use of a Michelson interferometer for performing Raman spectroscopy offers the benefits of superior spectral resolution, frequency accuracy, and relatively high energy throughput relative to a dispersion-based spectrometer. Since most biomolecules lack electronic transitions in the near-IR region, the near-IR FT-Raman technique is particularly attractive in the structural study of biological samples that are sensitive to visible or UV light.

ACKNOWLEDGEMENT

We are grateful to Miss Leigh Ann Lipscomb and Miss Carolina G. Castillo for obtaining parts of the data presented in this report. This work was supported in part by the National Institutes of Health grants GM18894.

6. REFERENCES

1. B. Chase, *J. Am. Chem. Soc.* **108**, 7485 (1986).
2. S. Nie, K. L. Bergbauer, J. J. Ho, J. F. R. Kuck, Jr. and N.-T. Yu, *Spectroscopy* **5**(7), 24 (1990).
3. I. W. Levin and E. N, Lewis, *Anal. Chem.* **62**, 1101A (1990).
4. F. J. Purcell, *Spectroscopy*, **4** (2), 24 (1989).
5. P. R. Griffiths and J. A. de Haseth, *Fourier Transform Infrared Spectroscopy* , Wiley and Sons, New York, 1986.
6. N. Bresciani-Pahor, M. Forcolin, L. G. Marzilli, L. Randaccio, M. F. Summers, and P. J. Toscano, *Coord. Chem. Rev.* **63**, 1 (1985).
7. M. Rossi, J. P. Glusker, L. Randaccio, M. F. Summers, P. J. Toscano, and L. G. Marzilli, *J. Am. Chem. Soc.* **107**, 1729 (1985).
8. J. Halpern, *Science* **227**, 869 (1985).
9. S. Salama and T. G. Spiro, *J. Raman Spectrosc.* **6**, 57 (1977).
10. S. Nie, L. G. Marzilli, and N.-T. Yu, *J. Am. Chem. Soc.* **111**, 9256 (1989).
11. S. Nie, P. A. Marzilli, L. G. Marzilli and N.-T. Yu, *J. Chem. Soc. Chem. Commun.* 770 (1990).
12. S. Nie, P. A. Marzilli, L. G. Marzilli and N.-T. Yu, *J. Am. Chem. Soc.* **112**, 6084 (1990).
13. L. Randaccio, N. Bresciani-Pahor, E. Zangrando, and L. G. Marzilli, *Chem. Soc. Rev.* **18**, 225 (1989).
14. H. A. O. Hill, J. M. Pratt, and R. J. P Williams, *Discussions Faraday Soc.* No.47, 165 (1969).

RAMAN MICROSPECTROSCOPY OF LIVING CELLS: NEW DEVELOPMENTS

J. Greve, G.J. Puppels and C. Otto

Department of Applied Physics
Biophysical Technology Group
University of Twente
P.O. Box 217, 7500 AE Enschede
The Netherlands

INTRODUCTION

Raman microspectroscopy is a technique which can yield very informative spectra of biological micromolecules. It was recently shown that it holds great potential for applications in the field of cellular biology[1,2]. By means of a sensitive confocal Raman microspectrometer single cell and single chiromosome measurements are possible (see also G.J. Puppels, et al. these proceedings).
In the present study we focus upon the application of such an instrument for the study of living cells. This may seem a difficult task: a cell contains many different molecules the spectra of which will be obtained all at the same time. Often an unraveling of the spectra will therefore be cumber some. Yet the desire to study macromolecules inside a living cell is clear: sometimes the systems are simply too complicated to design a reliable model system, also one would like to study cellular processes for which the cell has to be functioning. As a first part of an ongoing study we report here on the first results obtained with some leukocytes. The leukocytes were selected because of their importance for the immune system. Also it is well known that many granules, containing highly specific enzymes, are present inside the granulocytes which form a subpopulation of the leukocytes. It might therefore be expected that some specific spectra could be obtained.

MATERIALS AND METHODS

In this study we use the confocal Raman microspectrometer which has been described previously[1,2]. It was designed for a high throughput of collected Raman scattered photons by using a single dispersive grating and a dielectric filter for suppression of the exciting laser light. The liquid nitrogen cooled CCD detector ensures an essentially background free signal detection. By applying a confocal detection scheme we obtained a detection volume of $0.45 \times 0.45 \times 1.3$ μm^3. Raman spectra were obtained at different positions inside single cells.
The leukocytes were isolated from human peripheral blood. Neutrophilic (NG) and eosinophilic (EG) granulocytes were purified by density gradient centrifugation, basophilic granulocytes (BG) were isolated by fluorescence activated cell sorting after labeling with fluorescein conjugated IgE. During preparation and Raman measurements the BGs were kept at 4°C. Raman spectra were measured while the cells were attached to a poly-l-lysine coated fused silica substrate. The cells were kept in Hank's buffered salt solution.

RESULTS

Excellent Raman spectra of cells were first recorded at 514 nm. It appeared however that even at laser powers below 5 mW substantial damage occured to the cell. This resulted in a reduced survival as could be illustrated by a standard cell-viability test using the acridine orange-ethidiumbromide test. We therefore changed to 660 nm. At this wavelength no cell damage was apparent even at laserpowers up to 20 mW and an irradiation duration of 300 seconds. In all further Raman measurements we used only 6 mW laserpower and irradiation times of 30-150 seconds. Cell viability of irradiated cells was 90% both before and after the measurement.

Spectra with very good signal to noise ratio were obtained from the nuclei of the different granulocytes (Figure 1A). These spectra showed essentially only lines due to DNA and protein. Peak positions and intensities of most lines were very reproducible with the exception of a line at 1048 cm^{-1} which could not be assigned so far and which varies widely in intensity between different cells. The intensity ratio $I_{1094}:I_{1450}$ which we take as indicator of DNA protein ratio varies about 10% between cells. The spectra of NG, EG and BG nuclei were identical.

In contrast to this, wide variation were found for spectra obtained from the cytoplasm (Fig.1B–D). The spectra are clearly specific for each different granulocyte type. Variations in intensity of the different lines were observed for separate cells from the same subpopulation. This holds especially for EG. These variations are most probably caused by the fact that a different number of granules may be present in the measuring volume when changing from position in the cytoplasm of individual cells or to a different cell. Also if too high intensities and too long irradiation times (> 30 s at 6 mW) were used the intensities of lines due to the specific peroxidases EPO and MPO (see below) in the granules showed a slight decrease between repeated measurements at the same spot.

DISCUSSION

The results obtained show clearly that very high S/N ratio spectra of living granulocytes can be obtained. This opens up the possibility of detailed studies of cellular processes like cell division, cell-cell interaction or phagocytosis by means of Raman spectroscopy. It also illustrates that great care has to be taken to prevent the killing of cells by the laser irradiation and the photochemical degradation of specific components even at 660 nm. From the nucleic spectra it is clear that the DNA is predominantly in the B-form as follows from the strong line at 1094 cm^{-1} (DNA backbone); the line at 833 cm^{-1} (ribose phosphate) and the positions of the ring breathing modes of the different DNA bases which correspond to C_2'-endo-anti sugar puckering. The proteins in the nuclei possess mainly α-helical and random coil configurations. This follows from the maximum of the amide-I band at 1659 cm^{-1}, the absence of a strong line below 1240 cm^{-1} in the amide-III region and the presence of a high intensity C-C skeletal mode at 936 cm^{-1}. The DNA:protein ratio inside the nuclei was determined from the ratio of the 1094 cm^{-1} DNA line intensity and the intensity of the 1449 cm^{-1} protein line. A value of 1:2·3 was found. Assuming a DNA: histone protein ratio of 1:1 inside the cells it follows that the ratio histone:nonhistone protein is about 1:1·3.

The cytoplasmic spectrum of NG showed mainly contributions from myeloperoxidase (MPO) present inside the granules as could be shown by comparison with spectra of purified MPO. The EG cytoplasmic spectra are 3 times more intense than found for NG. Again the main contribution comes from peroxidases inside the granules as could be shown by comparison with literature data on eosinophilic

294

Granulocyte (Raman) spectra

*660 nm; 6 mW; 30 s
(cytoplasm); 150 s
(nucleus)   averages of 5
(NG, BG) or 10 (EG)
measurements*

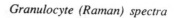

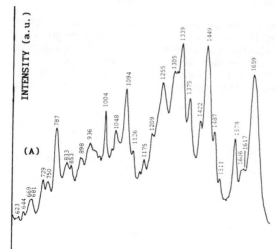

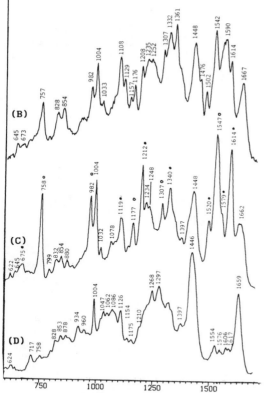

A *Nucleus NG*

B *Cytoplasm NG*

C *Cytoplasm EG*

D *Cytoplasm BG*

peroxidase (EPO) spectra[4], taking into account the different excitation wavelengths. The cytoplasmic spectra of BG varied more widely, which may be caused by the fact that the granules inside BG are bound to the cell membrane and the cytoplasmic spectra may therefore refer to more strongly varying compositions.

No evidence was found so far for the presence of heparin or histamin which form the main constituents of BG cytoplasmic granules. Some of the lines (1297, 1268) may be due to phospholipid contributions. Probably originating from dense multiple concentric membrane arrays in the BG granules[5]. This could also account for the CH_2 bending vibration, positioned at 1446 cm^{-1}.

In conclusion we can state that Raman spectroscopic studies of single living cells look promising.

REFERENCES

1. G.J. Puppels, Ph.D. Thesis, University of Twente, 1991.
2. G.J. Puppels, F.F.M. De Mul, C. Otto, J. Greve, M. Robert-Nicoud, D.J. Arndt-Jovin, T.M. Jovin, 1990, Nature, 347, 301-303.
3. G.J. Puppels, W. Colier, J.H.F. Olminkhof, C. Otto, F.F.M. De Mul, J. Greve, 1991, J.Raman Spectrosc. In press.
4. S.S. Sibbett, S.J. Klebanoff, J.K. Hurst, 1985, FEBS Letters 2911, 271-275.
5. A.M. Dvorak, S.J. Ackerman, 1989, Lab. Invest. 60 (4), 557-567.

CORRELATIONS BETWEEN VIBRATIONAL SPECTRA OF BIOMATERIALS AND BIOCOMPATIBILITY

A. Bertoluzza [1*], *P. Monti* [1] *and R. Simoni* [2]

Centro di Studio Interfacolta' sulla
Spettroscopia Raman, University of Bologna (Italy)

[1] Dipartimento di Biochimica, Sezione di Chimica e
Propedeutica Biochimica, via Selmi 2, Bologna
[2] Dipartimento di Chimica "G.Ciamician", via Selmi 2

INTRODUCTION

The characterization of the biocompatibility of a prosthesis in relation to the structure and reactivity of the component materials (biomaterials) constitutes at present an advanced structural research line.

Starting from the consideration that each living organism is predisposed to safeguard its biological integrity through both a non-specific response against a foreign body and a specific response (immunological reaction) which depends on the molecular structure of the biomaterial, biocompatibility is, from a biological and clinical aspect, a "biological limit condition of relative tolerance", deriving from the reciprocal interactions between the organism and the prosthesis.

From a structural point of view, biocompatibility is to be referred to the response of the organism to the prosthetic material, a response which is due to the molecular structure of the biomaterial and to the molecular interactions between tissues and the surface of the prosthesis. Therefore, structural investigations on the bulk and surface structure of biomaterials and on molecular interactions between biomaterials and tissues, conducted using non-destructive and non-invasive molecular methods such as Raman and ATR/FTIR spectroscopies, constitute a new and promising approach to the molecular characterization of biocompatibility. Some correlations between vibrational spectra of prosthetic materials and biocompatibility are here reported.

RESULTS AND DISCUSSION

The main phenomena influencing biocompatibility are of different origins, that are:

i) physical, which depends on the bulk structure of the biomaterial and the agreement between its physical properties and those of the host tissue or the tissues to which the biomaterial comes in contact;

ii) chemical, which depends on the structure and reactivity of the biomaterial surface;

iii) biochemical, which depends on the molecular interactions between the biomaterial surface and the surrounding tissue.

Vibrational Raman and ATR/FTIR spectroscopies are able to characterize in a non-destructive and non-invasive way all three of the different aspects of biocompatibility. There therefore exist close correlationships between vibrational spectra and biocompatibility, which have been verified by this research group for different classes of biomaterials: ceramics, bioactive glasses, polymers, etc..[1]

In this work examples of the correlationships between vibrational spectra and the three aspects of biocompatibility (physical, chemical, and biochemical) are discussed with reference to hydrogels which are biomaterials of ophthalmological interest used for the making of contact lenses.

The Physical Aspect of Biocompatibility.

The two hydrogels of relevance here are made of organic polymers based on 2-hydroxyethylmethacrylate (HEMA) or vinylpyrrolidone (VP), copolymerized with a small quantity of methacrylate co-monomers and generally cross-linked with ethyleneglycoldimethacrylate, which unites the linear polymeric chains and confers a suitable chemical stability to the polymer, as also shown by thermogravimetric measurements.[2,3] (In the following discussion the former hydrogel will be termed PHEMA and the latter PVP). Moreover, the presence of the ester and alcohol groups in the case of PHEMA, or ester and amide in PVP confers hydrophylic properties on the polymers: the water content is about 40% w/w and about 70% w/w respectively.

The hydrogels thus obtained possess suitable optic properties able to correct, by means of lens thickness, visual defects. Besides, their elastic modulus agrees with that of the cornea,[2,3] something which does not happen with the dry biomaterials, and the water content in the lens garantees the permeability to oxygen, necessary for the metabolism of the cornea. These prerequisits (optical properties, elastic modulus and permeability to oxygen) constitute the physical aspect of biocompatibility.

The Raman spectra give an account of the bulk structure of the two types of biomaterial and show a complete polymerization and homogenicity of such a structure. In particular the spectra show that the methacrylate group undergoes, by H-bonding with water, a decrease in the carbonylic frequency of about 8 cm^{-1} in both cases. Other than the C=O of the methacrylate group, PVP possesses a different hydrophilic group that is the C=O of the lactamic ring, which, by H-bond interaction with water lowers its own frequency by about 23 cm^{-1}.

In the bulk structure of hydrogels water becomes evenly distributed, interacting directly by H-bonds with the hydrophilic groups and successively with other water molecules which tend towards the typical distribution of liquid water, as shown by the differential scanning calorimetric curves.[2,3] In this respect the Raman spectra show that water, other than interacting with the hydrophilic groups of the polymer, also interacts with the hydrophobic ones in that the spectra show modifications of the C-H bending vibrations. The disturbance of the hydrophobic component of the hydrogels by water assumes an important role with regards to the other aspects of biocompatibility (chemical and biochemical).

The Chemical Aspect of Biocompatibility.

The surface structure of hydrogels is directly responsible for their chemical biocompatibility, which concerns the surface reactivity of the lens. This reactivity is due to the hydrophilic and hydrophobic groups of the biomaterials, and is mediated by the water molecules adsorbed on the surface, molecules which undergo a turnover once the lens is in contact with the surrounding biological fluids. Such a turnover has been verified by us by applying a PHEMA lens exclusively containing D_2O to rabbit eye. The Raman spectrum of the lens after application no longer contains the typical D_2O bands thus showing a complete exchange of lens water with that of the ocular tissue. Moreover, thermogravimetric measurements taken from PVP lenses applied to patients who have shown intollerance in a follow-up of 3 months have clearly shown an appreciable loss of water (an average about 8%) with respect to the nominal value.[4] We think that the loss of water is due to the turnover of water between lens and ocular tissue and is the most important parameter which influences the surface reactivity of the lens. In fact, lacking water molecules adsorbed, the surface hydrophilic groups can exercise greater molecular interactions regarding the tissues than in the ideal hydrated lens.

In order to study the chemical aspect of biocompatibility it is essential to evaluate the percentage of water not so much in the isolated lens as in the lens applied in vivo, where the turnover exists. Towards this end, a Raman method has been set up to directly evaluate the percentage of water contained in the contact lens.[5] This method has now been extended to the lenses applied in vivo. Preliminary measurements of multichannel Raman spectroscopy, directly obtained from a PHEMA lens applied to human eye, have shown a broad fluorescence band which masks the water bands. This fluorescence band is probably due to the formation of a protein layer at the lens-corneal tissue interface.

The Biochemical Aspects of Biocompatibility.

The increase in surface reactivity due to the turnover of water is directly responsible for the increase in surface interactions between lens and tissues. In particular the ATR/FTIR spectra of hydrophilic lenses taken from patients who have complained of intollerance and which were ascertained to have a reasonable decrease of water content with respect to the nominal value, showed the presence on the lens surface of the bands typical of proteins.[4] The intensity of such bands appears proportional to the degree of water lost through the lens-tissue turnover.

At the same time bacterial adhesion tests[6] have shown that Staphilococcus aureus adheres more easily onto PHEMA than onto PVP when the lenses are new, namely not yet applied. On lenses worn by patients and then treated with a suitable cleaner (regenerated lenses) no difference in bacterial adhesion was found between the two lens types. Moreover the ATR/FTIR spectra clearly show the presence of proteic substances on the regenerated lenses.

From this data, one can deduce that the difference in the quantity of bacteria adsorbed onto the new lenses is due to a different surface reactivity of the hydrogels, which depends on both the type of reactive surface centres and the different surface layer of water molecules. Viceversa, the fact that the amounts of bacteria adhered to both PVP and PHEMA regenerated lenses are the same could be due to the protein monolayer deposited on the lens surface which makes the surfaces of the two types of lenses similar. Thus the adsorption of Staphilococcus aureus is no longer influenced by the types of active surface centres and by the different water layer on the lens surfaces.

The singling out of the phenomena which regulate the various aspects of biocompatibility through correlations with the vibrational spectra, also allow the planning of prostheses with enhanced biocompatibility. As regards contact lenses such improvements are directed towards stabilizing the surface water layer of the hydrogels, thus improving the chemical aspect of biocompatibility, and making the biomaterial surface less reactive (by means of either physical and chemical treatments) in order to limit the proteic adsorption and increase the biochemical biocompatibility of the lens.

REFERENCES

1. A. Bertoluzza, C. Fagnano, P. Monti, R. Simoni, A. Tinti, M.R. Tosi and R. Caramazza, Clinical Materials, 1991, in press.
2. A. Bertoluzza, P. Monti, J.V. Garcia-Ramos, R. Simoni, R. Caramazza and A. Calzavara, J. Mol.Struct., 1986, 1434, 69.
3. A. Bertoluzza, P. Monti, R. Simoni, J.V. Garcia-Ramos, R. Caramazza, M. Cellini, L. De Martino and A. Calzavara, 'Stud. Phys. Theor. Chem.', vol.45 (Laser Scattering Spectroscopy of Biological Objects), Elsevier, Amsterdam, 1987, p. 595.
4. A. Bertoluzza, P. Monti and R. Simoni, 'Advances in Biomaterials', vol.7 (Biomaterials and Clinical Applications), Elsevier, Amsterdam, 1987, p. 315.
5. A. Bertoluzza, C. Fagnano, P. Monti and R. Simoni, 'Proc. XIth Int. Conf. on Raman Spectroscopy', Wiley, Chichester, 1988, p.723.
6. C. Arciola, P. Versura, G. Ciapetti, A. Pizzoferrato, R. Caramazza and P. Monti, 'Advances in Biomaterials', vol. 8 (Implant Materials in Biofunction), Elsevier, Amsterdam, 1988, p.349.

RAMAN MICROSPECTROSCOPIC INVESTIGATION OF CHROMOSOMAL BANDING PATTERNS

G.J. Puppels[o], C. Otto[o], J. Greve[o]
M. Robert-Nicoud[*], D.J. Arndt-Jovin[*], T.M. Jovin[*]

[o] Biophysical Technology Group, Faculty of Applied Physics, University of Twente, P.O. Box 217, 7500 AE Enschede, The Netherlands
[*] Max Planck Institut for Biophysical Chemistry, Department of Molecular Biology, PF 2841, D-3400 Göttingen, F.R. Germany

We have recently described a sensitive confocal Raman microspectrometer (CRM) which makes it possible to obtain high quality Raman spectra of single living cells and chromosomes[1,2]. The spatial resolution of the instrument is < 0.5 μm in lateral and ~ 1.3 μm in axial direction. This high spatial resolution allows the investigation of chromosomal banding patterns.

Polytene chromosomes are a strongly amplified form of the interphase chromosomes, found in cells of salivary glands and other tissues of Dipteran insects. They possess a characteristic pattern of dark bands and light interbands which can be observed under a light microscope. Not much is known about the molecular basis of this phenomenon.

In metaphase chromosomes banding can be induced by means of physico-chemical treatment(s) and staining. Although these banding patterns cannot be observed on untreated unstained chromosomes they must of course reflect one or more properties of the chromosomes in their natural form. Many theories about the origin of these metaphase chromosomes banding patterns have been put forward. But unequivocal experimental evidence for any of them is lacking.

Figure 1 shows Raman spectra obtained from bands (A) and interbands (B) of unfixed polytene chromosomes (Chironomus thummi thummi), as well as a difference spectrum (C). Strong signal contributions of DNA and proteins are present.

It can be seen that the predominant DNA structure is the B-form (phosphate vibrations at 1094 cm^{-1}, ~ 833 cm^{-1} (overlapped by tyrosine vibrations) ring breathing vibrations of the bases at 681 (Guanine), 729 (Adenine) and 749 cm^{-1} (Thymine) consistent with a C2'-endo-anti sugar pucker). DNA base composition was determined by fitting the doublet at 669 (T) and 681 (G) cm^{-1} with Lorentzians and comparing their intensity ratio with that found for calf-thymus DNA which contains 42% GC basepairs[3]. Ctt-polytene chromosomes were found to contain 31 ± 4% GC.

Information about protein secondary structure is obtained from the Amide I and Amide III bands. Their positions and intensities show the chromosomal proteins to contain mainly α-helical and random coil domains and little β-sheet structure.

Comparing the signal intensities of spectra A and B, it is evident that the chromatin concentration is higher in bands than in

interbands. The difference spectrum (C) makes clear that the DNA-protein ratio is higher in bands than in interbands and furthermore that the average protein secondary structure is different in bands and interbands (see positions Amide I bands). These type of Raman microspectroscopic measurements may lead to a better understanding of the cause and meaning of the polytene chromosome band-interband pattern.
For metaphase chromosomes a similar approach is feasible.

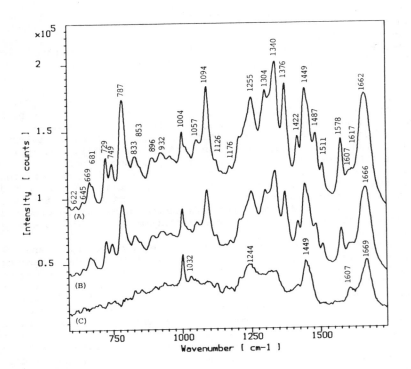

Figure 1
Raman spectra of unfixed Ctt-polytene chromosomes.
A) Average of measurements on 4 different bands
B) Average of measurements on 4 different interbands
C) Difference spectrum B-A after scaling of A and B on the 1094 cm^{-1} DNA line (shown 3 times enlarged).
In all measurements the laser power was 15 mW (660 nm) and the measuring time 10 minutes. The spectra have been shifted along the ordinate for clarity.

[1] G.J. Puppels, et al. (1990) Nature 347, 301-303.
[2] G.J. Puppels, et al. J. Raman Spectrosc.: in press.
[3] G.J. Puppels (1991) PhD thesis, University of Twente.

HIGH-SENSITIVE DETECTION OF INTRACELLULAR CAROTENOIDS IN SINGLE LIVING CANCER CELLS AS PROBED BY SURFACE-ENHANCED RAMAN MICROSPECTROSCOPY

M. Manfait[1], H. Morjani[1], R. Efremov[2], J.-F. Angiboust[1], M. Polissiou[3] and I. Nabiev[2]

1 Laboratoire de Spectroscopie Biomoléculaire, GIBSA, UFR de Pharmacie, Université de Reims, 51096 Reims cedex, France.
2 Shemyakin Institute of Bioorganic Chemistry, USSR Academy of Sciences, Moscow, USSR.
3 Agricultural University of Athens, Laboratory of General Chemistry, Athens, Greece.

INTRODUCTION

It has recently been demonstrated that the effect of very large (in some instances up to 10^9) enhancements of the Raman cross-section for molecules in the close vicinity of a metal surface can find many very important applications in life sciences[1-3] including antitumor drugs in vitro [4] and in vivo [5]. This paper deals with investigations on one of carotenoids as an agent of cell differentiation-dimethylcrocetin (DMCR)-as well as the study of its interaction with living cancer cells. DMCR chromophore has a very high extinction coefficient for the electronic transition in the visible region. We selectively detected the DMCR SERS signal in the cytoplasm and in the nucleus of single living cancer cell under the microscope. Significant spectral differences in the spectra between free DMCR and interacting with cytoplasm or nucleus have also been demonstrated.

EXPERIMENTAL

SERS spectroscopy

A vacuum disperse method was used to obtain silver films on specially prepared glass slides as described in [6]. Island films with absorption maxima varing from 450 to 660 nm were applied to detect SERS spectra which were obtained on a DILOR Omars-89 Raman spectrometer connected with an Olympus BH-2 microscope equipped with a 100X water immersion objective (Leitz Fluotar), using the 514.5 nm line of an argon laser (Spectra-Physics, Model 2020-03). Laser power on the sample was less than 3mW.

Chemicals and cells

Dimethylcrocetin (DMCR) is a lipopholic derivative of crocin which is the main coloring component of the extract from the natural pigment of Crocus sativus L (Saffron). DMCR was synthesized by alkaline hydrolysis of crocin in methanol.

K562 and Friend (leukemia cell lines) are incubated in presence of 10^{-6} M DMCR for 4 hours, washed twice with 0.9% NaCl and then deposited on the silver island films with an absorption maxima in the green.

RESULTS AND CONCLUSION

In the SERS spectrum of 10^{-6} M DMCR (Figure 1-a), the most intense bands at 1532 and 1161 cm^{-1} assigned to $-C=C-$ and $=C-C=$ stretching modes of the conjugated polyene chain can be detected. After incubation of K562 and Friend cells in 10^{-6} M DMCR we detect in the cytoplasm a strong SERS signal of DMCR displaying the two main bands at 1530 and 1160 cm^{-1} (Figure 1-b, 1-c). Comparing to free DMCR, new components at 1280 and 1150 cm^{-1} can be observed for the two treated cell strains. In addition for Friend cells the band at 1529 cm^{-1} was less intense than it of free DMCR or in K562 cells. In other hand, the SERS signal of DMCR was more intense in the K562 cytoplasm than in the nucleus (Figure 1-b, 1-d). Comparing to the measurements done on glass slide (Figure 1-e), we can conclude that the SERS active plate technique is a powerful tool for the study of microamounts of drugs or xenobiotics in a single living cell. Investigations on the interaction of DMCR or other carotenoids, performed on single living cells, are considered as important for elucidation of the biological activity concerning cytotoxicity and differentiation effects of these molecules.

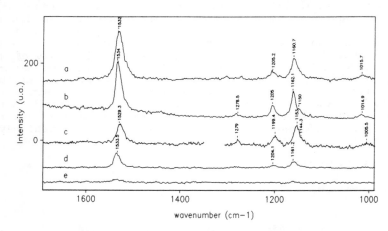

Figure 1: SERS spectra on silver island films. (a) free DMCR; (b) and (c) DMCR in the cytoplasm of K562 and Friend cells respectively; (d) DMCR in the K562 nucleus and (e) Raman spectra of treated K562 cells deposited on glass slide.

REFERENCES

1. I.R. Nabiev and R.G. Efremov, J.Raman Spectrosc., 1983, 14, 375.
2. N.G. Abdulaev, I.R. Nabiev, R.G. Efremov and G.D. Chumanov, FEBS Lett., 1987, 213, 113.
3. I.R. Nabiev, R.G. Efremov and G.D. Chumanov, Sov.Phys.Usp., 1988, 31, 241.
4. Y. Nonaka, M. Tsuboi, and K.-J. Nakamoto, J.Raman Spectrosc., 1990, 21, 133.
5. I.R. Nabiev, H. Morjani and M. Manfait, Eur.Biophys.J., 1991, in press.
6. R.S. Sennett and G.D. Scott, J.Opt.Soc.Amer., 1950, 40, 203.

MICRO-SERS ANALYSIS OF ELLIPTICINE INTERACTION IN LIVING CANCER CELLS

J.-M. Millot[1], H. Morjani[1], J. Aubard[2], J. Pantigny[2], I. Nabiev[3], M. Manfait[1]

1 : Laboratoire de Spectroscopie Biomoléculaire, UER de Pharmacie, 51096 Reims, France.
2 : Institut de Topologie et de Dynamique des Systèmes, 1 rue Guy de la Brosse, 75005 Paris, France.
3 : Shemyakin Institute of Bioorganic Chemistry, USSR Academy of Sciences, 117871 Moscow, USSR

INTRODUCTION

Ellipticine is an antitumor alkaloid whose physico-chemical studies have described its interaction mechanism with DNA. In vitro, the apparent pK of ellipticine is strongly affected when this weak base is bound to biological molecules : pK=7.4 (in aqueous solution), pK=9 (bound to the DNA). Thus, in biological systems, ellipticine may exist both in the cationic protonated form and in the neutral form in balance. Surface-enhanced Raman spectroscopy studies of ellipticine allow to characterize the protonation state of the drug in aqueous solutions [1].

MATERIAL AND METHODS

Ag hydrosol was prepared by the reduction of Ag nitrate with trisodium citrate and was aggregated by sodium perchlorate 0.05 M [2]. 10^6 leukemia K562 cells were exposed to 10 µM ellipticine for 1 hour, washed and resuspended in 100 µl of a phosphate buffered saline. This cell pellet was incubated into 400 µl of aggregated Ag hydrosol for 30 min. at 37°C [2]. The micro-SERS spectra were recorded by using a DILOR OMARS 89 micro-Raman spectrometer coupled with an optical microscope equipped with a 100X water immersion objective.

RESULTS

Ellipticine in Aqueous Solutions

The intensities and the profiles of the main lines 1588 and 1406 cm^{-1} prove to be very dependent on the protonation state of the molecule. More exactly, the value of the ratio $R=I_{1588}/I_{1406}$ is superior to 1 when the ellipticine is in the cationic form (spectrum 2) and is inferior to 1 in the neutral form (spectrum 1). In presence of DNA, the pK value of ellipticine shifts from 7.4 to 9 and spectrum 3 corresponds to spectrum 2 of the protonated form.

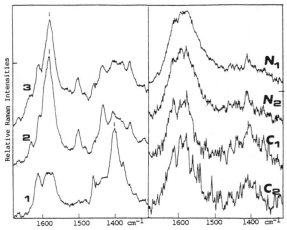

Figure 1 : Micro-SERS spectra of ellipticine (3 μM) : pH 8 (spectrum 1), pH 5.5 (spectrum 2), complexed to DNA pH 8 (spectrum 3). Micro-SERS spectra obtained by focusing a laser spot on treated cells (10 μM ellipticine) in nuclei (spectra N_1, N_2) and in cytoplasms (spectra C_1, C_2).

Ellipticine in Living Cells

From the nuclear spectra (N_1, N_2), the value of ratio R is superior to 1. This suggests that in the nucleus, the drug is intercalated between DNA base pairs under its cationic form, as expected from the pK of ellipticine bound to DNA in vitro. However, the noticeable lack of the 1500 cm⁻¹ line in the nuclear spectra is not expected, thus showing the limits of the modelisation with the ellipticine-DNA complex. From the cytoplasmic spectra (C_1, C_2), the value of ratio R superior to 1, corresponds to a cationic form of ellipticine. However, the intensities of the 1406 cm⁻¹ line are higher, as compared with the intranuclear spectra. This could result from a contribution of neutral ellipticine within cytoplasmic compartments such as mitochondria, as observed from previous microspectrofluorometric results[3].

In conclusion, micro-SERS appears to be a powerful tool for studying molecular events in single living cells (interaction with DNA, protonation states, structure modifications) which are associated with biological and antitumoral effects.

REFERENCES

1. J. Aubard, J. Pantigny, G. Levy, J.F. Marsault, G. Dodin, M.A. Schwaller. '4th Eur. Conf. Spectrosc. Biol. Mol.', Royal Soc. of Chem. Ed., York, 1991.
2. I. Nabiev, H. Morjani, M. Manfait. Eur. Biophys. J., 1991, in press.
3. M.A. Schwaller, J. Aubard, F. Sureau, F. Moreau, J.M. Millot, M. Manfait. Cancer Res., 1991, in press.

In vivo STUDY OF PHOTOSYNTHETIC BACTERIA BY INFRARED MICROSPECTROSCOPY

Y. Ozaki and K. Okada

Department of Chemistry, Kwansei Gakuin University, Nishinomiya 662, JAPAN

INTRODUCTION

Nondestructive structural analysis of biological materials are very important in medicine as well as in biology.[1] Recent rapid progress in FT-IR spectroscopy has enabled one to measure infrared spectra of cells, tissues, organs, and whole living bodies in vivo. The present paper demonstrates potential of infrared microspectroscopy[2] in probing the structure of constituents of complex biological materials. Two kinds of photosynthetic bacteria Rb. sphaeroides R26 and G1C are employed as examples.

EXPERIMENTAL

Infrared spectra of the bacteria were mesured with a JEOL JIR-100 FT-IR spectrometer equipped with a micro infrared attachment (IR-MAU 110). The spectra were obtained at 4 cm^{-1} resolution, with 100 scans.

RESULTS AND DISCUSSION

Figure 1 shows infrared spectra of Rb. sphaeroides R26 (a) and G1C (b) measured in vivo by infrared microspectroscopy. Note that spectral quality of Figure 1(a) and (b) is very high. The spectra are, in general, very similar to those of chromatophores of R26 and G1C, respectively, showing that they largely reflect them. Intense bands at 1655 and 1547 cm^{-1} are assignable to amide I and II modes of proteins. These amide frequencies indicate that the proteins constituting bacteria have mainly α-helix structure. Bands at 1739 cm^{-1} are probably due to C=O stretching modes of ester groups of bacteriochlorophyll a (BChl a) included in the bacteria. Particularly striking is that the relative intensity of the C=O stretching band is much stronger in Figure 1(a) than 1(b). Broad features in the 1100-1000 cm^{-1} region are assignable to C-O stretching modes of polysaccha-

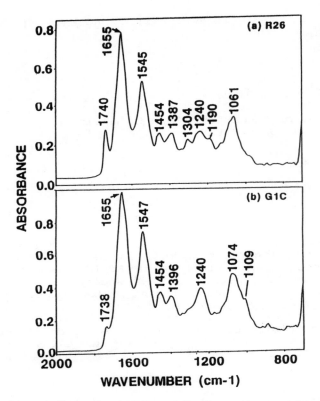

Figure 1. Infrared spectra of Rb. sphaeroides R26 (a)
and G1C (b) measured in vivo by infrared microspectroscopy.

rides.
      The two spectra resemble   each other, but the
calculation of difference spectra between Figure 1(a)
and 1(b) clearly shows that the intensities of 1740,
1304, and 1200 cm^{-1} bands are considerably stronger in
Figure 1(a) while that of 1109   cm^{-1} is more intense in Figure 1(b).   These results have not been well
understood yet, but it may be concluded from them that
the relative amounts of BChl a are more in R26 than in
G1C and structure and/or kinds of polysaccharides are
significantly different between them.

References
1. Y. Ozaki, Appl. Spectrosc. Rev., 1988, 24, 259.
2. R. G. Messerschmitt and M. A. Harthcock,   'Infrared
   Microspectroscopy', Marcel Dekker, 1988.

INTRACELLULAR DISTRIBUTION AND INTERACTIONS OF DOXORUBICINE IN SENSIBLE AND RESISTANT K562 CELLS STUDIED BY CONFOCAL FLUORESCENCE IMAGING AND MICROSPECTROFLUOREMETRY.

R. Colombo[1], S.M. Doglia[2], M. Manfait[3], H. Morjani[3] and A. Villa[2].

[1]Department of Biology, Zoology and Cytology and [2]Department of Physics, University of Milan, 20133 Milan, ITALY; [3]Laboratoire de Spectroscopie Biomoleculaire, Universite de Reims, 51096 Reims, FRANCE.

Recent advances in microspectrofluorometry have enabled the detection of fluorescence signals from microvolumes within a single living cell[1,2]. By this method the concentration of doxorubicine in the nuclei of sensible and resistant K562 tumor cells has been evaluated[3]. The higher nuclear concentration of doxorubicine in sensible cells suggested that the cytotoxic effect of the drug is related to the nuclear drug concentration. The lower concentration in the nuclei of resistant cells could instead be the consequence of the extranuclear interactions of the drug in presence of the complex processes of multidrug resistance.

Fluorescence emission spectra of doxorubicine in the cytoplasm of the K562 R and S cells has been studied by laser microspectrofluorometry (OMARS 89, Dilor, France). Preliminary results[4] indicated that the spectral response in the case of resistant cells is quite different from that of free drug in solution. Moreover, the fluorescent signal from different microvolumes within the cytoplasm is not homogeneous both in intensity and in spectral shape.

In order to understand these results, we have performed fluorescence imaging analysis of doxorubicine in K562 S and R cells by a confocal laser microscope (Lasersharp-MRC 500). K562 cells in exponential growth phase (at a concentration of 10^5 cells/ml) were incubated with 2 μM doxorubicine (Roger Bellon, France) for 1 hour in culture medium and washed twice; cell smears on a glass slide were employed for fluorescence imaging. Very different patterns have been recorded in the two cases: (1) in the K562 S cell the drug fluorescence is confined only to the nucleus and no appreciable fluorescence can be seen in the cytoplasm, in agreement with the spectral analysis; (2) in the K562 R cell the nucleus fluorescence is very weak, but an intense signal appears on the cell membrane and its proximity (in a layer of about 1μM) within the cytoplasm. This doxorubicine fluorescence on the membrane is not homogeneous, indicating that specific interactions occur in the resistent cells between the drug and the membrane components. These could include the P-glycoprotein and/or actin structures involved in clustering processes of the membrane proteins[5]. To explore the complex phenomenon of multidrug resistance, we have studied the doxorubicine fluorescence in K562 R in presence of agents known to reverse the resistance. The effect of verapamil, a calcium channel blocker, was studied by confocal fluorescence imaging and by microspectrofluorometry. K562 cells were incubated with 2 μM doxorubicine and with 5 μM verapamil (Bioserda, France) for 1 hour in culture medium and washed twice; again cell smears on glass slides were employed for fluorescence imaging measurements, whereas for microspectrofluorescence cells were studied in a PBS suspension on a petri dish. The doxorubicine fluorescence imaging of verapamil-treated K562 R cells indicates that in the cytoplasm the drug is completely absent, as in the case of sensitive cells. Instead, the nuclear fluorescence is increased, in agreement with the results of the spectral analysis. By microspectrofluorometry the nuclear concentration of doxorubicine in K562 R cells has been studied as a function of the extracellular concentration of verapamil, but without reaching the value observed in the K562 S case. This suggests that verapamil never reverses resistance completely.

Since doxorubicin interacts with actin[6], investigations are in progress to establish whether the fluorescent signals localized in the membrane area of K562 R cells are due to the drug interaction with a particular membrane protein or with actin filaments of the cortical web. Unfortunately, K562 cells are not the best model system to study doxorubicine cytoplasmic localization, because of their small cytoplasm volume. We plan to extend this work to larger cellular systems.

REFERENCES

1. L. Ginot, P. Jeannesson, J.-F. Angiboust, J.-C. Jardillier and M. Manfait., Studia Biophys., 1984, 47, 332.
2. M. Manfait, M. Gigli, J.-M. Millot, T. Rasoanaivo, C. Perchard, S. Nocentini, P.-Y. Turpin, P. Vigny and S. Doglia., "Spectroscopy of Biological Molecules - New Advances", E. D. Schmid, F. W. Schneider, F. Siebert (eds), John Wiley & Sons, Chichester, 1987, p.445.
3. M. Gigli, S. M. Doglia, J.-M. Millot, L. Valentini, and M. Manfait, BBA, 1988, 950, 13.
4. H. Morjani, DEA Thesis, University of Reims, 1989.
5. H. M. Coly, P. R. Twentyman, and P. Workman, Cancer Chemother. Pharmacol., 1989, 24, 284.
6. R. Colombo and A. Milzani, BBA, 1988, 968, 9.

THE COMBINED USE OF FT-IR MICROSCOPY AND MULTIVARIATE STATISTICS FOR THE CHARACTERIZATION OF BACTERIAL MICROCOLONIES

D. Naumann

Robert Koch-Institut des Bundesgesundheitsamtes, Nordufer 20, D-1000 Berlin 65

INTRODUCTION

The Fourier-Transform Infrared Spectroscopic patterns of intact bacterial cells can be used to rapidly differentiate a variety of bacterial strains and taxa (1-4). Encouraged by the recent advent of high-quality FT-IR microscopes, we have tried to establish some essential features of a technique which combines the advantages of light microscopy with the sensitivity and selectivity of FT-IR to detect, enumerate and differentiate/identifiy bacterial cells and to characterize bacterial growth.

A TYPICAL PROTOCOL

In order to measure bacterial microcolonies, the following procedure has been elaborated: An aliquot of a sufficiently diluted bacterial suspension was plated on an adequate solid growth medium (agar plates) and was incubated 6 to 8 hours at 37 °C. A round, IR-transparent BaF_2-plate (typically 25 mm in diameter) was then gently and rectangularly pressed onto the agar surface using a special, home-made device. Thus, small amounts (replica) of the bacterial microcolonies (aprox. 2 to 3 bacterial layers per spot) were transferred to the IR-plate, were dried to transparent films within seconds at ambient conditions and were subjected to FT-IR microscopy. The optical properties (size etc.) of the replica and the number of spots were estimated operator controlled or by imaging techniques.

RESULTS AND DISCUSSION

Fig. 1 gives the dendrogram of a cluster analysis performed on 10 FT-IR spectra of some selected bacterial spots from a mixed culture containing three different kinds of bacteria. Identification of the different bacterial strains was attained independently by conventional microbiological techniques. Data elaboration and cluster analysis was carried out as already described (3).

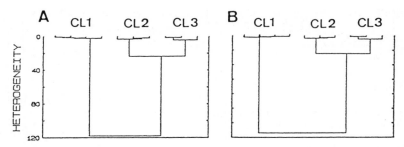

<u>Figure 1</u> Dendrogram of the classification of FT-IR measurements
on 10 microcolony spots obtained from a mixed culture.
A: FT-IR measurements with a 100 μm aperture.
B: FT-IR measurements with a 80 μm aperture.
CL1 = Escherichia coli; CL2, CL3 = two different strains
of Staphylococcus aureus. Cluster analysis was performed
using the first derivatives of the spectra considering
the spectral ranges 1200 - 1500 and 900 - 1200 cm^{-1}.
The two spectral ranges were equally weighted (3).
Ward's algorithm was applied. FT-IR instrumentation:
IFS-66 spectrometer and FT-IR microscope A-560 from
Bruker, Karlsruhe, FRG. Number of scans: 512,
apodization : triangular.

From the results depicted in Fig. 1 we conclude: (i) bacterial
microcolonies (40 to 80 μm in diamter) can be transferred from
solid culture plates onto IR-transparent plates by a special
replica (printing) technique for subsequent FT-IR measurements,
(ii) colony spots can be enumerated, (iii) the data accessible
from the light-microscopic images (numbers, size and different
shapes etc.) and from the FT-IR spectra of bacterial colonies
(cell composition, structural data etc.) can be succesfully used
to differentiate and classify bacteria and to characterize colony
growth. Furthermore, the application of multivariate statistic
image analysis to both, the 2D-images of colony prints (replica)
and the 1D-"images" of bacterial FT-IR spectra, strongly enhances
the capacity and the versatility of the technique (1, 2, 4, 5).

REFERENCES

1. D. Naumann, V. Fijala, H. Labischinski and P. Giesbrecht,
    J. Mol. Struct., 1988, <u>174</u>, 165.

2. D. Naumann, V. Fijala, H. Labischinski and P. Giesbrecht,
    Microchim. Acta, 1988, <u>I</u>, 373.

3. D. Helm, H. Labischinski, D. Naumann, J. Gen. Microbiol.,
    1991, <u>137</u>, 69.

4. D. Naumann, H. Labischinski and P. Giesbrecht, in: Modern
    Techniques for Rapid Microbiological Analysis, W.H. Nelson,
    Ed., VCH Verlag Chemie, New York (in press).

5. M. van Heel, Ultramicroscopy, 1984, <u>13</u>, 165.

CORRELATIONS BETWEEN THE MOLECULAR STRUCTURE OF OCULAR TISSUES AND PATHOLOGY BY RAMAN LASER SPECTROSCOPY

A. Bertoluzza, C. Fagnano

Centro Studi Interfacolta' sulla Spettroscopia
Raman, Dipartimento di Biochimica, Sezione
di Chimica e Propedeutica Biochimica
Via Selmi 2, Universita' di Bologna (Italy)

R. Caramazza, S. Mancini and E. Barbaresi

Istituto di Clinica Oculistica
Policlinico S. Orsola, Univ. Bologna

Traditional and multichannel Raman spectroscopies are, at present, techniques greatly suitable for correlating the molecular structure of biological tissues and their biological functions. The ocular tissues are well adapted for this purpose because of both their transparency and their accessibility for in vivo measurements.

Regarding this we have previously studied by Raman spectroscopy: isolated normal rabbit lenses; isolated normal and cataractous human lenses; isolated rabbit lenses in which cataract was induced by an antimetabolite added in the diet; normal rabbit lenses directly in a whole eye ball; normal animal and human lenses directly in situ and in vivo. Correlations between pathology (i.e. cataract) and structural modifications deduced from the Raman spectra have been found. These data constitute, at present, an useful tool for the molecular diagnosis of cataract (1).

In the ophthalmological field a new surgical technique of the cornea (computerized refractive surgery) is developing. By means of such a technique it is possible to remove consecutive layers of corneal tissue using an excimer laser in order to correct myopia. This technique has several other applications which are actually less developed (i.e. treatment of superficial opacities of the cornea, treatment of phterigium and lasertrabeculectomy in glaucoma).

Till now scientific research has been devoted to the physical aspects of this technique, while lacking studies about the effects of the laser radiations on corneal tissues. Concerning this we have planned an interdisciplinary spectroscopic-clinical research devoted to the Raman multichannel characterization of the corneal tissues directly in situ and in vivo, before and after the laser treatment. This is in order to point out the possible structural modifications caused by the excimer laser radiation.

In this work some preliminary results have been reported on the Raman spectrum of a fresh rabbit cornea immersed in physiological solution, of the same cornea after its dehydration in air and of the cornea treated with HCl aqueous solution 1N for 150 sec and successively dehydrated.

Figure 1 shows the Raman spectrum of the fresh rabbit cornea immersed in physiological solution. This spectrum substantially agrees with the one (feline cornea) found in the literature (2).

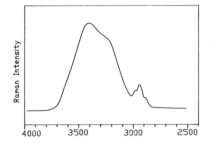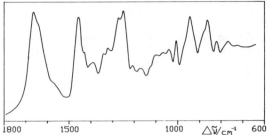

Fig. 1 - Raman spectrum of the fresh rabbit cornea.

In order to keep the cornea transparent in the dehydrated state it was compressed between two glass windows. In this way the spectra of both normal and HCl treated cornea were obtained.

By comparing the spectra obtained as above we can note:

- the disappearance, in the spectrum of the dehydrated cornea, of the broad bands centered at 3420 and 3250 cm^{-1} (and also at 1640 cm^{-1}), due to water of the fresh cornea;
- the appearance of two bands in the same region at 3450 and 3320 cm^{-1} in the spectra of the dehydrated cornea;
- the disappearance of the 3450 cm^{-1} band passing from the dehydrated cornea to dehydrated cornea treated with HCl;
- dehydration involves also a change in intensity of the band centered at 1270 cm^{-1};
- treatment with HCl involves also a change in intensity of the bands centered at 940 and 855 cm^{-1}.

These spectroscopic data show a correlation between the modifications in the Raman spectra of the cornea and phenomena such as caustication.

We are now studying the meaning of these modifications by the extension of the measurements on the main components of the system. At the same time, also on the basis of the results obtained, we are setting up the measurements for evaluating in situ and in vivo structural modifications of the cornea after treatments with an excimer laser.

REFERENCES

1. A. Bertoluzza, C. Fagnano, P. Monti, R. Simoni, J.V. Garcia-Ramos and R. Caramazza, Optica Pura y Aplicada, 1988, 21, 43 and references therein.
2. S.C. Goheen, L.J. Lis and J.W. Kaufmann, Biochim. Biophys. Acta, 1978, 536, 197.

RAMAN SPECTRA OF A NEW BONE CEMENT FOR ORTHOPAEDIC USE

A. Bertoluzza[1], M.A. Morelli[2], A. Tinti[1] and M.R. Tosi[3]

Centro Studi Interfacolta' sulla
Spettroscopia Raman,
Universita' di Bologna (Italy)

[1] Dipartimento di Biochimica, Sezione di
Chimica e Propedeutica Biochimica
via Selmi 2, Bologna

[2] Dipartimento di Chimica "G.Ciamician"
via Selmi 2, Bologna

[3] Istituto di Citomorfologia Normale e
Patologica del CNR
via di Barbiano 1/10, Bologna

In the surgery of hip prosthesis two techniques are used: the first involves the utilization of the acrylic bone cement self cured in situ. Its main function is to secure the prosthesis in its seat, fitting it to the bone modifying its size, volume, shape, and distributing the stresses by its modulus of elasticity. Polymethylmethacrylate(PMMA) is the cement with the highest follow-up of biocompatibility.

The second one involves the implant of a prosthesis without cement but able to originate bioactivity by the coating of the metallic stem.

Either bioactive glasses (the Hench "bioglass" is a typical example) or ceramics (hydroxylapatite) or mixed bioglass-ceramic materials are used as coatings. These materials have the function to elicit specific physiological responses including bonding to tissues.

In this work we study the Raman spectra of new bone cements, made of PMMA, obtained by self curing of the monomer and a mixture of 1:1 of PMMA and sodium or calcium, sodium and calcium or calcium and aluminium polyphosphate bioactive glasses with a precise molar ratio, R = basic oxide/acid oxide. These biomaterials have the function of integrating cement properties with those of bioactive materials showing both elastic properties and the fastening to the bone by physiological responses due also to the mineral component of the cement.

The Raman spectra of the bone cements with different compositions, obtained by us, show, as regards the mineral component, a certain trend of the stretching mode (at about 1170 cm^{-1}) $\nu_s PO_2^-$ of the polyphosphatic chain (with a molar ratio basic oxide/acid oxide =1) depending on the cation (Na or Ca). As R rises, a gradual intensity and frequency decrease as regards this band (with a shoulder increase in the region of the lower frequencies) is observed. Also a new band is observed, with increasing intensity, attributable to the stretching mode $\nu_s PO_3^{2-}$ (at about 1030 cm^{-1}), that for terminal phosphate groups in oligophosphates is due to the demolition of the polyphosphatic chains by the basic oxide during the glass melting.

The correlation observed previously,[1] between frequency and relative intensity of this terminal stretching mode band and average chain length, in its turn, depends on R.

As regards bioactive oligophosphate glasses with Ca and Al the intensity and frequency decrease is less marked; the new band in the previous case also at 1030 cm^{-1} is not so strong and is slightly shifted towards a lower wavenumber. In agreement with previous measurements [2] the spectroscopic trend suggests a partial covalent interaction Al-O as regards the prevailing ionic one of alkaline and alkaline earth ions.

Regarding the organic component, in comparison with the spectrum of PMMA, Raman spectra show some modifications of the relative intensities of the bands typical of the bending modes (scissoring, rocking, e tc.) in-plane and out-of-plane and also of stretching modes of σ bonds characteristic of lateral chain groups.

This vibrational study has characterized the new cements main structural aspects that correspond to their function as biomaterials; that is to integrate PMMA properties with those of bioactive coatings showing both elastic properties and the fastening to the bone by physiological responses.

ACKNOWLEDGEMENT

This work was supported by C.N.R. (Italy) progetto finalizzato "Chimica Fine" sottoprogetto "Chimica e Tecnologie dei Polimeri"

REFERENCES

1. A. Bertoluzza, M.A. Battaglia, R. Simoni and D.A. Long, J. Raman Spectr., 1983, 14, 178.
2. A.Bertoluzza, C.Fagnano, A.Marinanageli, R.Simoni, A.Tinti and M.A.Morelli, Biomaterials and Clinical Applications, Ed. A. Pizzoferrato, P.G.Marchetti, A.Ravaglioli and A.J.C. Lee,1987, p.511.

CORRELATIONS BETWEEN VIBRATIONAL SPECTRA (ATR-FTIR) OF VASCULAR BIOPROSTHESES AND BIOCOMPATIBILITY

A. Bertoluzza[1], A. Tinti[1] and M.R. Tosi[2]

Centro Studi Interfacolta'
sulla Spettroscopia Raman
Universita'di Bologna (Italy)

[1] Dipartimento di Biochimica,
Sezione di Chimica e
Propedeutica Biochimica

[2] Istituto di Citomorfologia
Normale e Patologica del CNR
Bologna (Italy)

The study of the behaviour of vascular prostheses made of Dacron coated with a biological substance (bioprostheses) is a very interesting and actual problem in vascular surgery.

The biological coating is mainly made for avoiding the preclotting which can cause bleeding and infection. Fibrine (1), elastine (2) , albumine (3) and collagen (4) have been used as biological coatings .

The functionality of the bioprostheses rises either from the kind of interaction which bind the biological coating to the polymeric tissue of the prosthesis, generally Dacron, or to the behaviour of the biological layer in the physiological environment. This problem has not yet been characterized at a molecular level for vascular prostheses with a biological coating in relation to their functionality.

In this connection, vibrational ATR-FTIR spectra of bioprosthesis surfaces before and after explantation and also in vitro, give useful information as regards the knowledge of vascular prosthesis functionality.

ATR-FTIR spectra (obtained by a Jasco FT-IR 7000 spectrometer) of a commercial sample (Gelseal Triaxial), before and after 15 days of implant in man, have shown the lack of the biological covering layer (5).

With the aim to characterize this release, we have planned an in vitro spectroscopic study regarding the biological coating of a prosthesis of this type. The samples have been stored, at 37°C, for different times, in a buffered saline solution at physiological pH.

By comparing the spectrum of Dacron (Fig. 1e), glycerine (Fig. 1a) and gelatin, the untreated prosthesis (Fig. 1b) shows the bands typical of glycerine and those typical of gelatin, at 1636 cm^{-1} (amide I) and at 1550 cm^{-1} (amide II). After 1h in buffer solution (fig1. c) the bands of the glycerine disappear and only those typical of the gelatin are present. A spectrum obtained after 24h of storage in buffered solution (fig.1d) does not show essential modification of the gelatin coating. The full release of the polypeptide from the prosthesis is instead deducible from the spectra after 48h in the buffer (fig 1e); in this case the spectrum is coincident with that of the surface of uncoated Dacron.

These spectroscopic measurements obtained by a non destructive and non invasive technique are meaningful to characterize the behaviour of the biological layer of the prosthesis in vitro because they can give a molecular explanation regarding the behaviour of the coating.

Now we are trying to evaluate the influence of the haemodynamic environment on the bioprosthesis surface and also to study the prosthesis behaviour in vivo.

REFERENCES
1. L.R. Sauvage, K. Berger, C.C. Davis et al., *Biologic and Synthetic Vascular Prostheses*, Ed. Grune and Stratton, 1982, p. 533
2. J.V. Bascom, *Surgery* , 1961, 50, 504
3. D.J. Liman, K.G. Klein, J.J. Brash, *Thromb. Diath. Haemor.*, 1970, 42, 109
4. A.W. Humphries, W.A. Hawk and A.M. Cuthberston, *Surgery*, 1961, 50, 947
5. M.R. Tosi, A. Tinti e P. Filippetti, *Atti del VI Congresso Nazionale S.I.A.Te.C.* Bologna 16-19 settembre 1990, p. 339

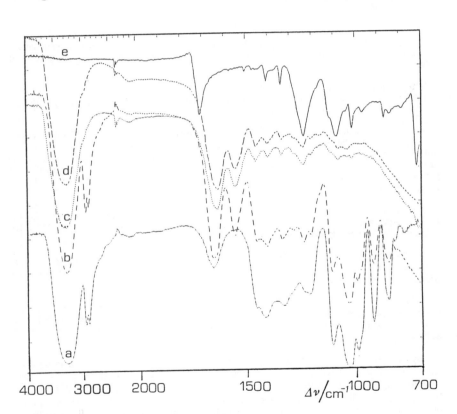

Fig. 1 - ATR-FTIR spectra of: a) glycerine
b) glycerine and gelatine coated v.p.
c) v.p. after 1h in buffer
d) v.p. after 24 h in buffer
e) v.p. after 48 h in buffer (Dacron)

RAMAN SPECTRA OF "PYROLITIC CARBON" DEPOSITED ON DIFFERENT MATRICES FOR PROSTHETIC USE

A. Bertoluzza[1], C. Fagnano[1], A. Tinti[1], R. Pietrabissa[2]

[1] Centro di Studio Interfacolta' sulla
Spettroscopia Raman, Dipartimento
di Biochimica, via Selmi 2,
Universita' di Bologna (Italy)

[2] Dipartimento di Bioingegneria,
Politecnico di Milano, P.za Leonardo
da Vinci 32, Milano (Italy)

Carbon coated surfaces display the best biocompatibility with blood and are extensively adopted in different devices (prosthetic heart valves, vascular grafts and others) in order to minimize trombus formation.

Among the different allotropic forms of carbon, isotropic pyrolitic one presents good mechanical properties such as high fatigue and wear resistance together with good blood biocompatibility.

The high temperature fluidized-bed process does not allow the deposition of isotropic pyrolitic carbons on matrices with a low melting point, i.e. polymers and also several metals.

In this work, Raman spectra of a "reference pyrolitic carbon" (a commercial prosthetic bileaflet heart valve) (fig. 1a) and of three different carbon coated materials (titanium fig. 1b, aluminium fig. 1c and NaCl fig. 1d) obtained with the ion beam process at room temperature are reported.

The Raman bands at 1590 and at 1360 cm^{-1} which are present in the spectrum of our "reference pyrolitic carbon" subtantially agree with the Raman bands of fluidized bed pyrolitic carbons (PC) deposited from propylene or methane onto glassy carbon beads or of glassy carbon (GC) processed at the lowest temperature (1) and also of carbon fibers obtained by pyrolizing a mixture of benzene and hydrogen at low temperature (2). In particular, the bands at 1590 (called G band) and at 1360 cm^{-1} (called D band) originate, in disordered carbons, from the G band at about 1580 cm^{-1} of the ordered crystalline hexagonal graphite. The additional line D is also observed in the spectrum of microcrystalline materials and increases in intensity with the inverse of the crystal planar domain size L_a; at the same time a new line (D') appears at about 1620 cm^{-1}(3). On the contrary, during the grafitization of disordered carbons by heat treatment, the disorder-induced line at 1360 cm^{-1} decreases in intensity while that at about 1590 cm^{-1} rises in intensity with a decrease of the line width. An additional strong line appears at about 2720 cm^{-1} and is also present, asymmetrically, in crystalline hexagonal graphite (1,3).

The Raman spectra of the three different carbon coated materials (fig. 1 b, c, d) obtained with the ion beam process at room temperature, do not agree with the spectral behaviour of the coating of pyrolitic carbon in the previous discussion. The intense and asymmetrical band at 1520 cm^{-1} (shoulder at about 1310 cm^{-1}) which appears in the spectra of the three samples agrees, on the contrary, with the spectrum of an amorphous carbon film prepared by ion beam sputtering deposition at room temperature (4) and, in a more recent work, with an amorphous hydrogenated carbon film obtained on a variety of substrates in a plasma chamber at different self-bias voltage (5). The agreement of the spectra obtained by us with those of the literature above mentioned (4,5) and the similar behaviour of the spectra of the literature during the annealing process (fig. 2, ref. 4; figs. 5 and 6, ref. 5) suggest that the coatings obtained by us are also made up of "a random network of covalently bonded carbon in some proportion of tetragonal to trigonal coordination with some bonds being terminated by hydrogen atoms" (5).

For prosthetic use an "annealing" pretreatment is therefore necessary to transform the amorphous "diamond like" carbon coating in an amorphous isotropic pyrolitic carbon coating. For this purpose Raman spectroscopy, also because of its non destructive and non invasive character, is a powerful tool in characterizing carbon biomaterials with good mechanical properties and high biocompatibility.

REFERENCES

1. R. Vidano, D.B. Fischbach, J. Am. Ceram. Soc., 1978, 61, 13.
2. T.C. Chieu, M.S. Dresselhaus, M. Endo, Phys. Rev. B, 1982, 26, 5867.
3. R.J. Nemanich, S.A. Solin, Phys. Rev. B,1979, 20, 392.
4. R.O. Dillon, J.A. Woollam, V. Katkanant, Phys. Rev. B, 1984, 29, 3482.
5. L.A. Farrow, B.J. Wilkens, A.S. Gozdz, D.L. Hart, Phys. Rev. B, 1990, 41, 10132.

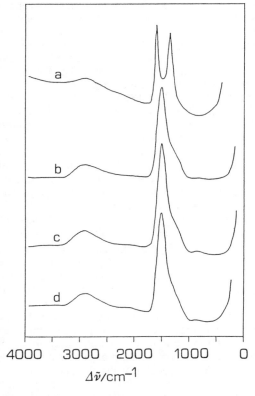

Fig. 1 - Raman spectra of: a) reference pyrolitic carbon, b) carbon coated titanium, c) carbon coated aluminium, d) carbon coated NaCl.

CHARACTERIZATION OF ADSORBED PROTEIN LAYERS ON THIN NYLON FILMS WITH FLUORESCENCE ENERGY TRANSFER BETWEEN PROTEIN AND NYLON: A METHOD FOR BIOSENSOR ANALYSIS

Edith S. Grabbe and Dennis J. Reeder

Biotechnology Division
National Institute of Standards and Technology
Gaithersburg, MD 20899
USA

INTRODUCTION

The field of biosensor fabrication has undergone a rapid expansion in recent years, with numbers of publications and patents increasing each year. However, the commercial implementation of biosensors has progressed far slower than desired. One significant problem that limits the production of reliable biosensors is the inability to immobilize biomaterials reproducibly onto the sensor substrate while retaining their activity. While considerable effort has been expended in developing methods of crosslinking proteins to various surfaces,[1-2] far less emphasis has been placed on analyzing the bound protein layers formed by these reactions. Determining these protein film structures would allow questions of bioinactivity, non-specific adsorption and evenness of coverage to be directly addressed. This information then would be used to refine immobilization procedures, resulting in systematic improvement of biosensor operation.

FLUORESCENCE EXPERIMENTS

Fluorescence spectroscopy, which has provided information concerning protein conformations in solution for decades, has recently been modified for thin film analysis using total internal reflection (TIR).[3-5] The goal of this study was to demonstrate the ability of TIR fluorescence (TIRF) to characterize layers of adsorbed IgG on thin nylon films. The use of nylon allows for comparison of protein immobilization techniques used in several commercially available bulk nylon enzyme or immunosensors. A method for depositing optically clear nylon films, covalently coupled to glass substrates has been developed. These film-coated glasses serve as both TIR elements and adsorption surfaces. Fluorescence techniques used to probe the IgG/nylon interface include: 1) energy transfer between adsorbed fluorescein isothiocyanate (FITC) tagged IgG and tetramethylrhodamine isothiocyanate (TRITC) tagged nylon, 2) binding activity, measured using TRITC la-

321

belled antigens, and 3) energy transfer between FITC labelled IgG and TRITC labelled antigen. These methods are
used to evaluate the structure of the adsorbed IgG layer
as a function of the pH of the surrounding buffer and adsorption and washing procedures. The two energy transfer
experiments characterize both the nylon/IgG and IgG/solu-
tion interfaces.

RESULTS

TIRF measurements of adsorption of IgG on bare nylon correlated well with simple staining experiments. Carbonate
buffer and TWEEN in phosphate buffered saline washes removed more nonspecifically bound IgG than did phosphate
buffer at pH 7. The TIRF system allowed for dynamic measurements, in which the wash rate, duration, and relative
amount of protein removed could be compared. Experiments
with TRITC tagged nylon and adsorbed IgGFITC revealed
that both quenching of the FITC and energy transfer to
the TRITC occurred, the former being the predominant loss
of fluorescence at low coverages of IgGFITC. At higher
coverage, quenching lessened and transfer remained constant, indicative of an IgG multilayer, where the upper
layers were no longer interacting with the TRITC labelled
surface. Antigen incubation studies also supported the
multilayer model. When a pH 7 buffer was used to wash
the surface, the antigen both bound and exchanged with
loosely adsorbed IgG, whereas after carbonate washes,
only binding occurred. In a pH 9 buffer, less energy
transfer took place between the FITC labelled IgG and the
TRITC tagged antigen than in pH 7 buffer, probably due to
a loss of binding between the antigen and antibody at
higher pH. However, the antigen did remain on the surface. These results show the utility of TIRF to monitor
protein film behavior. Future studies include use of
singularly labelled peptides to measure surface/protein
separation distances.

REFERENCES

1. P.V. Sundaram, Ann. NY Acad. Sci., 1988, 542, 165.
2. P.L. Domen, J.R. Nevens, A.K. Mallia, G.T. Hermanson,
   & D.C. Klenk, J. Chromatog., 1990, 510, 293.
3. B.K. Lok, Y-L. Cheng, & C.R. Robertson, J. Col. Int.
   Sci., 1983, 91, 87.
4. M.L. Pisarchick & N.L. Thompson, Biophys. J., 1990,
   58, 1235.
5. C.F. Schmidt, R.M. Zimmermann, & H.E. Gaub, Biophys.
   J., 1990, 57, 577.

NMR STUDY OF DOMOIC ACID, A MARINE TOXIN

M. Falk, J.A. Walter, P.F. Seto and D. M. Leek

Institute for Marine Biosciences
National Research Council of Canada,
1411 Oxford Street, Halifax, Nova Scotia, Canada B3H 3Z1

NRCC No. 31964

INTRODUCTION

Domoic acid is a neurotoxin of marine origin, recently found responsible for a serious episode of human poisoning[1]. This compound has been known for some 30 years to have anthelminthic and insecticidal properties[2] and is now known to be the agent binding most strongly to the kainate receptors of the central nervous system[3]. This has prompted renewed interest in the chemistry and spectroscopy of this compound. In the present work, which will be published in greater detail elsewhere, we have used NMR spectroscopy in order to derive information on the protonation equilibria which have a bearing on the physiological and neurological activity of this molecule. We have also used NMR spectroscopy to measure the solubility of domoic acid in water as a function of pH, in the presence of NaCl, and in methanol. In addition, we have used UV spectroscopy to measure the octanol-water partition coefficient of domoic acid, a parameter important in modelling its biological and environmental behaviour.

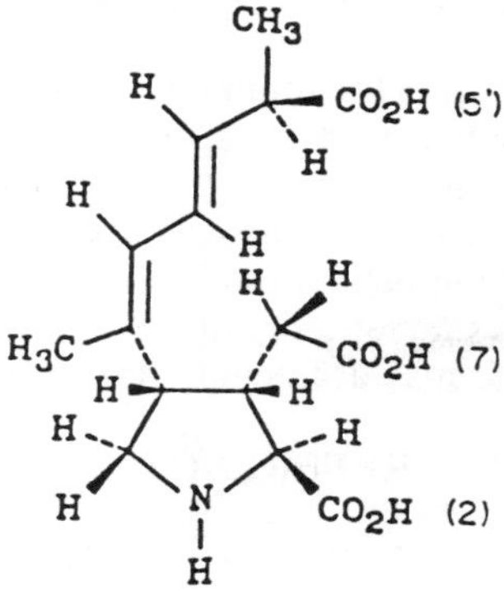

Fig. 1.
Molecular structure of neutral
domoic acid molecule

Domoic acid possesses one secondary amino group, >NH, and three carboxylate groups, -COO⁻, all capable of protonation. Therefore, domoic acid molecules in solution are a mixture of five distinct protonation states, their proportions being determined by the four pK_a values, and varying with pH.

Each protonation state has a characteristic NMR spectrum. From the variation of the experimental spectrum with pH one may derive the individual spectra of the five protonation states, obtain accurate pK_a values, and identify the site of protonation corresponding to each pK_a.

EXPERIMENTAL

The same stock of crystalline domoic acid dihydrate was used as in our earlier work.[5] NMR spectra in H_2O and D_2O were recorded at 20°C with a Bruker MSL-300 spectrometer at 300 MHz (^1H) and 75.5 MHz (^{13}C) with tetramethylsilane as a reference. The pH was varied by the addition of small amounts of NaOH (NaOD) or HCl (DCl) in an NMR tube, using a Fisher Accumet model 610A pH meter and an Ingold 6030 combination electrode. The pD values were obtained by adding 0.40 pH units to the pH meter readings in D_2O solution[5]. For solubility measurements, domoic acid crystals were added to the solution in an NMR tube until saturation was attained. ^1H NMR spectra of the saturated domoic acid solution and of a sucrose solution of known concentration were recorded under identical conditions and the domoic acid concentration calculated from the average integrated intensities of the proton resonances in the two spectra.

RESULTS AND DISCUSSION

Mathematical analysis of the variation of chemical shift of the ^1H and ^{13}C resonances with pH yielded the spectral parameters for the five protonation stages of the molecule and the four pK_a values in H_2O: 1.85, 4.47, 4.75, 10.60. The values of the pK_a's in D_2O are an average 0.40 units higher. An inspection of the variation of chemical shifts with pH established unequivocally the order of protonation: The C-2 carboxyl is the most acidic, followed by the C-5' carboxyl, then the C-7 carboxyl and finally the >NH group.

The solubility of domoic acid in water is markedly pH-dependent, passing through a minimum at the isoelectric point, pH 3.13. The increase toward higher and lower pH values indicates that the anionic and cationic forms of domoic acid are more soluble than the neutral form. The low value of the octanol-water partition coefficient, K_{ow} = 0.0037 at pH 5.32, indicates that aquatic organisms cannot take up domoic acid directly from the water[6] and may accumulate it only through dietary intake.

REFERENCES

1. J.L.C. Wright et al., Can. J. Chem. **67**, 481 (1989).
2. T. Takemoto and K. Daigo, Chem. Pharm. Bull. (Tokyo) **6**, 578 (1958).
3. G. Debonnel, L. Beauchesne, and C. de Montigny, Can. J. Physiol. Pharmacol. **67**, 29 (1989).
4. M. Falk, J.A. Walter and P.W. Wiseman, Can. J. Chem. **67**, 1421 (1989).
5. P.K. Glasoe and F.A. Long, J. Phys. Chem. **64**, 188 (1960).
6. D. Mackay, Environ. Sci. Technol. **16**, 274 (1982).

SURFACE-ENHANCED RESONANCE RAMAN SPECTROSCOPY OF HEMATOPORPHYRIN DERIVATIVE (HPD)

S. E. J. Bell*, J. J. McGarvey and B. Byrne

School of Chemistry
The Queens University of Belfast
Belfast BT9 5AG
U.K.

Hematoporphyrin derivative (HPD) is currently undergoing clinical trials as a sensitizer for photodynamic cancer treatment.[1] The material is prepared from hematoporphyrin by reactions which generate a mixture of dimers and higher oligimers which may be either ester- or ether-linked. The ratio of ester- to ether-linked oligimers depends both on the exact method of preparation and the conditions under which the material has been stored.[2] Since the *materia medica* is a very complex chemical mixture structural characterization has proved challenging. The most successful technique applied so far is mass spectroscopy although this suffers from the disadvantage that it cannot be used to directly distinguish between ester- and ether-linked materials since both are formed by loss of H_2O in a condensation reaction.

We have carried out preliminary surface-enhanced resonance Raman (SERRS) studies on HPD and simple monomeric model compounds in order to determine the feasibility of characterizing both the compositions of HPD preparations from different sources and, more importantly, to aid identification of materials which are extracted from *in vivo* tissue experiments. In the latter case the quantities of material which are available for analysis would be expected to be very limited since typical treatment doses are approximately 1 mg/kg body weight.

Figure 1 shows SERRS spectra of a commercial HPD preparation (Paisley Pharmaceuticals) and hematoporphyrin. Although the spectra show only slight differences in the region 1000 cm^{-1} - 1450 cm^{-1} marked differences are apparent in the 1450 cm^{-1} - 1600 cm^{-1} range with large intensity increases and moderate shifts in band positions. No silver ion incorporation occurs for porphyrins adsorbed on the surface of the sols used but the spectra of HPD/sol solutions do gradually change over the course of several hours to resemble those of the monomeric hematoporphyrin parent. The technique is extremely sensitive. Acceptable quality spectra can be readily obtained from 1μl of solution which contains 1 mg/litre of

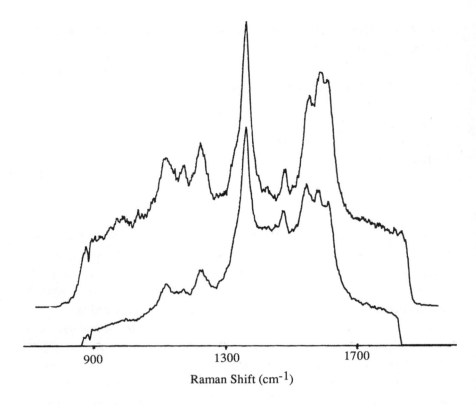

Raman Shift (cm^{-1})

Figure 1 SERRS spectra of hematoporphyrin (bottom) and HPD (top). Excitation wavelength 457.9 nm, concentration 0.1 g/litre, total accumulation time 50 seconds.

hematoporphyrin, this concentration is of the order which is expected for physiological extracts obtained in pharmacokinetic studies.

It has been established that prolonged storage of HPD at room temperature results in conversion of ester- to ether-linkages.[2] Significantly, this change is reflected in the SERRS spectra of fresh and aged samples of HPD recorded in the present work. This suggests that although detailed assignments are not yet available for the observed Raman bands the spectra are indeed sensitive to this interconversion despite the fact that the changes occur at the periphery of the porphyrin rather than within the core.

REFERENCES

1. C-W. Lei, SPIE, 1987, 847, 51.
2. B. Musselman and C.K. Chang, SPIE, 1987, 847, 96.

THE STUDY OF INTERACTIONS OF HPD MOLECULES AND PHOTOPROCESSES OCCURRING IN THEM FROM RRS SPECTRA

N. G. Golubeva,   Wang Lijun,   I. V. Shpak

Radiophysical Department
The University
Kiev
USSR

When applying hematoporphyrin and its derivatives (HPD) in tumor photodynamic therapy the state of HPD molecules is of great importance, as well as photoprocesses occurring with their participation. In this connection an investigation has been carried out of RRS spectra of some HPD in aqueous solutions and at their interaction with biologically active molecules. The influence of exciting laser radiation on these spectra was studied. To explain the effect of resonance processes on the HPD RRS we measured the dependence of line parameters under different exciting conditions.
For exciting the RRS spectra and irradiating a sample we used laser radiation of λ = 441.6 (He-Cd); 476.5, 488.0, 496.8, 514.5 (Ar) and 632.8 nm (He-Ne) wavelengths.
    The HPD RRS spectrum was investigated at its interaction with human serum albumin (HSA). It has been stated that the HSA influence is displayed in variations of intensities, halfwidths and degrees of polarization of several   HPD lines, and the appearance of extra low-frequency bands. RRS lines being most sensitive to the influence of HSA have been selected, and the interpretation of apparent variations carried out. The study has been performed of the dependence of line parameters on the HSA and HPD concentrations in solutions. With the aim of comparisons the variations observed in luminescence and absorption spectra have been analysed. From data obtained and results of the study of influence of HPD molecule aggregation on RRS spectrum in solutions some conclusions have been drawn on the existence of HPD molecules associated with HSA in different ways and the variations in aggregate formations of HPD molecules. It is known that under the action of laser radiation in aqueous solutions of HPD and at its interaction with biologically active systems there occur photoprocesses observed from luminescence and absorption spectra [1,2]. These processes are accompanied by the formation of photoproducts whose origin has not been studied well enough. The break of the pyrrol ring double bond when

retaining the inner 18-element chromophore [3], or the
porphyrin photo-reduction [4] may well be the reasons for
their appearance. To analyze these processes RRS spectra
were studied at varied excitation conditions and extra
irradiation in different spectral regions. Spectrum
structure variations observed under these circumstances
allow to suggest the existence of molecules with hydrated
pyrrol rings in a solution but they cannot be accounted
for by this effect only. The dependence of spectrum
structure on irradiation time enabled us to follow the
dynamics of photoprocesses occured.

REFERENCES

1. G. Jonushauskas, R. Kapochyute et al., J.Appl.Spectr.
    (USSR), 1989, 50, 929.
2. N.G. Golubeva, A.O. Velikanov et al.,
    Ukrainian J.Phys. (USSR), 1990, 35, 18.
3. K.N. Solovyev, L.L. Gladkov et al., Specroscopy of
    Porphyrins: Vibrational States, Nauka i Technika,
    Minsk, USSR, 1985.
4. J. Harel, G. Mahassen, H. Leranan, Photochem.& Photo-
    biol., 1974, 20, 1983.

STRUCTURAL INVESTIGATIONS OF MALARIAL PIGMENT.

C. BREMARD and **J.C. MERLIN** : Laboratoire de Spectrochimie Infrarouge et Raman, (CNRS UPR A 2631 L), UST Lille Flandres Artois, Bât. C5, 59655 Villeneuve d'Ascq cedex, France.
S. MOREAU : INSERM, Unité de Recherche 42, 369 rue Jules Guesde, 59650 Villeneuve d'Ascq, France.
J.J. GIRERD : Laboratoire de Chimie Inorganique (CNRS URA 420), Université de Paris Sud, 91405 Orsay cedex, France.

The growth of the intraerythrocytic malaria parasite *Plasmodium* is made at the expense of the host cell cytosol. Haemoglobin, as other proteins, is digested within food vacuole where the released amino acids are used for parasite feeding, and the porphyrin moiety is accumulated in large pseudo crystalline dark brown aggregates which form the malarial pigment. The investigation of the exact nature and structure of this pigment is of fundamental importance because it has been implicated in the therapeutic action of chloroquine, the main antimalarial drug.

Previous investigation performed on solvent extracts of infected erythrocytes[1] has shown that the pigment consists of a ferriprotoporphyrin IX and an apoprotein believed to be a partially digested globin. However, the exact structure of malarial pigment remains unknown. Data on aqueous iron porphyrin solutions lead us to postulate that porphyrins could exist in two forms, a hydroxo monomer and/or a dimeric mu-oxo compound. Spectroscopic investigations, based on resonance Raman (RR) spectroscopy and including electron paramagnetic resonance (EPR) spectroscopy and magnetic measurements can resolve this problem. The experiments are undertaken on intact malarial pigment, obtained from *in vitro* cultures of *Plasmodium falciparum* (FCR3 strain), the main human malaria species.

The recorded RR spectra of crude pigment, monomeric haematin and dimeric mu-oxo complex are shown in Figure 1. As can be readily observed, they exhibit closely positioned lines. The wavenumbers of the ν 4 and ν 10 marker bands of porphyrin core, confirm the presence of an iron III in high spin state inserted in a porphyrin ring. The relative intensities of the RR lines in the pigment spectrum are in full agreement with those obtained for haematin sample even if different exciting laser lines are used. The symmetric and antisymmetric stretching modes of the Fe-O-Fe bridge, observed at 416 and 890 cm^{-1} respectively in the mu-oxo spectrum, are not detected in the pigment spectrum even if the excitation wavelength used corresponds to the maximum enhancement[2]. A hydroxyde monomeric species can thus be postulated. According to previous studies on water-soluble sterically hindered porphyrin[3], the broad line observed near 600 cm^{-1} in the RR spectrum of the pigment could tentatively be assigned to a Fe-OH stretching mode.

A good way to detect the monomer is to look at the paramagnetic properties of the crude pigment. The EPR spectrum at 77 K revealed the presence of a paramagnetic iron; the g values observed (5.8 and 2.0) are typical of an iron III compound in high spin state engaged in square

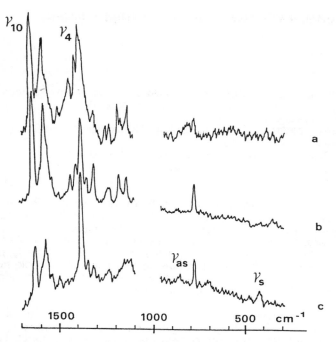

ν_{10} ν_4

ν_{as} ν_s

1500   1000   500   cm^{-1}

a

b

c

Figure 1 : RR spectra of crude malarial pigment (a), haematin in solid state (b) and mu-oxo ferriprotoporphyrin IX in solid state (c) recorded with 514.5 nm excitation.

pyramidal coordination. As the antiferromagnetically coupled mu-oxo compound cannot be detected in such experiments, quantitative information about the spin state of iron species is needed. By elemental analysis, using atomic spectrometry, only iron was detected (0.15% w/w). A known amount of pigment was submitted to a magnetic susceptibility analysis as a function of the temperature. The magnetic pattern obtained by plotting the magnetic susceptibility X vs 1/T is typical of a pure monomeric iron III compound in high spin state. The slope of the line is in full agreement with S = 5/2. The monomeric iron III protoporphyrin hydroxyde is the unique porphyrin species in the malarial pigment.

The formation of malaria pigment during haemoglobin degradation in the acidic food vacuole of the parasite leads to a monomeric compound. As dimerization could occur in a soft acidic medium, we must postulate that iron porphyrin moieties are maintained as isolated species during haemoglobin degradation. The recent determination of a specific protein elaborated by the parasite, which exhibits a very high affinity for ferriprotoporphyrin IX[4], could explain this phenomenon.

REFERENCES

1. T. Deegan and B.G. Maegraith, Ann. Trop. Med. Parasitol. 1956, 50, 194.
2. P. Kowalewski, Thèse Ingénieur Docteur, UST Lille Flandres Artois, 1989.
3. R.A. Reed, K.R. Rodgers, Y.O. Su, K. Kushmeider and T.G. Spiro, XII ICORS, J.R. Durig and J.F. Sullivan Eds, Wiley & Sons, New York, 1990.
4. J.O. Ashong, I.P. Blench and D.C. Warhurst ,Trans. R. Soc.Trop. Med. Hyg., 1989, 83(2),167.

SPECTROSCOPIC INVESTIGATION OF CONGO RED BINDING TO AMYLOID-LIKE PROTEINS

A. ELHADDAOUI[1], S. TURRELL[1], J.C. MERLIN[1], S. LAMOTTE[1]
G. VERGOTEN[2] and A.DELACOURTE[3]
[1] Laboratoire de Spectrochimie Infrarouge et Raman
(CNRS UPR A 2631 L), UST Lille Flandres Artois, Bât. C5,
59655 Villeneuve d'Ascq cedex, France.
[2] Laboratoire de Genie Biologique et Medical, Faculté de Pharmacie,
INSERM U 279, Université de droit et santé de Lille, France.
[3] INSERM, U 156, 59045 Lille, France.

Neuropathologically two of the most characteristic features of Alzheimer's disease are senile plaques and neurofibrillary tangles, both of which have peptide amyloid as a basis[1]. X-ray diffraction studies show a beta-pleated sheet conformation, and sequencing has revealed lysine residues occuring at strategic locations in the chain[2].

Congo red is one of the two histological dyes used for the identification of the presence of amyloid-type proteins in post-mortem tissue examinations for the diagnosis of Alzheimer's disease[3]. Although its interaction with these proteins, determined by a green birefringence when the tissues are examined with a polarisation microscope, is known to be specific, its nature is not understood. The present work was undertaken to gain information on the linkage of Congo red to amyloid and hence to better understand the structure and mechanisms of the action of the peptide. As poly L-lysine has a pleated sheet structure and gives a green birefringence under polarisation, it was chosen as a model for amyloid.

Raman spectra of congo red in basic, neutral and acidic 10^{-5} M solutions are presented in Figure 1. As the spectra were recorded using argon Laser lines varing from 488 nm to 568.2 nm, resonance conditions led to spectral enhancement of chromophore bands. It is observed that a change from the acid to the basic form of the dye affects predominantly the bands due to N=N (1442 cm^{-1} range) and the amino-phenyl (1350-1390 cm^{-1} range) stretching modes. Spectra recorded of solutions of two poly L-lysines (40 000 and 3650) with Congo red are shown in Figure 2. Because of the before-mentionned resonance conditions, the spectra present features characteristic uniquely of the dye. It is observed that , in the case of poly L- lysine 40 000, whether the dye-protein interaction is made in neutral solution (a), in a neutral solution which is then acidified (b) or in an acid solution (c), the resulting spectrum is one of the basic form of the dye. The same result is observed for poly L-lysine 3650 (in neutral solution (c) and in neutral solution which is then acidified (d). No interaction at all is observed between Congo red and L-lysine .

The unique and irreversible quality of the observed interaction implies first that the beta-pleat conformation is necessary, the linkage cannot be simply a matter of the dye being trapped in a beta protein.

A conformational study made using molecular mechanics techniques yielded an energy minimisation and stable conformations which, when visualised, support the hypothesis of Congo red binding on different peptide molecules in the beta pleat of the fibrils.

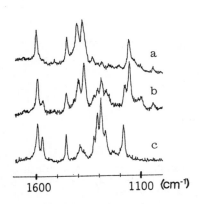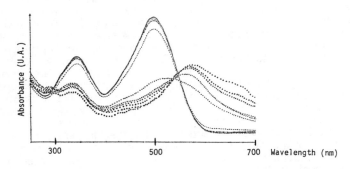

Figure 1: Raman spectra of
　　　Congo red in a) basic
　　　b) acidic pH 3,5 and
　　　c) acidic pH 3,5 (third conformer)

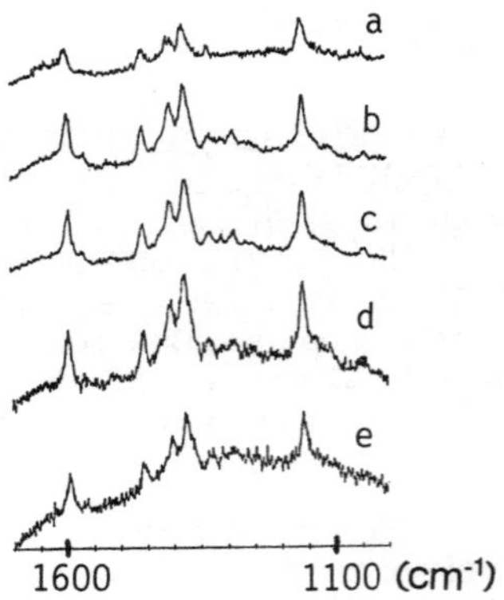

Figure 2: Raman spectra of Congo red
　　　with poly L-lysine 40 000 (a, b, c)
　　　and poly L-lysine 3650 (d, e)
　　　with varying pH values.

A routine check for the isobestic point of dye revealed a problem in the pH range of 3.5 to 4.9 (see Figure 3). Subsequent studies yielded evidence for the existence of third form for Congo red with an azure-blue coloration. This form possess unique visible (maximum at 750 nm) and Raman spectra (see Figure 1c). While the third form is stable, an increase in pH reverses it to the red, basic form and a following acidification leads back to the violet, acid form. It is hence, obvious that further investigations of nature of this form of the dye are necessary in order to understand the interaction between Congo red and poly L-lysine.

Figure 3: Visible absorption spectra of Congo red 10^{-5} M solutions of pH from 2.9 to 10.5

REFERENCES

1. R. Katzman (Ed) " Biological Aspects of Alzheimer's Disease" (Cold Spring Harbor Laboratory, New York), 1983.
2. J. Kang, H. Lemaire, et al. Nature, 1987, 325, 733.

UV-VIS-, FLUORESCENCE AND RAMAN SPECTROSCOPIC STUDIES OF THE LIGNAN (+)-PINORESINOL

Henrik Tylli

Department of Chemistry
University of Helsinki
E. Hesperiankatu 4
SF-00100 Helsiki, Finland

INTRODUCTION

A large group of naturally occurring lignans have the 3,7-dioxabicyclo[3.3.0]octane (tetrahydrofuro[3,4-c]-furan) ring as a common structural unit [1]. (+)-Pino-resinol or 2,6-diguaiacyl-cis-3,7-dioxabicyclo[3.3.0]-octane (I) is found in the resin of several species of Pinus and Picea [2] and it has also been isolated from the fir cambial sap. Pinoresinol is an efficient anti-fungal agent and it therefore plays an important role in the protection of the plant against diseases. The occurrence of pinoresinol type structural units in softwood lignin is well established [3] and the diaryl-tetrahydrofurofuran unit appears as a building block in certain naturally occurring antioxidants [4] and also in a number of compounds which are of significant pharmacological interest [5-7].

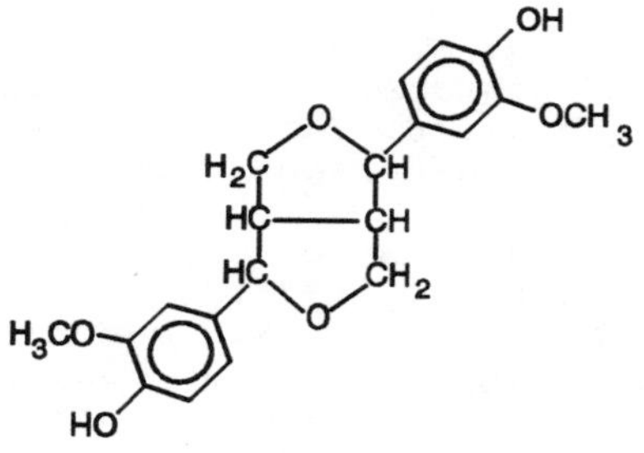

(I)

EXPERIMENTAL

Pinoresinol diacetate was isolated from the resin of spruce (Picea abies) according to the procedure given by Erdtman [8] and the sample of pinoresinol was prepared from the diacetate as described by Freudenberg and Dietrich [9]. White crystals with mp. 121°C were

obtained. UV-VIS- and fluorescence spectra were
recorded in hexafluoropropanol solutions with Shimadzu
UV-240 and RF-5001PC spectrometers. Raman spectra of
polycrystalline samples were measured as reported
previously for guaiacol [10].

RESULTS

The UV-VIS-spectrum of pinoresinol displays three broad
bands with maxima at 205, 227 and 280 nm, the last
absorption being a composite band. This is in close
agreement with previously reported results for other
diaryl tetrahydrofurofuran lignans [11]. Excitation at
235 nm yields an intense emission with a maximum at 310
nm and an emission tail extending to about 400 nm. This
is rather typical for a methoxysubstituted phenol. A
similar fluorescence spectrum with almost the same
intensity was obtained if an excitation wavelength of
277 nm was used. Excitation at longer wavelengths
yielded a very weak emission with a maximum at 440 nm,
which most probably is due to a minor impurity present
in the sample.
Low-frequency Raman spectra of polycrystalline samples
were recorded in order to gain information about the
internal dynamics of the molecule. The torsional
transitions of the OH, OCH_3 and phenyl groups were
obtained by comparison with the corresponding spectra
of guaiacol [10]. Low-frequency skeletal modes originating from the tetrahydrofurofuran ring complicate the
assignment.

REFERENCES

1. A. Pelter and R. S. Ward in C. B. S. Rao, (Ed.),
    'Chemistry of Lignans', Andhra University Press,
    Andhra Pradesh, 1978, Chapter 7, p. 227.
2. K. Weinges, F. Nader and K. Kunstler in ref. 1,
    Chapter 1, p. 1.
3. K. Lundquist and R. Strömberg,
    Holzforsch., 1988, 42, 375.
4. T. Osawa, M. Nagata, M. Namiki and Y. Fukuda,
    Agric. Biol. Chem., 1985, 49, 3351.
5. S. Ghosal, S. Banerjee and R. Srivastava,
    Phytochem., 1979, 18, 503.
6. S. Ghosal, S. Banerjee and D. Jaiswal,
    Phytochem., 1980, 19, 332.
7. H. Greger and O. Hofer,
    Tetrahedron, 1980, 36, 3551.
8. H. Erdtman, Svensk Kemisk Tidskr., 1934, 46, 229.
9. K. Freudenberg and H. Dietrich,
    Chem. Ber., 1953, 86, 4.
10. H. Tylli, H. Konschin and H. Tenhu,
    J. Mol. Struct., 1990, 220, 129.
11. O. Hofer and R. Schölm,
    Tetrahedron, 1981, 37, 1181.

VII

NUCLEIC ACIDS AND THEIR

INTERACTION WITH PROTEINS

AND DRUG MOLECULES

STRUCTURES OF PACKAGED DNA IN VIRUSES INVESTIGATED BY RAMAN SPECTROSCOPY

George J. Thomas, Jr.* and James M. Benevides
Division of Cell Biology and Biophysics
School of Basic Life Sciences
University of Missouri-Kansas City
Kansas City, Missouri 64110, U.S.A.

INTRODUCTION

In recent years the structures of DNA have been revealed in considerable detail by both spectroscopic[1] and crystallographic[2] studies of oligonucleotides of defined base sequence. These studies have helped to identify the distinguishing features of A- B- and Z-DNA; they also suggest substantial variability in the canonical B-DNA conformation, depending upon the particular base sequence and environmental factors such as pH and ionic composition. The structure of DNA in biological assemblies, however, may differ significantly from that of protein-free DNA in the crystal or in aqueous solution. Typically, the DNA packaged in a viral protein shell (capsid), or in a eukaryotic chromosome, is more highly condensed than unpackaged DNA in solution. Packaged DNA also may be intimately associated with specific nucleic acid-binding proteins or other factors which have the capability to limit or extend the configurations accessible to protein-free DNA. Raman spectroscopy represents a powerful technique for investigating nucleic acid structure in biological assemblies. Here we describe an empirical Raman approach to DNA structure determination in viruses and other biological assemblies. We apply the Raman method to two bacterial viruses: bacteriophage T7, which packages a double-stranded DNA (dsDNA) genome, and bacteriophage φX174 which packages single-stranded DNA (ssDNA). The present results illustrate the diversity in structure of packaged DNA, and reveal that protein-DNA interactions can be a significant determinant of DNA conformation in mature virions. The prospect of extending this methodology to gene regulatory complexes also is considered.[3]

METHODS

Our approach is based upon the empirical correlation of specific Raman vibrational bands ("Raman markers") with detailed structures or conformations of nucleotide residues in DNA, as established by X-ray crystallography.[1] In effect we have determined the Raman signatures of nucleotides in canonical A-, B- and Z-DNA oligonucleotides for which high resolution crystal structures have been determined. This primary database has

Table 1   Selected Markers (cm^{-1}) in Canonical DNA Structures.

Group	A-DNA	Z-DNA	B-DNA
O-P-O	706±5		790±3
	807±3	745±3	828±2(GC)
			835±2(GC/AT)
			839±2(AT)
PO$_2^-$	1099±1	1095±1	1092±1
CH$_2$	1418±2	1425±2	1422±2

Base	C3'endo/anti	C3'endo/syn	C2'endo/anti
G	664±2	625±3	682±2
A	1335±2	1310±5	1339±2
C	1252±2	1265±2	1255±5
T	662±2	650±2	669±2
	745±2	770±2	748±2

been augmented with additional structure-spectra correlations established from fiber-diffraction and related analyses.[4]   An abbreviated tabulation of Raman markers is given in Table 1.

Experimental procedures for the collection of Raman spectra from biological specimens, including nucleic acids and viruses, have been described in various papers from our laboratory. Further details are given in recent reviews.[4,5]

RESULTS, DISCUSSION AND CONCLUSIONS

Raman spectra (600-900 cm^{-1} region) of phage T7 and protein-free dsDNA are compared in Fig. 1.[6]   The computed difference spectrum clearly shows large spectral differences between packaged and unpackaged states of dsDNA. Although both states display only C2'endo/anti nucleoside markers (Table 1), the OPO backbone conformation marker is shifted dramatically --- from 825 cm^{-1} in the phage to 836 cm^{-1} in protein-free dsDNA.   This indicates a highly perturbed B-form backbone in packaged dsDNA.   The 825 cm^{-1} marker of T7 is close to that of "GC-DNA" (Table 1), i.e. dsDNA containing only GC pairs and manifesting a relatively wide minor-groove.   This perturbation affects essentially all of the packaged T7 dsDNA.   Other spectral changes attendant with packaging are enhanced intensities at 760, 772 and 805 cm^{-1} bands, which have been further discussed.[6]

Fig. 2 compares Raman data for packaged (Fig. 2b) and unpackaged (Fig. 2c) states of the ssDNA of bacteriophage φX174.[7]   Here, the data are strikingly different than for T7. Although again only C2'endo/anti nucleoside conformers are evident, the Raman intensity in the region of backbone conformation markers (810-860 cm^{-1}) is extraordinarily feeble.   In order to provide a quantitative interpretation to these results, we have calibrated the integrated intensity in the 810-860 cm^{-1}

interval in terms of the percentage of nucleotide residues which adopt the B-form backbone geometry, as illustrated in Fig. 3.

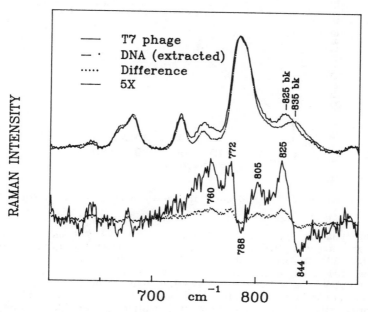

Fig. 1 Conformation change with T7 packaging of dsDNA.[6]

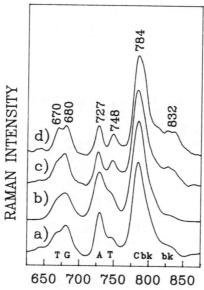

Fig. 2 (a) ssDNA model, O_r1+; (b) packaged φX174 ssDNA; (c) unpackaged φX174 ssDNA; (d) dsDNA model, RFIII φX174 dsDNA.[7]

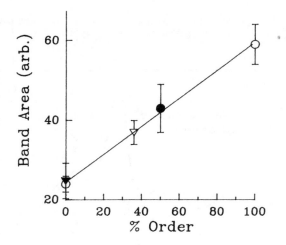

Fig. 3 Correlation of 810-860 cm^{-1} intensity with percent B-form.[7]

Our results show that the packaged ϕX174 genome contains no B-form geometry ($0\pm5\%$), despite the fact that the unpackaged genome has 36% B-form geometry. This result signals the strong structural influence of protein-DNA interactions in the capsid. A model for ssDNA packaging in ϕX174 has been proposed.[7]

The secondary structures of both ssDNA and dsDNA are dramatically altered by packaging in icosahedral capsids. These structural alterations, which are transparent to many biophysical and biochemical probes, are readily detected by laser Raman spectroscopy. Further, the Raman signatures of packaged DNA can be characterized both qualitatively and quantitatively, which leads, in favorable cases, to the elucidation of DNA-protein interactions and possible packaging mechanisms.

REFERENCES

1. G. J. Thomas, Jr. and A.H.-J. Wang, Nucl. Acids Mol. Biol., 1988, 2, 1;   S. Dijkstra, J.M. Benevides and G.J. Thomas, Jr., J. Mol. Struct., 1991, 242, 283.
2. O. Kennard and W.N. Hunter, Quart. Rev. Biophys., 1989, 22, 327.
3. J.M. Benevides, M.A. Weiss and G.J. Thomas, Jr., Biochemistry, 1991, 30, in press. (See also this volume.)
4. G.J. Thomas, Jr., J.M. Benevides and B. Prescott, Biomolec. Stereodynam., 1986, 4, 227.
5. G.J. Thomas, Jr., in Biological Applications of Raman Spectroscopy, 1987, Vol. 1, T.G.Spiro (Ed.), Wiley, NY, p. 135.
6. G.J. Thomas, Jr. and P. Serwer, J. Raman Spectrosc., 1990, 21, 569.
7. J.M. Benevides, P.L. Stow, L.L. Ilag, N.L. Incardona and G.J. Thomas, Jr., Biochemistry, 1991, 30, in press.

VIBRATIONAL SPECTRA AND ASSOCIATIVE INTERACTIONS OF POLY-L-(Asn, Asp) WITH POLYNUCLEOTIDES

P. Carmona[*1], M. Molina[1], P. Martinez [2] and A. Ben Altabef[3]

[1]Institute of Optics (CSIC), Madrid, Spain.
[2]Department of Physical Chemistry, University of Cádiz, Cádiz, Spain.
[3]Department of Physical Chemistry, University of Tucumán, Tucumán, Argentina

INTRODUCTION

Interactions between proteins and nucleic acids are of extraordinary importance in essential processes of the living cell. Apparantly the selectivity of protein-nucleic acid binding is determined in large part by specific interactions between amino acid side chains and nucleotide bases[1]. This is shown by observation of comparatively short sites of proteins which are responsible for specific binding to nucleic acids[2-3], m oreover, a recent work reveals the decisive role of even single point contacts in recognition[4]. High specificity of the point interactions may be provided by hydrogen bonding between interacting molecules. In this connection, it seems convenient to study polypeptides and polynucleotides with a low number of amino acids and nucleobases respectively. Protein side chain amide and carboxylic groups are, owing to the presence of proton donors and proton acceptors, very active groups in complex formation between some tRNAs and their respective aminoacyl synthetases[1]. That is why we study in this work a copolypeptide containing L-asparagine and L-aspartic acid. We have, thus, undertaken a study of the conformational structure of this polypeptide using infrared and Raman spectroscopy in order to determine whether structural changes accompany the binding of this biopolymer to poly(rA).poly(rU) duplex and which molecular subgroups may participitate.

METHODS

Conventional Jobin-Yvon instrumentation was used for Raman spectroscopy. Polypeptide and polynucleotide water solutions (0.1 M in monomers) were used in cacodylate buffer (pH 6.5). After using a fitting program for band resolution in polypeptide spectra infrared vCOO$^-$ bands were evaluated, whereby the percentage of Asp residues was found to be 31%. Infrared spectra were recorded in a Perkin-Elmer FTIR 1725X spectrometer using CaF$_2$ cells.

RESULTS AND DISCUSSION

The infrared spectrum of poly-L-(Asn, Asp) (PLAA) in heavy water solution (Figure 1) shows the absorption maximum of the amide I to occur at 1646 cm^{-1}. Polypeptide fragments in a non-ordered conformation are usually associated with an infrared band which in spectra measured in heavy water is located in the 1648-1640 cm^{-1} range[5]. As the 1639 cm^{-1} shoulder is very close to the vCO mode (1637 cm^{-1}) of acetamide in D$_2$O we assign it to the amide groups in the polypeptide side chains. Another amide I band component appears as a shoulder at 1652 cm^{-1} which could be assigned to alpha helices[5]. However, this is not consistent with the CD spectrum of this biopolymer in aqueous solution which does not show any signal of helical polypeptide arrangement. The presence of the amide I band component at 1652 cm^{-1} may, then, be caused by the asparagine side chain carbonyl groups that are buried to some extent in internal domains of the polypeptide and therefore are less solvated than the carbonyl groups causing the 1639 cm^{-1}

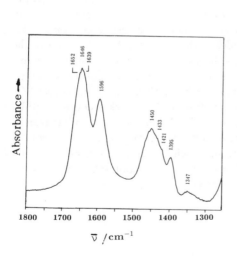

Figure 1. Infrared spectrum of PLAA in D$_2$O solution.

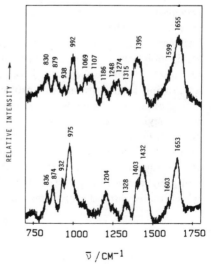

Figure 2. Raman spectra of (upper) PLAA in H$_2$O solution and (lower) in D$_2$O solution.

shoulder. Although the Raman spectrum of the polypeptide in aqueous solution (Figure 2) shows the 1274 cm^{-1} band in the amide III range usually assigned to helical secondary structure[6], N-deuteration shifts this band to 1328 cm^{-1} (Figure 2) which falls in the range of $\delta C_\alpha H$, twisting and wagging CH_2 motions. This result shows that the above vibrations are mixed with the in-plane bend of polypeptide backbone NH groups to some extent. CPK molecular models constructed with asparagine side chain NH_2 group hydrogen bonded to the peptide $C=O$ group in the same amino acid residue result in a cycle which makes the amide III, wagging and twisting CH_2 coordinates within the cycle nearly parallel and therefore strongly coupled. One should, then, be cautious in associating polypeptide backbone conformation with characteristic frequency ranges in the amide III region.

The conformational structure of PLAA remains unordered when bound to poly(rA).poly(rU) duplex. Although a small increase of the amide I infrared band component at 1652 cm^{-1} is detected, circular dichroism reveals a spectrum which is typical of unordered polypeptide backbone. The slight increase of absorbance at this frequency may, then, be caused by the rise of the amount of polypeptide carbonyl groups which become less solvated as a result from their interactions with nucleobases. Another cause that may explain this spectral change is the formation of certain types of loops which contribute at this frequency[6,7]. On the other hand, the poly(rA).poly(rU) duplex remains in the A form upon complex formation with PLAA (Figure 3). However, some Raman lines of nucleobases are the most sensitive to PLAA binding. Concerning adenine, the most significant increases in intensity take place at 1335 and 1446 cm^{-1}, and a decrease in intensity occurs at 1568 cm^{-1}. Finally, binding of PLAA also leads to a decreased intensity in

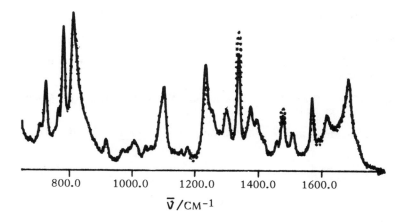

$$\bar{\nu}/\text{CM}^{-1}$$

Figure 3. Raman spectra of (—) poly(rA).poly(rU) and (···) the PLAA-poly(rA).poly(rU) complex uppon substraction of the PLAA spectrum.

the 1230 cm^{-1} uracil band. This vibration is assigned to the (C$_5$C$_6$ stretch + C$_5$H bending) mode[8]. It is expected that the intensity of this Raman line decreases when stacking interactions occur[9]. Stacking and/or hydrogen bonding may influence the relative intensities of these nucleobase vibrations. Stacking alone does not explain the simultaneous intensity decreases and increases of the above bands. Asparagine side chain amide group may, then, form hydrogen bonds with the adenine five-membered ring. In this connection, kinetics of deuteration of this adenine ring in the absence and presence of PLAA is in progress.

Acknowledgements. We thank the Dirección General de Investigación Científica y Técnica for financial support (PB87-0200).

REFERENCES

1. T. A. Steitz, Quart. Rev. Biophys., 1990, 23, 205.

2. L. Regan, J. Bowi and P. Schimmel, Science, 1987, 235, 1651.

3. W. F. Anderson, Y. Takeda, O. H. Ohlendorf and B. W. Matthews, J. Mol. Biol., 1982, 159, 745.

4. Y. M. Hou and P. R. Schimmel, Nature, 1988, 333, 140.

5. D. M. Byler and H. Susi, Biopolymers, 1986, 25, 469.

6. S. Krimm and J. Bandekar, Adv. Protein Chem., 1986, 38, 181.

7. S. J. Prestrelski, D. M. Byler and M. N. Liebman, Biochemistry, 1991, 30, 133.

8. S. Chin, I. Scott, K. Szczepaniak and W. B. Person, J. Am. Chem. Soc., 1984, 103, 3415.

9. E. W. Small and W. L. Peticolas, Biopolymers, 1971, 10, 1377.

SURFACE-ENHANCED RAMAN SPECTROSCOPY OF NUCLEIC ACIDS AND ANTITUMOUR DRUGS/NUCLEIC ACIDS COMPLEXES in vitro AND IN LIVING CELLS

I. Nabiev[+] , K. Sokolov[+] , H. Morjani[*]and M. Manfait[*]

+ Shemyakin Institute of Bioorganic Chemistry,
USSR Academy of Sciences, Moscow, USSR
* Laboratoire de Spectroscopie Biomoleculaire,
Universite de Reims, France

INTRODUCTION

It has recently been demonstrated that the effect of very large (in some instances up to 10^9) enhancements of the Raman cross-section for molecules in the close vicinity of a metal surface can find many very important applications in life sciences[1] Water-soluble proteins, peptides and amino acids, membrane-bound complexes, nucleic acids and their complexes with peptides and different low-molecular weight compounds, including antitumor drugs and their complexes with DNA in vitro[3,4] - this is only a very short list of biologically important molecules for the structural-functional investigations of which the new physical chemical technique of surface-enhanced Raman spectroscopy (SERS) has been successfully applied[1,2] So, the obvious prospects of this technique led us to go in for SERS applications to structural-functional investigations of living cells.

Experimental

Stock solutions of DOX, ACM and THP-DOX (Laboratoires Roger Bellon, Paris, France) and calf thymus DNA (Sigma Chemical Company, type I) were prepared in Dulbecco's PBS. K562 is a human leukaemia cell line, established from a patient with chronic myelogeneous leukaemia in blast transformation. In drug uptake studies, cells in exponential growth phase were incubated in RPMI containing the appropriate drug concentration, using 24-well multidishes in a moist air/CO_2 incubator at 37° C.

Silver hydrosol was prepared as described in R[2] and preaggregated by the addition of sodium perchlorate up to the final concentration 0.05 M. Cells were introduced into the silver hydrosol and after the 30 min of incubation they were precipitated, washed free of the hydrosol outside with PBS, and used for Raman or micro-Raman measurements.

SERS spectra were obtained on a DILOR Omars-89 Raman spectrometer connected with an Olympus BH-2 microscope, using the 514.5 nm line of an argon laser (Spectra-Physics, Model 2020-03). The laser power on the sample was 10 mW for the measurements in the

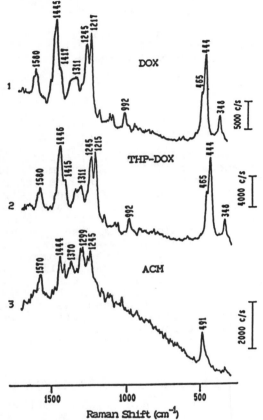

Figure 1 SERS spectra of doxorubicin DOX (1), THP-doxorubicin (THP-DOX,2), and aclacinomycin (ACM,3) adsorbed on silver hydrosol. Drug concentrations - 10^{-8} M.

macrochamber and less than 0.2 mW for the measurements under the microscope.

SERS spectra of the antitumor drugs ACM,DOX,THP and their complexes with calf thymus DNA in vitro. Normal resonance Raman spectra of DOX and THP can be obtained at concentrations not less than 10^{-4} M and with a very high contribution of the fluorescence background. The SERS spectra of these drugs (Fig.1) are similar to their resonance Raman spectra reported earlier by M.Manfait et al.[5] and correspond to the chromophoric part of the molecules. The resonance Raman spectrum of ACM was obtained earlier but with a very high background due to high quantum yield of fluorescence for this drug in the visible region[3,4].

SERS effect is accompanied by strong quenching

of fluorescence providing the possibility to extend the range of molecules that can be investigated by this technique. Moreover, the ultrahigh sensitivity of SERS for this kind of molecule permits to obtain spectra at concentrations up to 10^{-10} M and to develop an express method of detecting antitumor drugs in living cells.

Fig.2 shows SERS spectra of DOX, its complex with

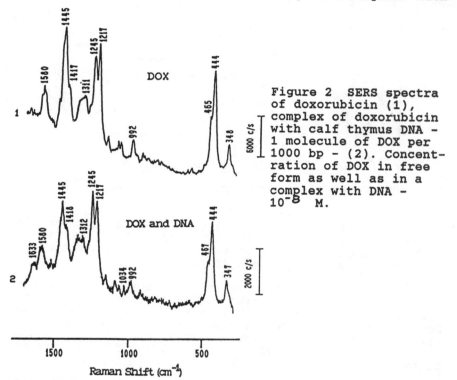

Figure 2 SERS spectra of doxorubicin (1), complex of doxorubicin with calf thymus DNA - 1 molecule of DOX per 1000 bp - (2). Concentration of DOX in free form as well as in a complex with DNA - 10^{-8} M.

calf thymus DNA in vitro adsorbed on silver hydrosol.

Living Erythroleukemic cancer cells,line K562. SERS spectra were detected from a population as well as from the single living cancer cells (Fig.3). The SERS spectrum of the cells treated with DOX closely corresponds to the spectrum of the in vitro complex of DOX with calf thymus DNA (Figs.2 and 3). Fig.3 shows that the signal of adenine moiety appears in the SERS spectrum of untreated cells (curve 3). This marker band was not revealed in Raman spectrum of living cells which were not incubated with hydrosol. It means that the hydrosol was introduced into the living cells through endocytosis during the incubation of cells with the hydrosol. The hydrosol used was more specific to adenine than to other nucleotides because it was activated by chloride ions contained in PBS[2].

The SERS spectrum of living cells with DOX in nuclei is very different from the spectrum of cells

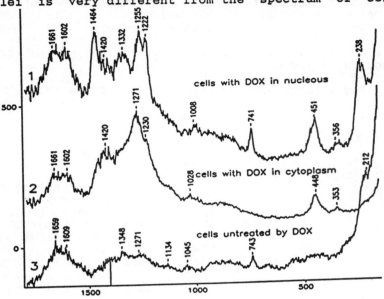

wavenumbers (cm−1)

Figure 3 SERS spectra of living cells with doxorubicin in nuclea incubated with hydrosol (1); cells with doxorubicin in cytoplasm (2) and untreated cells incubated with pure hydrosol (3).

with DOX in cytoplasm (Fig.3) but is closely correlated with the spectrum of the in vitro complex of DOX with calf thymus DNA. The spectrum of cells with DOX in cytoplasm is different from the spectrum of free DOX adsorbed on hydrosol (Fig.1 and 3). It means that the DOX in cytoplasm has some kind of target different from its target (DNA) in the cell nucleus.

Further investigations on the interaction of DOX and other anthracyclines with nucleus and cytoplasm performed on living cells could help to understand the drug resistance mechanism which leads to different responses in clinical chemotherapy.

REFERENCES

1. T.M.Cotton, 'Spectroscopy of Surfaces', Wiley & Sons, New York, (1988), 91.
2. I.R.Nabiev, R.G.Efremov, Sov.Phys.Usp., 1988, 31, 241
3. G.Smulevich, A.Feis, J.Phys.Chem., 1986, 90, 6388.
4. Y.Nonaka, M.Tsuboi, K.Nakamoto, J.Raman Spectrosc., 1990, 21, 133.
5. M.Manfait, L.Bernard, T.Theophanides, J.Raman Spectrosc., 1981, 11, 68.

THE STRUCTURE OF DNA IN A DNA-COPPER PORPHYRIN COMPLEX: A UV RESONANCE RAMAN APPROACH

L. Chinsky* , P. Mojzes and P.Y. Turpin.

L. P. C. B. (CNRS UA 198), Institut Curie and Université Pierre et Marie Curie, 75231 Paris Cedex 05, France,

INTRODUCTION

Nucleic acids interact with a wide variety of compounds including metal complexes, antibiotics, chemical oncogenes and molecules of biological interest. There are basically three modes of interaction involved in these interactions : intercalation, groove binding and covalent bonding, all being often reinforced by hydrogen bonding and/or coulombic interactions. Binding of the water-soluble porphyrin H2(TMpy-P4) and its metal derivatives to nucleic acids has been extensively studied with a variety of spectroscopic methods (UV-visible and CD [1], linear dichroism in a flow gradient [2], fluorescence [3], ESR [4] and NMR [5-7]). These studies demonstrated that planar four-coordinate porphyrins [H2, M= Cu(II) and Ni(II)] are selectively intercalated at G-C sites, whereas porphyrins having axial coordination [M= Zn(II), Co(III), Fe(III) and Mn(III)] form "outside bound" or "groove bound" complexes at the A-T sites of DNA.[1,8]

The interest of studying such a kind porphyrins is that H2(TMpy-P4) and its metal derivatives are regarded as attractive alternatives for hematoporphyrin derivatives (HPDs), which are used for photoradiation therapy applications. Hence, studies on porphyrin-DNA interactions may provide structural information essential in understanding the roles of HPDs in photodynamic therapy.

RESULTS AND DISCUSSION

Excitation in the Visible Region : Resonance Raman of the porphyrins

Resonance Raman (RR) spectra of ground state Cu(II)-, Ni(II)- and Co(III)(TMpy-P4) and their complexes with nucleic acids, obtained with exciting lines located in the Soret region (visible excitation) of these metalloporphyrins, have shown different marker bands for the intercalation and groove binding processes.[9,10] However for Cu(TMpy-P4), which was thought to interact preferentially with G-C sites of DNA, it has been recently shown, by using high power pulses of excitation in the Soret band, that an electronically excited Cu(TMpy-P4) may be trapped selectively at the A-T sites of DNA, in forming an exciplex which is specific to this combination.[11] Through a second Raman experiment made with a two colour pulsed excitation, it has been possible to determine the resonance Raman excitation profiles (RREP) of marker lines of both the exciplex and the ground state complex, and subsequently to propose a characterization of this exciplex.[12] In addition this experiment confirmed the very short lifetime of the exciplex (previously estimated at ca. 20 picoseconds [11]).

Excitation in the UV region : Resonance Raman of the DNA

Complementary information can be obtained on the nucleic acid moieties of the complexes by exciting them in the UV range, i.e. under resonance conditions for the nucleic bases. We essentially investigated, with 223 , 257 and 280 nm excitation wavelengths, the complexes of Cu(TMpy-P4) with poly rG, poly dG - poly dC, poly [(dG-dC)]$_2$, poly dA - poly dT and poly [(dA-dT)]$_2$, whose native structures in solution are very different from each other.[13]

Two types of information can be obtained from the UV RR spectra of the porphyrin - polynucleotide complexes : i) the nature of the porphyrin interaction and the target molecule - intercalation being monitored by a hypochromic effect (decrease in intensity) of the whole spectrum, and specific base targeting being reflected in shifts and intensity variations of some lines of the base in question, and ii) DNA structural changes, induced by the formation of the complexes, being monitored by other variations of the nucleic base Raman features.

G-C containing polymers. Poly[(dG - dC)]$_2$ in low salt aqueous solution is known to adopt a right-handed B form helix. In interaction with Cu(TMpy-P4), the polynucleotide yields a very hypochromic RR spectrum (taken at 257 nm) as compared to that of the free species : this confirms that an intercalation process of the porphyrin in the duplex is involved, which was already known from other experimental approaches and from visible RR results on the porphyrin moiety in the same complex.[9] In addition to that, the UV RR spectrum shows an important shift of the guanine 1353 cm^{-1} line to 1361 cm^{-1}, along with a sharp intensity decrease of the cytosine 1640 cm^{-1} line : this means that the polynucleotide in interaction undergoes a structural change, which makes it resembling the poly dG - poly dC structure, i.e. close to a A form.

In contrast the 257 nm RR spectrum of poly dG - poly dC, whose structure has been ranked by Benevides and Thomas as A/B or A-type form,[13] does not yield any spectral change when the polymer interacts with the porphyrin : surprisingly the spectra of the free and bound species are almost superimposable, which means that the 257 nm excitation is actually not favorable to monitor the interaction in this case.

However the 257 nm spectrum of poly rG, known to be under a A form in aqueous solution, shows a quite significant hypochromism when the polymer is complexed with the porphyrin : this seems to be in favor of some intercalation process of the latter. In addition, the observation of a +7 cm^{-1} upshift of the 1354 cm^{-1} line (C8N7, N1C6 and N7C5 motions [14]) might reveal some kind of specific local interaction of the porphyrin with the most reactive part of the guanine base in the polymer.

A-T containing polymers. All RR experiments made with A-T containing polynucleotides did not yield any hypochromism upon complexation with Cu(TMpy-P4), whatever the excitation wavelength and the structure of the polymer : this clearly confirms that the interaction process is likely not an intercalation. Instead, these observations are compatible with a groove-binding process of interaction.

The structure of Poly[(dA-dT)]$_2$ definitely differs from that of a conventional B-DNA. Benevides and Thomas ranked it in the A/B-type helices, or even in the A-type structure (see[R]3 for discussion), while other experimental evidence suggested an alternation of C3'-endo and C2'-endo puckers of purine and pyrimidine nucleosides, respectively.[15] With a 257 nm excitation, the spectrum of this alternate copolymer does not show any change when it complexes with the porphyrin, like did poly dG - poly dC in the previous case : again this excitation wavelength is not favorable to monitor the interaction. In contrast, the same polymer excited at 223 nm yield very interesting RR feature changes upon porphyrin complexation. A group of adenine lines located in the 1310-1370 cm^{-1} region, all involving the N9-C8-N7-C5 positions of the base [14], and the 1484 cm-1 thymine line (prominent in the spectrum taken at this wavelength of excitation), involving the N1-C2-N3 positions, are strongly modified upon porphyrin complexation. Both modifications involve specific chemical groups located in the narrow groove of the duplex.

It is now well recognized that the poly dA - poly dT structure is a heteronomous form in which the purine nucleosides all have the conventional A-helix geometry (C3'-endo/anti) while the residues of the dT strand retain the usual B geometry (C2'-endo/anti).[15-17] For this homopolymer duplex, both the 257 and 223 nm excitation wavelengths lead to RR spectral changes, upon porphyrin-polymer complexation, which can be interpreted in terms of structural evolution towards a A-type helix, as featured by the alternate copolymer poly[(dA-dT)]$_2$.

The polymer structure in the exciplex state. We have been able to collect information on the structure of the polynucleotide involved in the exciplex, with a two-beam mixing experiment : one beam at 560 nm pumped the porphyrin in its excited state, thus allowing the formation of the exciplex (see above), the second beam from the same laser selectively probed in the same shot the polymer in the UV. This was possible, in spite of the very short lifetime of the exciplex (ca. 10-20 ps), because of the 20 ns duration of the laser pulse and the high power of excitation continuously recycling the excited state. To date we only investigated at

223 nm the better candidate for the exciplex formation, i.e. poly[(dA-dT)]$_2$. Two kinds of spectral changes are observed on the polymer involved in the exciplex : i) the adenine group of lines in the 1310-1370 cm^{-1} region and the thymine 1484 cm^{-1} line behave in about the same manner as for the ground state porphyrin-polymer complex, thus likely reflecting the location of the interaction in the narrow groove, and ii) a drastic decrease of the thymine 1650 cm^{-1} line, mainly involving the C6=C5-C4=O moiety of the nucleic base. It is not yet clear to unequivocally assign these spectral changes either to global structure changes (such as an increase of the A-type character) or to local interactions only.

Finally, the influence of the polymer sequence and of its structure on the process of the porphyrin specific interaction will be discussed in the light of these RR studies, along with the biomedical significance of the formation of such a drug-DNA exciplex.

REFERENCES

1. R.F. Pasternak, E.J. Gibbs and J.J. Villafranca, Biochemistry 22, 2406, 1983.
2. N.E. Geacintov, V. Ibanez, M. Rougée and R.V. Bensasson, Biochemistry 26, 3087, 1987.
3. J.M. Kelly, M.J. Murphy, D.J. McConnell and C. OhUigin, Nucleic Acid Res. 13, 167, 1985.
4. G. Dougherty, J.R. Pilbrow, A. Skorobogaty and T.D. Smith, J. Chem. Soc. Faraday Trans. 2, 81, 1739, 1985.
5. D.L. Banville, L.G. Marzilli, J.A. Strickland and W.D. Wilson, Biopolymers 25, 1837, 1986.
6. L.G. Marzilli, D.L. Banville, G. Zon and W.D. Wilson, J. Am. Chem. Soc. 108, 4188, 1986.
7. J.A. Strickland, D.L. Banville, W.D. Wilson and L.G. Marzilli, Inorg. Chem. 26, 3398, 1987.
8. R.F. Pasternak, A. Antebi, B. Ehrlich, D. Sidney, E.J. Gibbs, S.L. Bassner and L.M. Depoy, J. Mol. Catal. 23, 235,1984.
9. N. Blom, J. Odo, K. Nakamoto and D. Strommen, J. Phys. Chem. 90, 2847, 1986.
10. J.H. Schneider, J. Odo and K. Nakamoto, Nucl. Acid Res. 16, 10323, 1988.
11. P.Y. Turpin, L. Chinsky, A. Laigle, M. Tsuboi, J.R. Kincaid and K. Nakamoto, Photochem. Photobiol. 51, 519, 1990.
12. L. Chinsky, P.Y. Turpin, A.H.R. Al-Obaidi, S.E.J. Bell and R.E. Hester, J. Phys. Chem., in press.
13. J.M. Benevides et G.J. Thomas Jr., Biopolymers 24, 667, 1985.
14. M. Tsuboi, S. Takahushi and I. Harada, in "Physico-chemical properties of Nucleic Acids", Duchesne J. Ed., Academic Press, New York, vol. 2., 91, 1973.
15. G.A. Thomas and W.L. Peticolas, J. Am. Chem. Soc. 105, 993, 1983.
16. B. Jolles, A. Laigle, L. Chinsky and P.Y. Turpin, Nucl. Acids Res. 13, 2075, 1985.
17. S. Arnott, R. Chandrasekaran, I.H. Hall and L.C. Puigjaner, Nucleic Acids Res. 11, 4141, 1983.

THE NUCLEATION PROPAGATION MODEL FOR THE B TO Z JUNCTION IN DNA

W. L. Peticolas, Z. Dai, M. Dauchez and G. A Thomas

Department of Chemistry
University of Oregon
Eugene, Oregon, USA 97403

The conformation of DNA in living cells is known to be overwhelmingly in the B form. But the Z-form of DNA may exist in living cells (1). If this be so, then the Z-form DNA must occur as a relatively short sequence of longitudinal CG base pairs that exits within a much longer DNA sequence in the B conformation (2). Longitudinal base pairs are called code words to distinguish them from Watson-Crick base pairs (2). Self-complementary sequences of the form, $d[(TT)_N(CG)_M(AA)_N]$ that start and end with N [TT.AA] code words have been synthesized. The conformations of the oligomers were determined in aqueous salt solutions using laser Raman spectroscopy. In certain cases part of the CG portion of the oligomers go into the Z-form. To rule out the possibility of hairpin formation, the melting points of the double helices were measured as a function of oligomer concentration both in solutions of low and high salt concentration. The marked dependence of the melting point on oligomer concentration showed that a duplex structure was present. From a plot of the melting temperature vs concentration the enthalpy of dimer formation was obtained in good agreement with duplex enthalpies found in the literature. Oligomers were synthesized with N = 1, 2, 3, 5 and varying sequence lengths of M CG code words. From a measurement of the intensities of the B- and Z-form guanine marker bands at 681 cm^{-1} and 628 cm^{-1} respectively, one can determine the fraction of the CG base pairs in the Z form from the ratio $I_{628}/(I_{628} + I_{681})$ (3).The Raman spectra show that for N = 1 the Z form is flanked by a single CG code word (2 CG base pairs) in the B form on each side while for N = 2 to 5 the Z form is flanked by 2 B-form CG code words on each side. The number of CG code words in the Z form is the excess of CG code words over the flanking B-form CG code words. Consider the following sequences:

BBBBBBBBBBBBBBBB	BBBBZZZZBBBB
5'-d(TTTTCGCGCGCGAAAA)-3'	5'-d(TTCGCGCGCGAA)-3'
3'-d(AAAAGCGCGCGCTTTT)-5'	3'-d(AAGCGCGCGCTT)-5
BBBBBBBBBBBBBBBB	BBBZZZZZZBBB
I	II

```
BBBBBBBBZZZZZZZZBBBBBBBB
5'-d(TTTTCGCGCGCGCGCGCGAAAA)-3'
3'-d(AAAAGCGCGCGCGCGCGCTTTT)-5'
BBBBBBBBZZZZZZZZBBBBBBBB
```

In the above diagrams, the conformation is given by the letters B for a base pair in the B form, and Z for a base pair in the Z form. Note that there are no bases pairs in the junction. This is the result of minimum potential energy calculations that show an abrupt transition from B to Z exists between two base pairs. This occurs when the torsional angles in the furanose-phosphate-furanose groups at the end of the B-form sequence connect a B double helix with a Z double helix. There appear to be no base pairs in the junction, but only base pairs on each side of the junction. The B form is nucleated by the TT code words at each end and the B-form propagates into the CG sequences. When N≥5 the length of the CG section in the B form remains constant at 3-4 base pairs on each side. This is also true for single B-Z junctions. Using the program SYBL, the potential energy of each of the oligomers containing junctions has been minimized using the AMBER all atom force field. A plot of the minimized potential energy of the B-Z junction versus its placement shows that the minimum occurs between the third and fourth CG base pair as is observed experimentally. The junction does not appear to occur between the TT and CG code words because the B-form nucleated in the TT section is propagated into the CG section for 1.5-2 code words [3-4 base pairs]. It is apparent from these measurements that there is a clash between the Z-forming and the B-maintaining code words. The oligomer (I) above stays in the B form because the 8 Z forming CG base pairs [or 4 CG code words] are too few and the N = 2 TT code words on each end are too many. Shortening the 2 TT code words on each end to 1 weakens their B-forming potential and permits the formation of junctions. Lengthening the CG portion from M = 4 to 8 code words strengthens the Z forming potential. It is apparent that length and base sequence determine the ability of a DNA to contain a Z-form section in the midst of a long B-form section. It seems certain that the value of M must be larger than 4 in order to support the Z DNA conformation in the midst of a long double helix in the B form. It is found that a minium value of M = 6 will stabilize the Z form in the midst of a long B-form segment. In such a duplex, the four central CG base pairs are in the Z form and flanked on each side by four CG base pairs in the B form and 10 or more AT base pairs in the B form. It is concluded that 12 sequential CG base pairs must exist in B DNA before they can be induced into the Z form.

1. Rich, A., Nordheim, A., & Azorin, F. J. Biomolecular Struct. and Dyn. 1, 1-19 (1983).
2. Peticolas, W. L. Wang, Y. and Thomas, G. A. Proc. Natl. Acad. Sci. USA 85, 2579 (1988).
3. Dai, Z, Thomas, G.A., Evertsz, E. & Peticolas, W.L. Biochemistry 28 6991-6996 (1989).

FTIR STUDY OF DNA TRIPLE HELICES

E.TAILLANDIER, M.FIRON and J.LIQUIER

Laboratoire C.S.S.B., U.R.A. C.N.R.S.1430
Université Paris XIII
74 rue Marcel Cachin F93012 Bobigny Cedex
FRANCE

An increasing interest in the formation of triple helices comes from the assumed existence of in-vivo intramolecular triple helical structures which may be formed by homopurine-homopyrimidine sequences. Such sequences are often encountered in the eukaryotic genome especially within putative regulatory regions of genes and hot spots of recombination. A possible role played by the triple helices in gene expression regulation has been proposed.

Few structural studies concerning the conformation of the sugars in the triple helices have been undertaken. The most currently used model considers A type (C3'endo) sugar conformations such as found by the X-ray fiber diffraction study of polyU.polyA.polyU (1). However two recent NMR studies show that in triplexes containing desoxyriboses all sugars are not in a C3'endo geometry (2,3). We have therefore studied by FTIR spectroscopy the conformation of sugars in triple stranded helices using IR marker bands of South (C2'endo) and North (C3'endo) geometries located, respectively, around 835 cm^{-1} for the South geometry and around 886, 860 and 805 cm^{-1} for the North geometry. The nucleic acids were studied in solution (H2O and D2O) and in films in different hydration conditions (in the case of most DNA double stranded helices the decrease of the hydration of a film induces a C2'endo->C3'endo transition of the sugars). The triple helical structures studied contain homopurine and homopyrimidine strands formed by adenines and thymines (or uracyls) bound to deoxyribose or ribose sugars. The triplets formed were dT-dA-dT, dT-rA-rU, rU-rA-dT and rU-rA-rU. The formation of the triple helixes was checked in the 1800-1500 cm^{-1} region in which modifications due to the formation of Hoogsteen base pairing are detected.

An example of the caracterization of the sugar geometries in triple stranded helices is given in figure 1 in the case of poly dT. poly dA. poly dT. The spectra presented are those of a film exposed to decreasing relative humidities from 93% to 47%. We detect,

355

when the RH is decreased, the emergence of an absorption located at 860 cm⁻¹, which coexists with a band around 835 cm⁻¹, showing the simultaneous presence of North and South sugars.

Studies concerning other triple helical structures containing isomorphous triplexes C-G-C⁺ and T-A-T are under progress. No structural studies dealing with purine-purine-pyrimidine triplexes have yet been published and we shall present results concerning the sugar conformation in such triplets.

REFERENCES
1. S. ARNOTT and P.J. BOND, Nature, (1973), 244, 99.
2. P. RAJAGOPAL and J. FEIGON, Biochemistry, (1989), 28, 7859.
3. C. de los SANTOS, M.ROSEN and D. PATEL, Biochemistry, (1989),28, 7282.

Figure 1: FTIR spectra of poly dT.poly dA.poly dT films between 750 and 1000 cm⁻¹ at decreasing relative humidities.

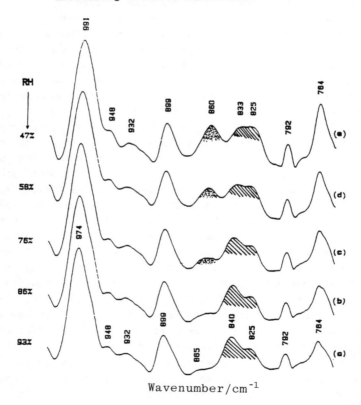

Wavenumber/cm⁻¹

356

B-Z TRANSITION OF POLY(dG-m⁵dC)·POLY(dG-m⁵dC). A FOURTH DERIVATIVE SPECTROPHOTOMETRIC STUDY.

P. Garriga*, D. Garcia-Quintana and J. Manyosa.

Unitat de Biofísica, Departament de Bioquímica i de Biologia Molecular, Facultat de Medicina, Universitat Autònoma de Barcelona. 08193-Bellaterra, Barcelona, Catalonia/Spain.

Fourth derivative (4D) spectrophotometry has been successfully applied to the analysis of conformational isomerizations of polynucleotides[1,2]. This technique provides further insights in the detailed study of electronic spectra of polynucleotides undergoing transition, allowing the association of changes in the derivative spectra to conformational features of the nucleic acid double-helices.

In this work, the 4D technique has been applied to the study of the B-to-Z isomerization of poly(dG-m⁵dC)·poly(dG-m⁵dC). The transition has been obtained using Mg^{2+} ion in ethanolic solution and the Z-conformation of the methylated polymer has been characterized, in this case by the appearance of a new 4D peak at 299.0 nm. This peak is specific of the Z-form of this polymer and has not been previously observed in Z-forms of other synthetic polynucleotides[1,3]. We attribute this 4D peak to interstrand stacking interactions of 5-methyl cytosines in view of the lately proposed model based on energetic calculations[4]. An increase of the temperature to 90°C results in an important reduction of the intensity of this peak that can be interpreted in terms of a weakening in the stacking interactions of 5-methyl cytosines.

On the other hand, the addition of Tb^{3+} to an ethanolic solution containing the Z-form induces the A-conformation of the polynucleotide[5]. The absorption spectrum obtained in the presence of Tb^{3+} shows an increased absorption in the 300-320 nm region which would arise from a new electronic band as already observed in other synthetic polynucleotides[2]. The 4D spectrum of this Tb^{3+}-induced A-conformation is significantly different from that of the Z-conformation, mainly in the long wavelength region. In particular, the most striking feature is the disappearance of the 299.0 nm 4D peak characteristic of the Z-conformation. The presumed band in the 300-320 nm region of the absorption spectrum is very poorly resolved, if at all, in the 4D spectrum.

This fact can be explained by the low intensity and relative broadness of this band. Deconvolution and curve-fitting of the absorption spectra have been carried out in order to obtain further information about the spectral features of such band in the long wavelength region of the spectrum above 300 nm. The appearance of this band can be related to a more stacked situation in the A-form with regard to the Z-form of the polynucleotide in accordance with a previous study of the A-form of poly(dA-dT)·poly(dA-dT) in ethanolic solution[2].

The structural parallelism among conformations of different synthetic polynucleotides, in view of the detailed analysis of the aforementioned long wavelength region, is discussed.

REFERENCES

1. P. Garriga and J. Manyosa, XII Congreso de la Sociedad Española de Bioquímica, València 1985.
2. P. Garriga, J. Sági, D. Garcia-Quintana, M. Sabés and J. Manyosa, J. Biomol. Struc. Dyn., 1990, 7, 1061.
3. P. Garriga, D. Garcia-Quintana and J. Manyosa, 4th Sitges International Conference on Protein-Nucleic Acids Interactions, Sitges/Barcelona 1988.
4. D.A. Pearlman and P.A. Kollman, Biopolymers, 1990, 29, 1993.
5. D. Chatterji, Biopolymers, 1988, 27, 1183.

SOLUTION CONFORMATION OF DNA AS PROBED BY ETHIDIUM FLUORESCENCE

M. Alloisio, M. Guenza and C.Cuniberti*

Istituto di Chimica Industriale, Università di Genova, e Centro Studi Chimico-fisici di Macromolecole Sintetiche e Naturali, C.N.R., Corso Europa 30, 16132 Genoa, Italia.

INTRODUCTION

The cationic dye ethidium bromide (3,8- diamino-5-phenylphenanthridinium bromide) is generally considered as the prototype of drugs that bind to double-stranded DNA by a non-specific intercalation mechanism in agreement with the nearest-neighbors exclusion binding model. This behavior along with the large fluorescence enhancement following the intercalation event has determined the extensive utilization of ethidium as a fluorescent probe of the topological and dynamical properties of DNA. The model proposed for the intercalation complex of ethidium with the standard B-form of DNA duplex (1) sees the planar phenanthridinium ring sandwiched between the base pairs in such a way that the two amino substituents at positions 3 and 8 can extablish hydrogen bonds with the phosphate groups across the two DNA strands. These hydrogen bonds appear to play a relevant role in maintaining the dye in a specific orientation with respect to the polymer frame. At the same time, a connection between these hydrogen bonds and the complete insertion of the dye within the hydrophobic pocket of the base pairs with the degree of fluorescence enhancement seems proven by the behavior of phenanthridinium derivatives lacking the 3-amino group (2).

Little is known about the structure of the ethidium-DNA complex prevailing in solution. From linear dichroism and fluorescence anisotropy decay measurements (3,4) it has however been inferred that the dye is tilted at 70° relative to the helix axis and undergoes limited orientational fluctuations around this position. In this paper we will report the results of a fluorescence study of ethidium intercalation in native DNA from calf thymus (CT-DNA) and in the synthetic alternating polydeoxiribonucleotide poly [(dA-dT)] (AT-DNA).

RESULTS AND DISCUSSION

Figures 1 and 2 show how the fluorescence quantum yield (ϕ_b) of the ethidium cation intercalated in CT-DNA and AT-DNA, respectively, depends on the solution ionic strength and temperature. It may be observed that in both cases the increase of Na^+ concentration determines the reduction of ϕ_b, although no salt effect was observed on the quantum yields of the free dye (ϕ_f). Furthermore the effect is larger with AT-DNA, which also gives rise to higher values of ϕ_b relative to the complexes with CT-DNA. Shielding of the phosphate charges by the counterions seems thus to influence the details of the complex geometry, either acting directly on the strength of the hydrogen bonding to the dye or through subtle variations in the double helix conformation. The major role of the DNA conformation on the positioning of ethidium finds strong support in the temperature dependence of ϕ_b which may be related to the well known "premelting" effects observed in the double helical structures (5). Indeed, ϕ_f undergoes a gradual reduction upon increasing the temperature which amounts to 10% in the range from 20° to 80°C. Thus the corresponding increase of ϕ_b with CT-DNA and its substantial reduction with AT-DNA indicate that the intercalation site may somewhat differ according to the local conformation and the conformational fluctuations of the double helix.

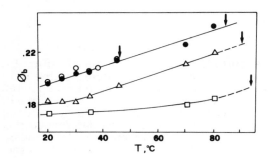

<u>Figure 1</u>
Quantum yield of EB/CT-DNA against temperature at : (o) 10^{-3}, (●) 0.15, (△) 0.5, (□) 1 M [Na$^+$]. Arrows show CT-DNA melting points.

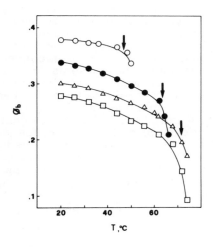

<u>Figure 2</u>
Quantum yield of EB/AT-DNA against temperature at : (o) 0.01, (●) 0.15, (△) 0.5, (□) 1 M [Na$^+$]. Arrows show AT-DNA melting points.

REFERENCES

1. C. C. Tsai, S. C. Jain & H. M. Sobell, <u>Proc. Natl. Acad. Sci.</u>, 1975, <u>72</u>, 628
2. A. Kindelis & S. Aktipis, <u>Biopolymers</u>, 1978, <u>17</u>, 1469
3. M.Hogan, N.Dattagupta & D.M.Crothers, <u>Biochemistry</u>, 1979, <u>18</u>,280
4. J.C. Thomas, S.A. Allison, C.J. Appellof & J.M. Shurr, <u>Biophys. Chem.</u>, 1980, <u>12</u>, 177
5. E.Pelacek, <u>Progr. Nucleic Ac. Res. Mol. Biol.</u>, 1976, <u>18</u>, 151

COMPARATIVE STRUCTURAL ANALYSIS OF DNA OLIGOMERS BY VIBRATIONAL SPECTROSCOPY AND ENERGY MINIMIZATION PROCEDURES

H. Fabian[1], W. Hölzer[2], U. Heinemann[3], H. Sklenar[1] and H. Welfle[1]

[1]Central Institute of Molecular Biology, Robert-Rössle-Str. 10, D-1115 Berlin-Buch, FRG
[2]Pedagogic University Erfurt, Nordhäuser Str. 63, D-5010 Erfurt, FRG
[3]Institute of Crystallography, Freie Universität Berlin, Abteilung Saenger, Takustr. 6, D-1000 Berlin 33, FRG

X-ray analyses of DNA oligomers have shown a high sequence dependent variability of the DNA double helix structure. The structure of a given oligonucleotide in the crystalline state, however, is not necessarily the same as in aqueous solution. Vibrational spectroscopy is very useful in studying the extent of preservation of the crystal structure in solution and on the manner in which the solution and crystal structure may differ (e.g.R1,2) We have analyzed the solution structures of DNA duplexes containing consecutive guanines (GGGGCCCC-I; GGGTACCC-II; GGGATCCC-III) and alternating GC sequences (CGCGCGCG-IV; CGCTAGCG-V) by vibrational spectroscopy. The spectroscopic studies show that in solution at low ionic strength the overall conformation of oligomers containing consecutive guanines is predominantly B-like in contrast to A-like structures found in the crystalline state. For the self-complementary DNA octamer GGGATCCC this was verified directly by comparison of the Raman spectra obtained from solutions and single crystals (Figure 1). The spectrum of the solution reflects a B-like conformation as indicated by the B-family marker bands, e.g. at 685, 837, 1094 and 1421 cm^{-1}. Remarkably, the crystal spectrum is quite different from the solution spectrum. The A-family marker bands at 664, 705, 807, 1101 cm^{-1} and the spectral characteristics in the 1200-1450 cm^{-1} region clearly indicate the A-form as the dominant conformation of GGGATCCC in the crystal in agreement with X-ray data[3]. However, a detailed inspection of the crystal spectrum reveals a weak Raman band near 691 cm^{-1}. The band near 691 cm^{-1}, although at slightly higher frequency than in typical B-form structures, may be assigned to guanosine residues within the C2'-endo/anti family conformation (B-family). Therefore, the presence of a weak Raman band near 691 cm^{-1} indicates that a small part of the molecules is in a B-like conformation in a sample which is predominantly in the A-form.
Evidence for the coexistence of two helical structures in a crystal have been obtained very recently for

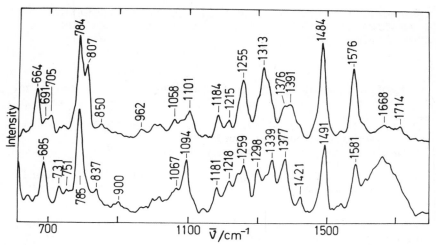

Figure 1 Raman spectra of the octamer d(GGGATCCC) in the
crystalline state (top) and in solution (bottom)

d(GG^{Br}UA^{Br}UACC) by X-ray diffraction[4] and for d(GGTATACC)
by infrared and Raman spectroscopy[2]. These two oligomers,
as well as d(GGGATCCC), crystallize in the space group P6$_1$.
Crystals of this space group have a solvent channel in
the center large enough to accommodate oligomer molecules
in the B-form.
Raman spectra of the oligomers (I) -(V) in solution indicate differences in the detailed geometries of the backbone and the furanose pucker and/or glycosyl orientation
of dG residues in homo (dG)·(dC) tracts in comparison to
those in alternating GC sequences[5]. In order to obtain
indications concerning the nature of the conformational
variants within the B-family responsible for the observed
spectral changes, model structures of the sequences studied were derived by using optimized force fields and
advanced energy minimization procedures[6]. Differences in
global parameters (width of the minor groove) and in
local conformations have been found.

REFERENCES
1. J.M. Benevides, A.H.J. Wang, G.A. van der Marel, J.H.
   van Boom and G.J. Thomas, Biochemistry, 1988, 27, 931.
2. J. Liquiers, E. Taillandier, W. Peticolas and G.A.
   Thomas, J. Biomol. Struct. & Dyn., 1990, 8, 295.
3. H. Lauble, R. Frank, H. Blöcker and U. Heinemann,
   Nucl. Acid Res., 1988, 16, 7799.
4. J. Douchet, J.-P. Benoit, W.B.T. Cruse, T. Prange
   and O. Kennard, Nature, 1989, 337, 190.
5. H. Fabian, W. Hölzer, G. Herrmann, O. Ristau,
   H. Sklenar and H. Welfle, J. Mol. Str., 1990, 217, 99.
6. H. Sklenar, R. Lavery and B. Pullmann, J. Biomol.
   Struct. & Dyn., 1986, 3, 967.

LOW FREQUENCY RAMAN, NMR AND IR INVESTIGATION OF METHYLATED NUCLEIC ACID BASES: SELF-ASSOCIATION AND COMPLEX FORMATION WITH AMINO ACID CARBOXYLIC GROUPS

N.V.Zheltovsky,   S.A.Samoilenko,   D.N.Govorun,
I.N.Kolomiets,   I.V.Kondratyuk,   Ya.R.Mishchuk,
A.V.Stepanyugin

Department of Molecular Biophysics, Institute of Molecular Biology and Genetics, Ukrainian SSR Academy of Sciences, 150 Zabolotny Str., Kiev, 252143, USSR

A study of the complex formation of nucleic acid bases with amino acids and their derivatives is important to understand the elementary mechanisms of the protein - nucleic acid recognition.

To determine the contribution of various atomic groups of nucleic acid bases to their interactions with amino acid carboxylic group, we investigated a complex formation of a number of the bases methyl derivatives with amino acid N-acyl derivatives containing neutral carboxylic group and sodium acetate modelling ionized carboxylic group by UV, IR and NMR spectroscopies.

Earlier it was established that all the bases except cytosine form the complexes with carboxylate-ion and on the contrary, only cytosine interacts with the neutral carboxylic group. In this work unlike its disubstitution, monomethylation of purine base amino group was shown not to prevent the interactions with carboxylate-ion. In contrast to analogous guanine derivatives, methylation of adenine at N1 and N3 atoms gives rise to a significant ability to interact with neutral carboxylic group. Cytosine derivatives methylated at N1 and C5 atoms remain capable to interact with neutral carboxylic group. Uracil methyl derivatives containing at least one unsubstituted imino group retain an ability to interact with carboxylate-ion. Results obtained are of interest for understanding of the role of the bases' methyl derivatives which are minors of nucleic acids, in protein - nucleic acid recognition and are useful for elucidation of muta- and carcinogenesis mechanisms.

The investigation of the low-frequency phonon spectra of nucleotide bases and their complexes with amino acids is one of the stages to elucidate the low

-energy vibration nature in nucleic acid and their complexes with proteins, which are evidently of great biological singificance in biological system. Low-frequency (15-200 cm^{-1}) Raman spectra of nucleotide bases methylated at glycoside nitrogen atoms in a polycrystal state and cocrystallizates of cytosine with glycine derivatives were studied and interpreted. The bases internal and external vibrations have been separated. Twisting vibrations of methyl groups have been assigned. H-bond translations were distinguished from librations in the cocrystalliates' spectra. Intensities of intermolecular vibrations in the Raman spectra appeared to be comparable with the intensities of intramolecular ones that shows good prospects for using them to diagnose structure and dynamics of biomological molecules and their complexes.

RAMAN SPECTROSCOPIC STUDIES OF COMPLEXES OF m-AMSACRINE WITH CALF-THYMUS DNA

C. A. Butler, T. F. Barton, R. P. Cooney* and W.A. Denny

Department of Chemistry and Cancer Research Laboratory
University of Auckland
New Zealand

Resonance Raman (RR) spectroscopy and surface enhanced Raman spectroscopy (SERS) were used to record spectra of the anti-tumour agent meta-amsacrine (m-AMSA, see Fig. 1.) in aqueous solution. The effects of solution pH, drug concentration, exciting line and potential were investigated. The aim of this study was to detect spectroscopically the intercalation of m-AMSA into calf-thymus DNA, and then to apply this technique to determining the DNA binding mechanisms of other drug molecules.

RR spectroscopy results in an enhancement of the spectral intensity of chromophoric groups within a molecule, enabling study of drug solutions at low, but biologically relevant, concentrations. This technique has been used to study other drug-DNA complexes such as those formed by adriamycin,[1] distamycin[2] and mitoxantrone.[3]

The RR studies were recorded on a Jobin-Yvon U1000 spectrometer, using a drug concentration of 10^{-3} molL^{-1}. Excitation profiles of m-AMSA were collected using argon ion laser excitation over the wavelength range 457.9 to 514.5 nm. The experiments were carried out over a pH range of 6.00 to 10.45. The SERS experiments were carried out at 10^{-5} molL^{-1}. Due to solubility problems of m-AMSA at high chloride concentration the supporting electrolyte used for SERS was a combination of 0.05 molL^{-1} KF and 0.01 molL^{-1} KCl. All potentials are quoted relative to a saturated calomel electrode reference. SERS spectra were collected from a polished silver electrode at potentials between -0.60V and 0.00V, following an oxidation-reduction cycle from -0.60V to +0.26V.

The excitation profiles for m-AMSA indicate that four classes of vibrational mode are involved in the visible electronic transition centred around 440 nm. The crystal structure of m-AMSA[4] suggests that the substituted phenyl group lies at an angle of 51° to the planar

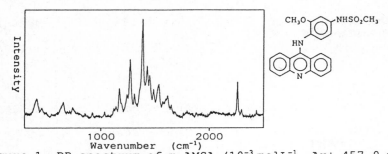

Figure 1. RR spectrum of m-AMSA (10^{-3} molL^{-1}, Ar+ 457.9 nm)

acridine group, and therefore is unlikely to be conjugated with the acridine group. Also a combination of visible spectroscopic data for various related molecules and the Raman excitation profiles suggests that the acridine group is dominant in the RR spectrum of m-AMSA.

In the investigation of the effect of solution pH on the RR spectrum of m-AMSA, vibrational bands showed intensity changes with pH. The bands at 1306, 1353, 1381 and 1528 cm^{-1} decreased in intensity while the bands at 1267, 1423, 1568 and 1610 cm^{-1} increased in intensity with increasing pH. Changes in the appearance of the SERS spectra with changing electrode potential appear to correlate with changing spectral patterns in the RR spectra of m-AMSA with varying pH. For example, the SERS spectrum obtained at -0.21V is very similar to the RR spectrum taken at pH 7.5. Although the pH of the bulk SERS solution remained at pH 6.4, cathodic adjustment of potential resulted in changes in the spectra which are similar to changes in the RR spectra of m-AMSA as pH is adjusted through the alkaline range. This may be due to elimination of protons in the thin-layer associated with cathodic interfacial processes.

SERS and resonance Raman spectra of calf-thymus DNA/m-AMSA complexes have also been recorded at neutral pH for drug:base pair ratios of 1:2 to 1:20. The changes detected in the spectra are indicative of hydrogen-bonding changes occurring upon drug-DNA complex formation. To assist interpretation of the these spectra, the acridine unit has been the subject of a molecular orbital calculation and a normal coordinate analysis. A final analysis of the spectra using the results from these calculations is in progress.

1. M. Manfait, L. Bernard and T. Theophanides, J. Raman Spectrosc., 1981, 11, 68.
2. D.S. Lu, Y. Nonaka, M. Tsuboi and K. Nakamoto, J. Raman Spectrosc., 1990, 21, 321.
3. B.S. Lee and P.K. Dutta, J. Phys. Chem., 1989, 93, 5665.
4. J.S. Buckleton and T.N. Waters, Acta Cryst., 1984, C40, 1587.

RECA–DNA COMPLEXES STUDIED BY LINEAR DICHROISM AND FLUORESCENCE SPECTROSCOPY

Per Hagmar and Bengt Nordén

Department of Physical Chemistry, Chalmers University of Technology
S-412 96 Göteborg, Sweden

Masayuki Takahashi

Institut de Biologie Moléculaire et Cellulaire, CNRS
F-670 84 Strasbourg Cedex, France

The recombination protein A (RecA) plays a central role in the DNA repair system of *Escherichia coli* by promoting general genetic recombination and induction of the SOS system[1,2]. Electron microscopy and spectroscopic studies such as linear dichroism and fluorescence have provided information about binding stoichiometry and structure of the RecA-DNA system[3,4]. RecA binds to single-stranded (ss) DNA and, in presence of the cofactor ATP(or the non-hydrolyzable analog ATPγS), also to double-stranded (ds) DNA in a cooperative manner forming a stiff fiber. The RecA fiber may accommodate up to three strands of DNA.

The main contribution to the light absorption at wavelengths longer than 280 nm of the RecA protein comes from two tryptophan residues (Trp 291 and Trp 309). So far the spectroscopic information has been limited by the superposition of the two tryptophans since only average properties are observed. We have studied orientation and surrounding environment of the tryptophan chromophores by comparing the linear dichroism and the fluorescence quenching

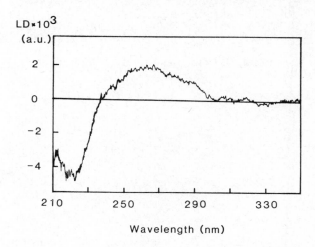

Figure 1: Linear dichroism of Trp 291 in RecA bound to ds-DNA. 1 µM protein, 3 µM bp ds-DNA, 5 uM ATPγS, shear gradient 90 s⁻¹.

properties of RecA with a genetically engineered RecA protein, both bound to DNA. In the modified protein Trp 291 has been replaced by a near-UV inactive amino acid (threonine). Both tryptophans are accessible to quenching by acrylamide whereas Trp 309 is protected from quenching by iodine ion, suggesting an anionic environment around Trp 309. The quenching properties are not altered by the activation of RecA for ds-DNA binding by its cofactor ATPγS. Binding to DNA does not affect the tryptophan fluorescence but the tryptophans become protected from acrylamide quenching. The tryptophans are similarly protected upon RecA binding to ss-DNA and to ds-DNA, indicating the protection to be due to protein-protein interaction rather than groove binding.

The RecA-ds DNA fiber is readily oriented in a flow field. The linear dichroism (LD), defined as the difference in absorption of light polarized parallel and perpendicular to the orientation direction (the fiber axis), shows that the DNA base pairs are oriented perpendicular to the fiber axis as it is for B-form DNA.The difference in LD of a complex of the wild-type protein and the modified protein bound to DNA shows the linear dichroism of the removed Trp 291 as indicated in figure 1. The L_a and L_b transitions of the tryptophan chromophore (250-300 nm) exhibit a positive LD and are thus oriented preferentially parallel to the fiber axis while the following absorption bands (220 nm), the B-bands, exhibit a negative LD and thus correspond to orientation perpendicular to the fiber axis. This, together with the directions of the transition moments within the tryptophan[5] enable us to conclude an orientation of Trp 291 with the pseudo-long axis of the indole moiety virtually perpendicular to the fiber axis and with the aromatic plane parallel to the fiber axis.

References

1. M. M. Cox and I. R. Lehman, Ann. Rev. Biochem., 1987, 56, 229-262
2. C. M. Radding, Biochimica et Biophysica Acta, 1989, 131-145
3. M. Takahashi, M. Kubista and B. Nordén, J. Mol. Biol., 1989, 205, 137-147
4. E. Di Capua, A. Engel, A. Stasiak and Th. Koller, J. Mol. Biol., 1982, 157, 87-103
5. B. Albinsson, M. Kubista, B. Nordén and E. W. Thulstrup, J. Phys. Chem., 1989, 93, 6646-6654

DNA RECOGNITION BY THE HELIX-TURN-HELIX MOTIF: INVESTIGATION BY RAMAN SPECTROSCOPY OF THE PHAGE LAMBDA REPRESSOR AND ITS INTERACTION WITH SPECIFIC OPERATOR SITES

James M. Benevides* and George J. Thomas, Jr.
Division of Cell Biology and Biophysics
School of Basic Life Sciences
University of Missouri-Kansas City
Kansas City, Missouri 64110, U.S.A.

INTRODUCTION AND METHODS

The major role of the helix-turn-helix (HTH) motif in DNA recognition, as deduced from high resolution X-ray diffraction studies of repressor-operator co-crystals,[1] is to provide a scaffold for support of the protein recognition surface. The HTH does not dictate how this recognition surface is to be aligned with DNA. Actual alignment is achieved through direct contacts between the protein and the edges of DNA base pairs in the major groove.[2] To determine whether a similar recognition mechanism applies also to HTH-DNA complexes in solution and to assess the importance of protein flexibility in DNA recognition, we have undertaken a comparative laser-Raman study of complexes of lambda operators O_L1 and O_R3 with both wild-type and mutant repressors. In this study we employ the HTH domain (N-terminal residues 1-102) of lambda cI repressor. In the wild-type protein, dimerization is mediated by hydrophobic packing of the dyad-related helix 5. The mutant repressor (Tyr88→Cys) contains an intersubunit disulfide bond and is constrained to function only as a covalent dimer. Although exhibiting enhanced operator binding, dimer-specific affinity is reduced tenfold relative to that calculated for the wild-type repressor. This suggests non-local effects of the intersubunit disulfide on HTH recognition.[3]

Wild-type (WT) and mutant (C88) repressors were overexpressed in E. coli and purified to 98%[3] Lambda operator O_L1 was synthesized by standard solid-support methods. Results were obtained on complexes containing a 1:1 ratio of protein dimer to DNA duplex dissolved in 0.2 M NaCl at 12 °C. Raman spectra were recorded on a Spex Ramalog V/VI at 1cm^{-1} increments with 1.5s integration time and 8 cm^{-1} slits. Further details are provided elsewhere.[4]

RESULTS, DISCUSSION AND CONCLUSIONS

Wild-type and mutant dimers exhibit similar secondary structures as indicated by comparisons of Raman amide I and amide III bands. The mutant dimer, however, lacks rigorous configurational symmetry since the disulfide exhibits mainly the gauche/gauche/trans rotamer. Important differences are also observed in multiple side-chain configurations.

Wild-type binding to O_r1 induces changes in the signatures of both DNA and protein, with significant perturbations to the local B-form conformation of DNA. This result provides direct evidence for the similarity of crystal and solution structures of lambda cI repressor-operator complexes.[5] Fig. 1 shows that wild-type and mutant repressors produce very different perturbations to the O_r1 DNA conformation upon binding, implying that specific DNA-protein contacts differ in the two complexes. These differences in the observed pattern of interaction suggest that flexibility at the dimer interface is critical for optimizing specific local interactions between the repressor and operator in an induced fit mechanism.

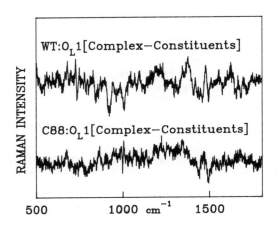

Figure 1   Difference spectra of repressor-operator complexes.

REFERENCES

1. T.A. Steitz, Quart. Rev. Biophys., 1990, 23, 205.
2. C. Wolberger, Y. Dong, M. Ptashne and S.C. Harrison, Nature, 1988, 335, 327.
3. R.T. Sauer, K. Hehir, R.S. Stearman, M.A. Weiss, A. Jeitler-Nilsson, E. G. Suchanek and C.O. Pabo, Biochemistry, 1986, 25, 5992.
4. J.M. Benevides, M.A. Weiss and G.J. Thomas, Jr., Biochemistry, 1991, 30, in press.
5. S.R. Jordan and C.O. Pabo, Science, 1988, 242, 893.

ACKNOWLEDGMENT

This research was supported by grants AI11855 and AI18758 from the U.S. National Institutes of Health to GJT.

PROTEIN/NUCLEIC ACID RATIO INFLUENCE UPON THE NUCLEOPROTEIN ELECTRONIC AND VIBRATIONAL SPECTRA

G.S. Litvinov, Y.L. Belokur, G.I. Dovbeshko

Scientific Research Center "Vidguk", Vladimirskaya 61-b, Kiev-33, 252033, USSR

INTRODUCTION

One of the urgent problems in biochemistry consists in the necessity of control over the protein and nucleic acid content in ribosomes, nucleosomes, viruses and other nucleoproteins. The information about the interrelation between protein and nucleic acid ratios in vibrational and electron spectra of viral nucleoproteins has been known for a certain period [1,2]. Along with it the valuable data could be obtained from the vibrational spectroscopy. Despite electron spectra, in vibrational spectra of nucleoproteins, viral ones in particular, there are non-overlapping bands of protein and nucleic acid [3].

METHODS AND MATERIALS

In view of this a comparative analysis of experimental and model electronic ultraviolet absorption (200 - 400 nm) and vibrational infrared absorption (5000-400 cm^{-1}) spectra of nucleoproteins with different ratio between protein and nucleic acid was fulfilled. The spectra of DNA and RNA viruses (bacteriophages T4, λ , C_q , E. coli spp, tobacco mosaic virus, carnation mottle virus, potatoe X and Y viruses; densonucleosevirus, some enthomoiridoviruses) and model protein - DNA mixtures with relative content from 1:1 to 20:1 were investigated. The experimental spectra of biopolymers and their complexes in the range of 400-5000 cm^{-1} were registered. The specimens were prepared in the form of dry films on the transparent substrates in IR spectra.
Model (calculative and experimental) spectra were obtained by additive summing up of molecular components spectra with the account of their relative content in nucleoproteins.

RESULTS AND DISCUSSION

Spectral intervals where the spectrum of viral nucleoprotein is adequately described by the superposition of

IR absorption band of separate components of viruses
are determined. Within the intervals of 1080-1170 cm^{-1}
and 1200-1300 cm^{-1} the absorption determined by PO_2^-
band of nucleic acid almost fails to overlap the protein absorption (fig. 1). On the contrary, the protein
absorption in the area of Amid I (1650 cm^{-1}) is strongly
overlapping NH deformational and C=O valent vibrations

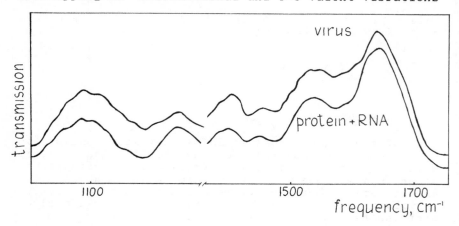

Figure 1 The illustration of identity of absorption
spectrum of the native nucleoprotein of carnation
mottle virus with composite spectra, obtained by
summing up of the viral protein and ribonucleic acid
spectra in equimolar relationship

of nucleic acid. But the Amid II protein vibration
(1540 cm^{-1}) undergoes the perturbation by nucleic acid
vibrations to a lesser degree. Therefore, the relation
of 1100/1540 cm^{-1} band intensities was chosen as an indicator of protein/nucleic acid ratio. The relation of
1540 cm^{-1} band maximum intensity to the absorption minimum intensity between Amid I and Amid II was also used
for this purpose. The calculations fulfilled by method
of differential moments well agree with the experimental data.
    Experimental and computing data on mixtures of
protein and nucleic acids with relative content from
1:1 to 20:1 showed the efficacy of a given approach to
the determination of molecular components´ content in
the living matter.

REFERENCES

1. H. L. Paul. Z. Naturforsch, 1959, 14 B, 427.
2. O. P. Sehgal at all. Phytopathology, 1970, 60, 177.
3. G.S. Litvinov, G.I. Dovbeshko, V.I. Strouk. ´Present
   developments in molecular spectroscopy´, Singapore,
   1989, 674.

CONFORMATIONAL TRANSITIONS OF E.coli RNA POLYMERASE INTERACTING WITH DNA AND REGULATORY FACTOR ppGpp

O. N. Ozoline, T. A. Uteshev, A. S. Solomatina, I. S. Masulis, S. G. Kamzolova

USSR Academy of Sciences
Institute of Biological Physics
Pushchino, Moscow region, 142292, USSR

Although the molecular mechanisms of DNA dependent RNA polymerase functioning have only been formulated in general terms,it is now well established that for this enzyme as well as for many others the limiting stages of the enzymatic reaction are complex conformational changes occuring at different steps of the transcriptional process.

To investigate these conformational transitions, RNA polymerase was specifically modified with fluoresceinmercuriacetate (FMMA),a specific reagent for SH-groups of proteins. It was previously shown that the modification has no effect on the kinetic parameters of initiation of RNA synthesis,nor does it lead to inhibition of the total RNA polymerase activity[1].That means,that FMMA-labeled RNA polymerase can be used to investigate functional conformational transitions of the enzyme.

As it is known from kinetic and footprinting data specific interaction of RNA polymerase with promoters is multistep and temperature-dependent reaction.Some inter - mediates of the process can be revealed at equilibrium conditions by temperature changing[2]. To investigate them, temperature-induced alteration of spectral parameters of FMMA-RNA polymerase in different complexes were compara - tively studied. To characterize the fluorescence spectra excited at 436nm,the intensity and the coefficient of fluorescence polarization (P) were registered at emission maximum.The polarization coefficient was determined by the standard procedure[1].

The data on temperature dependance of P are presented on Fig.1.It is seen that the values of P of the free label and FMMA-RNA polymerase complex monotonously decrease when temperature rises.The bend at 44 ± 1 C on curve 2A reflects denaturation-induced alteration of RNA polymerase polymerase[1].According to the data presented on the Fig.1B RNA polymerase undergoes temperature-induced conformational transitions at 13° during unspecific interaction with promoterless DNA fragments. These alterations are identical for the two different fragments used, indicating that conformational transitions followed by unspecific complex formation with DNA appeared to be sequence independent.

Temperature induced conformational transitions of

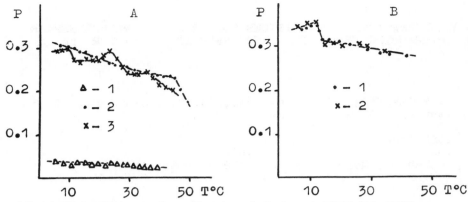

Fig.1 Temperature dependence of P for A:FMMA(1),FMMA-
RNA polymerase(2),FMMA-RNA polymerase+T7ΔD111-DNA (3);
B:FMMA-RNA polymerase+nonpromoter fragments.The sample −
(200 µl)contained 0.01 M Tris-HCl,pH 7.9,100mM NaCl,10mM
MgCl₂.Concentration of enzyme was 38mM(2,3A),111nM(1B),
17nM (2B).Fragments concentration was 17nM(1B) and 1.5nM
(2B) for 1151 and 782 b.p.Hae III fragments of T7-DNA
respectively.FMMA:RNA polymerase ratio in all cases was
1:1.In the case of B 0.01M Hepes pH 7.9 was also added.

the complex of FMMA-RNA polymerase with T7ΔD111-DNA are
more complicated.Four additional bends can be identified
in the temperature range 20-35°C.This DNA has 5 promoters
(A1,C,D,E,F) recognized by bacterial RNA polymerase. Most
of them form open promoter complexes at 20- 35°C.It seems
quite reasonable that the label attached is sensitive to
the conformational changes of the enzyme interacting with
different promoters.To verify this suggestion the fragments containing individual promoters were used and it
was proved that it is the case.Conformational transitions
of the enzyme in specific complexes with promoters A1 and
D appeared to be different for each other and differed
from those of nonspecific complexes.Therefore the pathway
of productive interaction with DNA and likely the final
configuration of the enzyme active center are promoter
and sequence dependent.
　　　The same label was used to reveal conformational
transitions of RNA-polymerase interacting with guanosine-
tetraphosphate-regulatory factor,changig the promoter selectivity of the enzyme. The fact means that the enzyme
structure preceding the interaction with DNA is also
essential for the proper promoter recognition.
　　　The data obtained demonstrate that the method of
fluorescent label is one of a few presently available efficient approaches for direct analysis of individual stages of RNA polymerase productive complex formation with
DNA.

1.　O.N. Ozoline, T.A. Uteshev, S.G. Kamzolova, BBA, 1991,
　　in press.
2.　D.W. Cowing, J. Mecsas, M. Record, C.A. Gross, J.Mol.
　　Biol., 1989, 210, 521.

ORIENTATION OF GROOVE BINDERS WITH RESPECT TO DNA LONG AXIS AS PROBED BY SERS.

J. AUBARD[*](1), J. PANTIGNY (1), G. LEVI(1), J.P.MARSAULT (1), M.A. SCHWALLER (1), E.W. THOMAS (2) and J.C. MERLIN (2).

- (1) Institut de Topologie et de Dynamique des Systèmes, (CNRS, URA 34), Université Paris 7, 1, rue Guy dela Brosse, 75005 Paris, France.
- (2) Laboratoire de Spectrochimie Infrarouge et Raman, (CNRS, UPR 2631), UST Lille Flandres Artois, 59655 Villeneuve-d'Asq, France.

DNA binding of small organic ligands may occur according to various mechanisms: external binding by electrostatic interaction with phosphate groups, outside binding to either the major or the minor groove by H-bonding, intercalation between base pairs due to hydrophobic interactions.

In order to gain more information concerning these different DNA binding modes, Surface Enhanced Raman Spectroscopy (SERS) of ellipticine, a pure intercalator, and berenil, a typical groove binder (1), was investigated.

Ellipticine   Berenil

The acido-basic properties of berenil, were recently established and showed that this molecule presents two pKs, respectively 10.5 for the diazoamino link and ca. 13 for the amidino groups. As expected, in the presence of hydrophobic macromolecules, such as DNA, these pKs are displaced toward higher values (2). We must thus consider that in pure water as well as in the presence of DNA, berenil essentially exists on the protonated dicationic resonant form presented above.

The SERS spectra obtained on partially aggregated silver colloids (30 mM NaCl), at pH 5, of berenil free and bound to DNA are shown in figure 1. These spectra compare well with resonance Raman spectra obtained recently in solution (3) but are much more intense and detailed. Very important intensity changes are observed in the SERS spectrum of the free drug with respect to the complexed drug, essentially the lines at 1429 cm^{-1}, 1393 cm^{-1}, 1308 cm^{-1}, 1182 cm^{-1} and 944 cm^{-1}. Clearly some lines are not perturbed, particularly the very intense ring streching line at ca. 1600 cm^{-1}.

The SERS spectra of ellipticines were reported elsewhere (4). Briefly, it was observed that the spectral perturbations detected, near the neutrality, arise essentially from the shift of the protonation equilibrium at N(2) nitrogen atom (see chem formula) on going from free to bound (to DNA) ellipticines. On the other

375

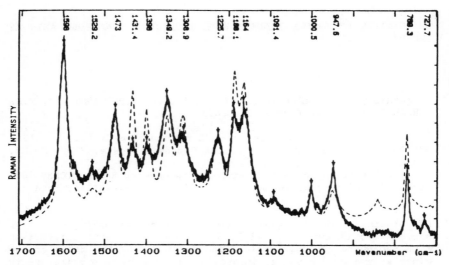

Figure 1. SERS spectra on Ag sols of berenil free (10^{-6}M, dotted line) and bound to DNA (drug/DNA ratio = 10^{-2}, solid line).

hand, for a given protonation state (i.e. the neutral or the cationic protonated forms), only sligth differences were observed between SERS spectra of free and bound molecules.

All these observations led to the conclusion that whereas no geometry change are observed for intercalators (ellipticines and ellipticiniums), in the case of berenil an orientational rearrangement takes place at the silver surface between the free and bound molecule. Indeed, the binding within the DNA minor groove fixes the berenil chromophore plane orientation at ca. 45° with respect to the helix axis (1). This orientational rearrangement of berenil, monitored by the groove binding mode, produces a concomitant variation of the geometry of adsorption at the metal surface, thus giving rise to the large intensity changes observed in the SERS spectrum.

Further experiments on various other groove binders (netropsin, stilbamidine, Hoechst 33258...) are currently underway to get more insight on the molecular mechanism involves at the Ag surface for this class of compounds.

REFERENCES

1. Ch. Zimmer and U. Wahnert, Prog. Biophys. Mod. Biol., 1986, 47, 31.
2. G. Dodin, J. Pantigny, M.A. Schwaller and J. Aubard, Anti-Cancer Drug Design, 1990, 5, 129.
3. J.C. Merlin, P. Clay, G. Petit, E.W. Thomas, J. Pantigny and J. Aubard, Proceeding of the 4th Eur. Conf. Spectrosc. Biol. Mol., 1991, (this book).
4. J. Aubard, G. Levi, J. Pantigny, J.P. Marsault, G. Dodin and M.A. Schwaller, Proceeding of the 4th Eur. Conf. Spectrosc. Biol. Mol., 1991 (this book).

COMBINED ANALYSIS OF ELECTRONIC AND RESONANCE RAMAN SPECTRA OF ANTITUMOR ANTHRACYCLINES AND THEIR COMPLEXES WITH DNA

G. Smulevich, A.R. Mantini and M.P. Marzocchi[*]

Dipartimento di Chimica, Universita' di Firenze, Via G. Capponi 9, 50121 Firenze, Italy

The antitumor anthracyclines, whose structural formula includes a substituted hydroxyanthraquinone chromophore, form intercalation complexes with DNA. The stacking interaction between the chromophore and the base-pairs of DNA gives rise to bathochromic and hypochromic effects in the absorption spectrum and to variations of the vibrational structure in the second-derivative absorption spectrum. Noticeable intensity changes are also observed in the resonance Raman spectrum [1-4]. The interpretation of these effects is complicated by the possible occurrence of intramolecular excited state proton transfer which affects the resonance Raman spectra. Dual excitation and emission have been observed in fact for the chromophore models 1,5- and 1,8-dihydroxyanthraquinone[5,6]. The rise of dual excitation and emission has been explained in terms of a Lippincott-Schroeder double minimum potential along the proton coordinate with reversed asymmetry at the excited state[7]. As a consequence the expected effects of deuterium substitution of the hydroxy-protons must be noticeable.

We measured the absorption, second-derivative, fluorescence and resonance Raman spectra of 1,4-, 1,5-, 1,8-dihydroxyanthraquinone, 1-hydroxyanthraquinone and their deuterated analogs in polar and non-polar solvents, and of idarubicin, epirubicin, deoxycarminomycin in H_2O and D_2O. For 1-hydroxyanthraquinone-d_0 and -d_1, being only one proton involved in the intramolecular hydrogen bond, the fluorescence spectra in n.alkane matrix were particularly useful.

The aim of this research is to determine the shifts of the potential curves between the ground and the excited states for all the modes, as follows:

i. to assign the spectral features due to proton motions at both the ground and excited states and to obtain the corresponding eigenvalues as function of the $R_{o..o}$ hydrogen bonded distance.

ii. to perform a combined analysis of the absorption,

second-derivative and resonance Raman spectra, via transform theory and model summing calculations, according to a method previously tested for β-carotene[8].

At first we performed the analysis for idarubicin and its model 1,4-dihydroxyanthraquinone which do not show the spectral complications due to proton transfer. The resulting shift parameters are comparable with the previous results obtained from the fluorescence excitation spectra in supersonic expansion[9]. The corresponding analysis for the idarubicin/ DNA complex indicates that the transition moment of the chromophore rotates upon interaction with the base-pairs.

The results for 1-HAQ indicate that the occurrence of dual fluorescence as well as the appearence of a remarkable vibrational progression at the excited state are due to the proton motion which acts also as a helping mode for all the resonance Raman active modes.

The analysis for the other compounds is in progress.

REFERENCES

1. G. Smulevich, A.R. Mantini and M.P. Marzocchi, J. Phys. Chem., 1990, 94, 2540.
2. G. Smulevich, A. Feis, A.R. Mantini and M.P. Marzocchi, Indian J. Pure Appl. Phys., 1988, 26, 207. Diamond Raman Jubilee.
3. Y. Nonaka, M. Tsuboi and K. Nakamoto, J. Raman Spectrosc., 1990 21, 133.
4. M. Manfait, A.J.P. Alix, P. Jeannesson, J.P. Jardillier and T. Theophanides, Nucleic Acids Res., 1982, 10, 3803.
5. M.H. Van Benthem and G.D. Gillispie, J. Phys. Chem., 1984, 88, 2954.
6. G. Smulevich, J. Chem. Phys., 1985, 82, 14.
7. G. Smulevich, P. Foggi, A. Feis and M.P. Marzocchi, J. Chem. Phys., 1987, 87, 5664.
8. A.R. Mantini, G. Smulevich and M.P. Marzocchi, J. Chem. Phys., 1989, 91, 85.
9. G. Smulevich, A. Amirav, U. Even and J. Jortner, Chem. Phys., 1982, 73, 1.

NUCLEIC ACIDS - DRUG INTERACTIONS STUDIED BY INFRARED LINEAR DICHROISM

Hartmut Fritzsche[*] and Allan Rupprecht

Institute of Microbiology and Experimental Therapy Jena Germany[*]

University of Stockholm
Arrhenius Laboratory
Stockholm
Sweden

INTRODUCTION

Deoxyribonucleic acid (DNA) changes its conformation as a function of the water activity. Both in solution and in films, the B-DNA undergoes a reversible transition to A-DNA when the water activity is reduced. We studied wet-spun oriented films of DNA (sodium salt) by infrared linear dichroism. By this technique, the conformation of a DNA sample can be analyzed in terms of B- and A-DNA.

RESULTS

The addition of drugs restricts the conformational flexibility of DNA. Dependent of the drug and its input concentration ratio - expressed in drug per DNA base pair - a fraction of DNA is frozen in the B conformation even under conditions of low water activity[1]. The freezing index F_i, which is given in DNA base pairs per drug molecule, describes this effect and is a new parameter of DNA-drug interaction which varies significantly among intercalating as well as non-intercalating drugs (Table 1)[2]. The extreme F_i values are 5-7fold higher than the number of DNA base pairs occupied by a drug molecule. The values of F_i can be correlated to the binding affinity. Among the non-intercalating drugs included in this study, a relation between the freezing index, F_i, and the flexibility of the drugs was found. These group of drugs, represented by netropsin and distamycin A, are known to bind preferentially to AT (adenine + thymine) sequences of B-DNA in the minor groove. The presence of GC (guanine + cytosine) base pairs reduces the binding tendency of these drugs. We obtained a significant reduction of the freezing index of these two drugs when we replaced the calf thymus DNA (42% G+C) by DNA from *Micrococcus lysodeikticus* with 72% G+C (Table 2). By contrast, the freezing index of intercalating drugs like the anticancer drug aclacinomycin A and violamycin BI is not significantly influenced by the binding to DNA with different G+C content (Table 2)[3].

379

Table 1

Freezing index, F_i, of DNA-drug complexes. The quantity F_i, expressed in DNA base pairs (bp) per drug molecule, describes the ability to inhibit the B-A conformational transition of DNA by "freezing" a part of the DNA in the B conformation. All experiments were done with wet-spun films of NaDNA from calf thymus. Accuracy is better than 10%

Drug	F_i bp/drug	Drug	F_i bp/drug
Intercalators		*Groove-binding drugs*	
Ethidium	4	SN-18071	2
Anthrapyrazole	5	NSC-101327	2
Mitoxanthrone	6	SN-6999	9
ß-Rhodomycin I	7	Pentamidine	12
Iremycin	9	Distamycin A	22
ß-Rhodomycin II	15	Netropsin	24
Aclacinomycin A	15		
Violamycin BI	18		
Daunomycin	28		
Adriamycin	28		

Table 2

Dependence of the freezing index, F_i, of DNA-drug complexes on the base composition of DNA as obtained from IR linear dichroism of wet-spun oriented films. DNA with 42% (G+C) was from calf thymus, DNA with 72% (G+C) was from *Micrococcus lysodeikticus*

Drug	F_i (bp/drug)	
	42% (G+C)	72% (G+C)
Netropsin	24	4.1
Distamycin A	21	4.4
Aclacinomycin A	15	14
Violamycin BI	18	12

REFERENCES

1. H. Fritzsche, A. Rupprecht and M. Richter, Nucleic Acids Res., 1984, 12, 9165.
2. H. Fritzsche and A. Rupprecht, Progr. Ind. Microorg., 1989, 27, 387.
3. H. Fritzsche and A. Rupprecht, J. Biomol. Struct. Dynamics, 1990, 7, 1135.

LOW-FREQUENCY RAMAN STUDIES OF SOME MOLECULES OF INTEREST FOR DRUG NUCLEIC ACID INTERACTION

O. Faurskov Nielsen,[a*] J. M. Espinosa,[a] D. H. Christensen,[a] J. Aubard,[b] M. Schwaller[b] and G. Dodin[b]

[a]Chem. Inst., University of Copenhagen, 5, Universitetsparken, DK-2100 Copenhagen, Denmark. [b]ITODYS, Université Paris VII, 1, rue Guy de la Brosse, 75005 Paris, France

For some years we have been interested in low-frequency studies of biomolecules.[1-3] However, the limiting factor in Raman spectroscopy by ordinary Ar-ion laser excitation is fluorescence. We have assigned a band at 100-115 cm^{-1} to a mode involving atoms in a hydrogen bond, and we have proposed this band to be important for understanding the interaction in biomolecules .[1-4]

The aim of the present paper is to investigate the possibility of low-frequency Raman studies with excitation by the 1064 nm line of a Nd/YAG-laser. Spectra were obtained on a Bruker FRA-106-Raman module coupled to an IFS-66 interferometer.

Figure. 1 shows Raman spectra of solid cytidine obtained by 514.5 nm and 1064 nm laser-lines, respectively. Both spectra are given in the $R(\hat{v})$-representation, as defined in the following

$$R(\hat{v}) \propto (\hat{v}_L-\hat{v})^{-4}\hat{v}[1-\exp(-h\hat{v}c/kT)]I(\hat{v}) \qquad (1)$$

where $\hat{v}$ is the Raman shift, and the other symbols have their usual meaning.[5,6] Especially in the low-frequency region this representation has several advantages as compared to the ordinary Raman spectrum.[1-6] Several bands are observed in Figure. 1. Small changes in band shapes between the spectra are caused by a difference in spectral resolution. A comparison between the two curves in Figure 1 shows that real bands are observed in the NIR-FT-Raman spectrum even to 100 cm^{-1} and below.

In a similar way the natural plant alkaloid ellipticine and 4 different substituted ellipticines were investigated by the NIR-FT-Raman technique. These compounds are all strongly coloured, and show high fluorescence emission when excited in the near UV and visible part of the spectrum. Even the NIR-FT-Raman technique revealed broad fluorescence-like bands for substituted ellipticines containing hydroxyl-groups.

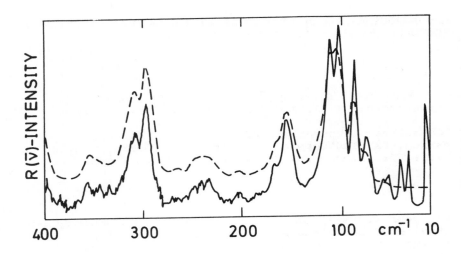

400   300   200   100   cm^{-1} 10

Figure 1. Raman spectra in the R($\tilde{\upsilon}$)-representation (see text) obtained of solid cytidine in a 180°-scattering configuration. The solid curve was obtained by excitation by an Ar-ion laser (514.5 nm) and the broken by excitation by a Nd/YAG laser (1064 nm).

NIR-FT-Raman spectra were obtained of two natural DNAs, Calf-Thymus and Sperm Herring. Low-frequency bands in the region 100 to 150 cm^{-1} could be assigned to hydrogen bonds between the base pairs. The modes might dynamically interact with an out of plane mode of ellipticine, and this dynamical coupling might be important for the drug/DNA interaction.

REFERENCES

1. O. Faurskov Nielsen, D.H. Christensen and E. Praestgaard, Biochem., Life Sci. Adv., 1988, 7, 57.
2. O. Faurskov Nielsen, D.H. Christensen and O. Have Rasmussen, J. Mol. Struct., 1991, 242, 273.
3. O. Faurskov Nielsen and G. Dodin, Acta Chem. Scand., 1990, 44, 1080.
4. O. Faurskov Nielsen, P.-A. Lund, L. S. Nielsen and E. Praestgaard, Biochem. Biophys. Res. Com., 1983, 111, 120.
5. M.H. Brooker, O. Faurskov Nielsen and E. Praestgaard, J. Raman Spectrosc., 1988, 19, 71.
6. W.F. Murphy, M.H. Brooker, O. Faurskov Nielsen, E. Praestgaard and J. E. Bertie, J. Raman Spectrosc., 1989, 20, 695.

RESONANCE RAMAN INVESTIGATION OF THE NON INTERCALATIVE BINDING OF BERENIL ON DNA

P. CLAY, J.C. MERLIN and **G.PETIT**, Laboratoire de Spectrochimie Infrarouge et Raman (CNRS UPR A 2631 L), UST Lille Flandres Artois, Bât. C5, 59655 Villeneuve d'ascq cedex, France.
E. THOMAS, Department of Biological Sciences, Salford University, Salford, M5 4WT, England.
J. PANTIGNY and **J. AUBARD**, Institut de Topologie et de Dynamique des systèmes (CNRS URA 34), Université de Paris VII, 1 rue Guy de la Brosse, 75005 Paris, France.

Berenil, a diarylamidine derivative, has veterinary applications as an anti-trypanosomal agent[1] which affects the replication of DNA by a non intercalating binding. Addition of DNA to a solution of Berenil produces a shift of the absorption spectrum (370 to 380 nm). An affinity for the minor groove of the double helix and a preferential binding on dA-dT rich regions were evidenced by previous studies[2]. In addition to the potential of such compounds as selective chemotherapeutic agents, the recognition of specific DNA sequences by small molecules is of considerable interest.

In order to collect information on the new conformation of the bound Berenil molecule, we have undertaken a spectroscopic study using the resonance Raman (RR) labelling technique and surface enhancement Raman spectroscopy (SERS). We report here preliminary experiments with RR spectroscopy. Figure 1 presents the typical RR spectra of Berenil at neutral pH obtained with and without DNA. As shown on this figure, the main spectral modifications observed are located on two lines (1188 and 1331 cm^{-1}).

Previous molecular modelling studies[3] indicate that the optimal fitting of Berenil in the groove is assisted by a change in the conformation of the diazoamino moiety together with a slight tilting of the amidimium groups which form hydrogen bonds with the adenine bases. Starting from this hypothesis, we focus our attention for the assignment of modes for which the vibration of these two parts are mainly involved. This can be performed on the basis of ^{15}N isotopic shift and literature data on related compounds[4]. By substituting the central nitrogen, it is possible to assign the N=N stretching mode which appears strongly coupled with a benzene ring vibration (1406 and 1443 cm^{-1}). The comparison with 4,4'-diamidino azobenzene, indicates that the double bond character is preserved in the diazoamidino bridge. The N-N stretching mode appears near 1188 cm^{-1}. The substitution of the four nitrogens of the amidinium groups produces different effects depending on whether the spectrum is recorded under resonance or off resonance (FT-Raman on solid form) conditions. If no shift is observed on the RR spectrum, the 1199 cm^{-1} line (1188 cm^{-1} in solution) is backshifted. This phenomenon clearly indicates that the N-N and the Ph-C(NH$_2$)$_2$$^+$ stretches appear at the same position, but only the last is observed under resonance conditions. It is known that ring-substituent stretches can be coupled with ring vibrations; two lines, at 1265 and 1331 cm^{-1}, can be assigned to the coupled modes related to the ring-amino stretching mode.

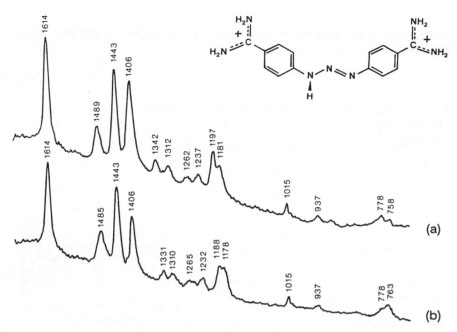

Figure 1 : Structure and RR spectra of Berenil (10^{-4} M, Tris pH 7.0, 457.9 nm), free (a) and bound (b) to Calf thymus DNA (4.5 10^{-3} M). The Raman lines from water and buffer was substracted.

From these assignments, it is apparent that the main perturbations arise for the amino bridge. As the NH function, which is placed on the convex side of the molecule cannot be involved in a direct interaction with the groove, a twisting around the amino bridge can be postulated. It is not possible, in the state of this study, to define exactly the conformational change but the increase of intensity and wavenumber of the 1188 cm⁻¹ line should be consistent with an increase of the N-N bond order and a more planar conformation. However the wavenumber of the N=N group is not perturbed upon binding.

In SERS experiments similar shifts in wavenumbers are observed[5], but significant changes in relative intensities occur, mainly for the lines assigned to the N=N stretching mode. This effect can be explained by an orientational effect with respect to the surface. Complementary information can thus be obtained on the new conformation and orientation of Berenil in the minor groove of DNA.

REFERENCES

1. L. BENNET Jr, Prog. Exp. Tumor Res. 1982, 7, 259.
2. B. BAGULEY, Mol. Cell. Biochem. 1982, 43, 167.
3. D.G. BROWN, M.R. SANDERSON, J.V. SKELLY, T.C. JENKINS, T. BROWN, E. GARMAN, D.I. STUART and S. NEIDLE, EMBO J. 1990, 9(4), 1329.
4. J.L. LORRIAUX, J.C. MERLIN, A. DUPAIX and E.W. THOMAS, J. Raman Spectrosc., 1979, 8(2), 81.
5. J. AUBARD, G. LEVI, J. PANTIGNY, J.P. MARSAULT, M.A. SCHWALLER, E.W. THOMAS and J.C. MERLIN, This proceeding, 1991.

SERS OF ELLIPTICINE AND DERIVATIVES FREE AND BOUND TO DNA.

J. AUBARD [*], G. LEVI, J. PANTIGNY, J.P. MARSAULT, G. DODIN and
M.A. SCHWALLER.

Institut de Topologie et de Dynamique des Systèmes (CNRS, URA 34)
Université Paris 7, 1 rue Guy de la Brosse, 75005 Paris, France.

Ellipticines are natural plant alkaloids which display some
antitumor properties (1,2). These molecules are planar conjugated
polycyclic aromatic dyes and interact with DNA by intercalation
between base pairs.

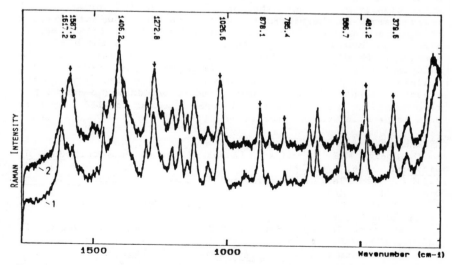

In this contribution we report the main results of a SERS
study of ellipticine and derivatives either free or complexed to
DNA. The SERS spectra obtained on partially aggregated Ag colloids
(30 mM NaCl), at pH 8.2, of ellipticine free and bound to DNA, are
presented in Fig.1.

Figure 1. SERS spectra on Ag sols of ellipticine free (8×10^{-7} M,
spectrum 1) and complexed to DNA (drug/DNA ratio = 10^{-2}, spectrum
2).

Upon complexation to DNA important relative intensity changes are observed, principally the lines at ca. 1617 cm^{-1}, 1588 cm^{-1}, 878cm^{-1}, 785cm^{-1} and 481 cm^{-1}. On the other hand certain Raman lines such as, 1273cm^{-1}, the triplet between 1100 cm^{-1} and 1200 cm^{-1}, the lines at 690 cm^{-1} and 566cm^{-1}, are not perturbed and can be used as internal standards.

The changes observed in SERS Raman spectra, between free and bound ellipticines, could arise either from a shift of the protonation state (3) and/or from a change in the orientation of the drug with respect to the metal surface.
First, recording at various pH, the SERS spectra of ellipticines free and bound to DNA, allowed us to estimate the protonation pKs of the N(2) nitrogen atom (see chem. formula). Indeed for each spectrum, plotting the ratios $I878/I566$ versus pH, leads to pK values of 7.5 for the water (colloid) solutions and ca.9 in the presence of DNA. These results are in total agreement with previous data obtained from fluorescence spectroscopy (3,4).
Secondly, the same experiments were carried out with the permanent cationic derivatives, 2-N—methyl- ellipticinium (NME; see chem. formula), whose structure does not depend on the pH and which interacts with DNA also by intercalation. In this case we observed that SERS spectra recorded in the absence and in the presence of DNA were similar and closely resemble those obtained with protonated cationic ellipticine.

From all these experiments it is clear that the main intensity changes observed between SERS spectra of ellipticines (free or bound), arise from the variation of the protonation equilibrium in the presence of DNA (3). Moreover the present experiments provide some insight on the geometry of ellipticines adsorbed on Ag sols. Our results on the quaternarized derivative NME, clearly rule out the possibility of an orientational rearrangement at the silver surface of the free drug with respect to the intercalated drug. Such an orientational effect should give large intensity changes in the SERS spectrum which are not observed. This leads us to consider that the interaction of the ellipticine (and ellipticinium) chromophore plane with silver colloids takes place perpendicularly to the Ag surface even in the presence of DNA. This necessarily indicates, in agreement with a recent binding model for intercalators (5), that the DNA long axis must be parallel to the surface.

REFERENCES
1. G. Dodin, M.A. Schwaller, J. Aubard and C. Paoletti, Eur. J. Biochem., 1988, 176, 371.
2. M.A. Schwaller, J. Aubard and G. Dodin, J. Biomol. Struct. Dyn., 1988, 6, 443.
3. G. Dodin, J. Pantigny, M.A. Schwaller and J. aubard, Anti-Cancer Drug Design, 1990, 5, 129.
4. M.A. Schwaller, F. Sureau, P.Y. Turpin and J. Aubard, J. Lumin., 1991, 48,
5. S. Fermandjan et al. J. Biol. Chem., 1991, 266,

CHANGES OF SOME MONONUCLEOTIDES STRUCTURES INDUCED BY DRUG INTERACTIONS. INFRARED STUDIES.

A. Hernanz and R.Navarro

Departamento de Química Física, Facultad de Ciencias, UNED
Senda del Rey, s/n,   28040-Madrid, Spain.

INTRODUCTION

It is known that the cytosine-β-D-arabinofuranoside (Ara-C) inhibits the growth of tumor cells and DNA synthesis.Some spectroscopic studies have been published about this antibiotic[1] although the mechanism of inhibition of DNA replication involves the direct interaction of Ara-C-TP with DNA polymerase, we do not know any spectroscopic work concerning phosphorilated derivatives of Ara-C. For this reason, we are investigating the interaction process between 5'-Ara-C-MP and 5'-GMP compared with 5'-GMP self-association and 5'-GMP+5'-CMP heteroassociation processes.In order to observe the modification of the mononucleotide structure in presence of the antibiotic, FTIR spectra of the following systems have been recorded, in deuterium oxide solutions, at neutral pH and room temperature:5'-GMP, 5'-CMP, 5'-Ara-C-MP, 5'-GMP+5'-CMP and 5'-GMP+5'-Ara-C-MP.High concentrated solutions (0.54M) have been used to make sure that the self-association and heteroassociation processes take place[2].

EXPERIMENTAL

Na_2(5'-GMP), Na_2(5'-CMP) (Boehringer-Mannheim) and the free acid of Ara-C-MP (Sigma) were used as received.Nucleotide solutions were prepared by dissolving the solid in ^2H_2O (99.95 atom %-Scharlau). Ara-C-MP neutral solution was made by reacting the antibiotic with a ^2H_2O solution of the sodium deuteroxide (Fluka), until the pH was 6.8 .This was done using a Schott-Gerat(CG822) pH-meter and a combination glass electrode. The 5'-GMP+5'-CMP and 5'-GMP+5'-Ara-C-MP samples were obtained mixing the corresponding solutions in equimolar amount. Infrared spectra were recorded at a apodized resolution of 2 cm^{-1} in a Bomem DA3 interferometer. A MCT detector was employed.

RESULTS AND DISCUSSION

In accord with their chemical structures similar Ara-C-MP and 5'-CMP behaviour will be expected. Consequently, we might hope that a heteroassociation process between the 5'-GMP and Ara-C-MP, like between the two mononucleotides 5'-GMP and 5'-CMP,takes place. The FTIR spectrum of one equimolar mixture of the Ara-C-MP and 5'-GMP (Figure 1e) agrees with this assumption. In any heteroassociating system,it is necessary to first understand the behaviour of the components , especially when they are capable of self-associating. For this reason, we have recorded and analyzed the spectra

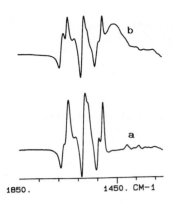

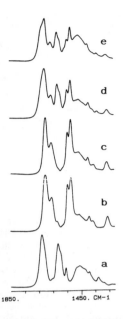

Figure 1 FTIR spectra of 0.54M solutions of the systems: a) 5'-GMP; b) 5'-CMP; c) Ara-C-MP; d) equimolar mixture of 5'-GMP and 5'-CMP; e) equimolar mixture of 5'-GMP and Ara-C-MP.

Figure 2 Difference spectra between: a) Fig.1d minus (Fig.1a + Fig.1b)/2 ; b) Fig.1e minus (Fig.1a + Fig.1c)/2

of the individual components in the same conditions as the mixture sample (Figures 1a and 1c). In order to detect possible differences between the antibiotic and 5'-CMP, this mononucleotide and the system 5'-CMP+5'-GMP were also studied (Figures 1b and 1d). The analysis of the heteroassociation processes were obtained by performing the comparison of the infrared bands proceeding from the equimolar mixtures above quoted and the infrared bands arising from the isolated components. The subtraction technique was used to make the differences more apparent (Figure 2). As occurs with 5'-CMP, a competitive heteroassociation process appears in the case of the antibiotic-mononucleotide against the self-association 5'-GMP process. The following spectral features are noted:(1) the region where the $\nu(C=O)$ and $\delta(NH_2)$ vibration bands appear is very sensitive to the drug-mononucleotide interaction. In particular, shifts to lower frequencies and intensity changes are observed for the complex 1670 cm^{-1} 5'-GMP band, in the spectrum of the mixture. This fact suggests that the 5'-GMP self-aggregation decreases while the heteroaggregation drug-mononucleotide arises. We resolved the complex 1670 cm^{-1} 5'-GMP band[3].(2) Some modifications of the intensities have been observed,too,in the bands at 1616 cm^{-1}(Ara-C-MP), 1578 cm^{-1}(5'-GMP),1737 cm^{-1}(5'-GMP) and 1504 cm^{-1}(Ara-C-MP).(3) The same trend in band shifts and intensity changes has been detected in the two heteroassociation process,5'-GMP+5'-CMP and 5'-GMP+Ara-C-MP,but both are not identical (see Figures 1d-1e and Figure 2).A more detailed study of this process needs, at a first step, to resolve the complex bands profiles. Further works are now in progress in this way.

REFERENCES

1. T.Theophanides,S.Hanessian,M.Manfait and M.Berjot,J.Raman Spectrosc. 1985,16,32.
2. R.Navarro,F.Peral,E.Gallego and M.Morcillo,J.Mol.Struct.,1986,163,357.
3. R. Navarro and A. Hernanz, in Chemistry and Properties of Biomolecular Systems,E.Rizzarelli and T.Theophanides (eds.),Kluwer, Dordrecht,1991.

A MONTE CARLO SIMULATION OF THE HYDRATION OF PURINE ALKALOIDS

V. I. Danilov and O. N. Slyusarchuk

Department of Quantum Biophysics, Institute of
Molecular Biology and Genetics, Academy of
Sciences of the Ukrainian SSR, 150 Zabolotny
Street, Kiev, 252143, USSR

To elucidate molecular mechanisms of the biological activity of purine series alkaloids (caffeine, theophylline, theobromine), it is necessary to study their physical and chemical properties in detail and, in particular, the nature of their stacking formation in aqueous solutions detected in many experimental investigations.

To study the systems of caffeine-water, theophylline-water and theobromine-water we applied a Monte Carlo technique. For this purpose, a Metropolis algorithm (1) was used for computer simulation of the behaviour of the purine series alkaloids mentioned above and their stacked associates in water cluster containing 200 and 400 molecules.

Computations of the energetic and structural characteristics of monomers and also of four main possible caffeine, theophylline and theobromine dimers were performed. Let us consider principal results for caffeine molecule.

The values obtained for the potential energy of the isolated water cluster and the system of caffeine molecule-200 water molecules allow one to compute the enthalpy of caffeine hydration from the experimental data available for this substance. According to the data obtained it is equal to -18.0 kcal/mol. Results from experiments (2,3) devoted to the determination of the caffeine dissolution (+3.4 kcal/mol) and sublimation (+23.9 kcal/mol) enthalpies indicate that the enthalpy of caffeine hydration amounts to -20.5 kcal/mol. Thus, the values of caffeine hydration enthalpies calculated on the basis of the theoretical and experimental data are in agreement.

Using the results obtained it is possible to compute the change in the association reaction of the molecules belonging to the alkaloids under study during the stacked dimer formation. The calculated values show that for caffeine the formation of two dimers different by location in water is energetically favourable. The main factor stabilizing the stacking is the change in the interaction of water molecules with each other connected with the change in water structure around monomers during their association. The base-base interaction contributes definitely to the stabilization of this configuration in water. The water-base interaction leads to the destabilization of the dimer configuration mentioned above.

Formation of two more possible caffeine dimers is energetically unfavourable, that is caused by the destabilizing effects of the water-water and water-base interactions. It is worth mentioning that in the present investigation, as well as in our works (4,5), the formation of the dimer, in which two molecules are located one above another ("right sandwich") in water, is shown favourable.

There are two energetically favourable dimers for theophylline. In the case of theobromine, the formation of all its four dimers is favourable. The fact that the formation of these dimers is also favourable is mainly due to the change in water structure around monomers during their association.

Caffeine molecules in the stack tend to arrange such that methyl groups draw nearer. This leads to the displacement of some of the water molecules localized near each methyl group and to their inclusion in the general net of H-bonds. Thus, the number of contacts of methyl groups with water decreases, in agreement with the known hydrophobicity properties of these groups.

It is necessary to note that our results showing the different character of the interaction of caffeine molecule and its stacked associate with water support the assumption regarding the different influence of caffeine monomer and dimers not only on the solvent but also on the DNA hydration shell.

Summing up the aforesaid, to study the dependence of the conclusions obtained on the size of water cluster, we carried out analogous investigation for the behaviour of the possible caffeine dimers in the cluster of 400 water molecules. There is only one dimer here favourable energetically and coinciding with that obtained in the cluster of 200 water molecules.

REFERENCES

1. N. Metropolis, A. W. Rosenbluth, M. N. Rosenbluth, A. N. Teller and E. Teller, J. Chem. Phys., 1953, 21, 1087.
2. A. Cesaro, E. Russo and V. Crescenzi, J. Phys. Chem.,1976, 80, 335.
3. H. Bothe and H. K. Cammenga, J. Therm. Anal., 1979, 16, 267.
4. V. I. Danilov and A. V. Shestopalova, Int. J. Quant. Chem., 1989, 35, 104.
5. A. V. Shestopalova, V. I. Danilov and V. Ya. Maleev, Dokl. Acad. Nauk SSSR (Russian), 1985, 282, 1000.

A SPECTROSCOPIC VIEW ON HOW DNA CONFORMATION IS REGULATED BY HYDRATION

W. Pohle

Department of Biophysical Chemistry,
Institute of Microbiology and Experimental Therapy,
Beutenbergstraße 11
Jena, O-6900, FRG

Infrared spectroscopy (IR) has been demonstrated to be[1-3] a valuable tool to study conformational changes in DNA and, furthermore, subtle environmental influencing of the phosphate groups via the sensitive PO_2^- vibrations, above all the antisymmetric stretching mode[4,5]. It is well-known for many years that one of the parameters determining the actual DNA conformation in a given sample is hydration, or, that is to say, water activity in the specimen[6,7]. More recently, an interesting concept was presented which explains the adoption of a certain DNA conformation by the economy of hydration of the phosphate groups: lowering the water activity in a DNA sample induces a B — A or B — Z (for alternating G - C sequences) transition in terms of diminishing the water binding to the free anionic phosphate oxygen atoms from "individual hydration" towards a more economic interphosphate bridging by water molecules[8]. Thus, shortage of water in the sample forces DNA to adopt A (or Z) conformation with interphosphate O - O distances small enough to allow "economic" water bridging [8].

IR investigations of oriented NaDNA films including the linear dichroism reveal that the cooperative B — A transition (cf. refs. 1 and 2) occurs at rather high relative humidity surrounding these films, namely at about 86%. Using the correlations derived from the results of conjugated gravimetric and IR experiments previously done[9,10], this value corresponds to the substantial number of about 12 water molecules per DNA nucleotide existing in the sample during the B — A transition. Thus, there seems to be no need of an economic interphosphate bridging by water molecules if, e.g., 10 or 11 H_2O per nucleotide are present. Regarding this amount of water (there is properly no shortage) in the pre-transition humidity region (around 80%), individual phosphate hydration might be considered more probable also for A-DNA. The spectroscopic finding is rather in favour of the mechanism proposed by Alden and Kim[11] who ascribed the A to B transition to be caused via the concomitant widening of the DNA grooves enforced by a critical amount of about 11 water molecules per nucleotide.

Moreover, it is strongly contradicting to the hydration-economy concept that the wavenumbers of the band due to the antisymmetric PO_2^- stretch for A and Z forms of DNA diverge substantially referred to the B-DNA value. Along with the B — A transition the wavenumber of this band increases drastically, which is in accord with the knowledge obtained from the results of quantum-chemical calculations that hydrogen bonding of water to phosphate leads to a partial electron withdrawal from the P $\doteq$ O bond[5]. According to the hydration-economy concept, this consideration should be valid also for the B — Z transition. In this case, however, not only no increase of the wavenumber of the antisymmetric stretching vibration band as for the B — A transition but even a drop is observed[3]. This fact suggests rather quite different hydration modes for A and Z conformations to exist.

REFERENCES

1. J. Pilet and J. Brahms, Biopolym., 1973, 12, 387.
2. W. Pohle, V.B. Zhurkin and H. Fritzsche, Ibid., 1984, 23, 2603.
3. E. Taillandier, J. Liquier and J.A. Taboury, in: Advances in Infrared and Raman Spectroscopy (R.J. Clark and R.E. Hester, Eds.), Wiley Heyden, 1985, p. 65.
4. W. Pohle and J. Flemming, J.Biomolec. Struct. Dynam., 1986, 4, 243.
5. W. Pohle, M. Bohl and H. Böhlig, J.Mol.Struct., 1991, 242, 333.
6. R.E. Franklin and R.G. Gosling, Acta Crystallogr., 1953, 6, 673.
7. V.I. Ivanov, L.E. Minchenkova, A.K. Schyolkina and A.I. Poletayev, Biopolym., 1973, 12, 89.
8. W. Sänger, W.N. Hunter and O. Kennard, Nature, 1986, 324, 385.
9. M. Falk, K.A. Hartman and R.C. Lord, J.Amer.Chem., Soc., 1962, 84, 3743.
10. S.J. Webb and M.D. Dumasia, Can. J. Microbiol., 1968, 14, 841.
11. C.J. Alden and S.H. Kim, J.Mol.Biol., 1979, 132, 411.

FORMATION OF D N A HELIX STRUCTURE AND ITS HYDRATION ENVIRONMENT AS REVEALED BY INFRARED SPECTROSCOPY AND OTHER PHYSICAL METHODS

V. Ya. Maleev, M. A. Semenov and A. I. Gasan

Department of Biophysics
Institute of Radiophysics and Electronics
Ukrainian Academy of Sciences
Kharkov, 310085, USSR

The present paper is aimed at studying the DNA-water system with various content of water to pick out the initial stages of the formation of DNA ordered structure and to evaluate the contribution of water into the double helix stabilization energy[1]. The infrared spectroscopy in the 900-3800 cm^{-1} region was chosen as the main experimental method for the finding the DNA hydration sites and their affinity to water molecules. To determine the energy parameters describing the interaction between DNA and water molecules we used also the differential scanning calorimetry and the gravimetry methods. The materials studied here were DNA samples from various sources.

The drastic changes in the frequency and intensity of absorption bands due to both the endocyclic (1577 and 1626 cm^{-1}) and exocyclic (1674, 1693, 1711 cm^{-1}) vibrations in DNA bases, and due to deoxyribose (982, 1053 cm^{-1}) were found already at lower humidities, $n < 10$, where n is the number of water molecules bound to the nucleotide. It was shown that these changes correlate with structural transitions of DNA from unordered state to A-conformation (at n=0-6) and from A- to B-form (at n=6-12). We found the hydration energy excess $\Delta E(n_i) = E_i - E_o$, where E_i is the energy of water binding to i-th type of hydration sites, n_i stands for the number of water molecules bound to the sites of this type, E_o

is the averaged energy of water-water interaction within a liquid phase (ca. 9.5 kcal/mole). The ΔE values have been estimated by 1) the frequency shift of OH stretching vibrational band due to water sorbed on DNA, 2) the analysis of gravimetry data, 3) the direct calorimetric measurement of the water evaporation enthalpy. Assuming that the hydration monolayer capacity ($n < 6$) is independent of the DNA conformation one can estimate the water contribution to the A- and B-DNA formation. The energy of DNA helical structure stabilization per one base pair appears to be equal to ca. 6 kcal/mole. This is the 60-70% of the total enthalpy of DNA melting.

The same conclusion may be drawn from the analysis of temperature plots of heat capacity obtained by the calorimetry for the DNA-water samples within a wide range of temperature and water content. In particular, the isothermal section of these curves (at $T=30^{\circ}C$) allows to derive a plot of bound water heat capacity C_p^w vs n. This dependence displays a "peak" at $n=4-10$ corresponding to the region of n, where the A-DNA is forming. This "peak" may be connected with the formation of ordered "crystalline" phase of water sorbed on DNA. The energy of disordering such "water structure" estimated with the square covered by the "peak" of the C_p^w vs n is of 6-7 cal/g or 4-5 kcal/mole of base pairs. This value agrees with the above-mentioned estimate of hydration energy contribution to the stabilization of a DNA helix.

The obtained results enable us to conclude that the principal energy contribution due to hydration is paid for the A-DNA formation, while the A- to B-DNA transition does not "profit at the expense" of enthalpy, thus having presumably the entropy sources.

REFERENCES

1. A. I. Gasan, V. Ya. Maleev, M. A.Semenov, Stud.biophys., 1990, 130, 171

MODELING CYCLIC AMP AND GMP CONFORMATIONS

J.Anastassopoulou and T.Theophanides

National Technical University of Athens,
Chemical Engineering, Laboratory of Radiation
Chemistry and Biospectroscopy, Zografou Campus,
Zografou 157 73, Athens, Greece

INTRODUCTION

Cyclic adenosine monophosphate (cAMP) and cyclic guanosine monophosphate (cGMP) play important roles in hormonal and cellular regulatory mechanisms. They are also used by cells in immunological functions, as secondary messengers, and for the transmission of cellular and intercellular signals that are Ca^{++} ion dependent[1].

Since the triggering process that discriminates the mediating role to be activated is thought to be intimately associated with their conformation, and since conformational properties are also believed to influence the preferential binding of platinum (in cis-platinum-diamine-dichloride (cisplatin), the most widely prescribed anti-cancer drug) to the N7 position of GMP subunits of DNA segments, a study to determine the conformational profiles of cAMP and cGMP in various force-fields was undertaken.

RESULTS

In this study with cAMP, 11 distinct conformers within a 5 Kcal/mol energy window were found in the AMBER force-field and with cGMP 14, respectively (Table 1). In both cases, the 6-membered phosphate ring in each of the distinct conformers adopts either a chair or a twist-boat conformation and the orientation of the base about the ribose sugar fragment is syn or anti. We calculated the relative stabilities due to the solvent effects, in AMBER and MACROMODEL force-field, using the water solvent model supported by the program. The relative values obtained are shown in brackets in table 1 and do not change the relative order found by the gas phase calculation. The AMBER calculations by MODEL[2] give excellent agreement with that found by X-ray. The global minimum conformer of cGMP superimposes on X-ray structure from the Cambridge data file with an RMS value of 0.142 Å (Fig.1) whereas, with cAMP the agreement is not as good with an RMS value of 0.337 Å (Fig.2). For molecular modeling of cAMP and cGMP the AMBER or MM2 force-fields, implemented in MODEL or MACROMODEL, and the MMX force-field in PCMODEL appear to give more consistent results with the X-ray findings than that using the AM1 Hamiltonian of AMPAC[3].

TABLE I. RELATIVE ENERGIES (KCAL/MOL) of cAMP and cGMP CONFORMERS

cAMP	MODEL	AMBER (H₂O)		MM2		MMX	AMPAC AM1	
	MODEL	MMODEL	MODEL	MMODEL	PCMODEL	1SCF	OPTM
Chair (*syn*)	0.0	2.2 (3.3)	0.5	1.8	1.7	2.9	3.7
Chair (*anti*)	0.1	0.0 (0.0)	0.0	0.0	0.0	0.6	0.0
TB (*syn*)	2.9	4.8 (7.0)	7.5	4.6	0.9	2.0	2.2
TB (*anti*)	3.3	3.0 (4.9)	7.8	3.3	3.5	0.0	0.2
cGMP							
Chair (*syn*)	0.0	0.4 (1.7)	0.2	0.5	0.0	0.0	0.0
Chair (*anti*)	1.0	0.0 (0.0)	0.0	0.0	1.3	1.7	1.5
TB (*syn*)	3.3	3.7 (5.8)	7.3	3.8	4.0	0.7	0.9
TB (*anti*)	4.2	2.9 (4.6)	7.5	3.2	1.6	1.2	1.2

Distance (Angstroms)
NH--O=P for cGMP Chair (*syn*)

X-ray (4.37)	3.92	4.13	4.03	3.86	4.13	3.92	3.65

Figure 1. Stereoview of cGMP calculated (- - -) and X-ray (——) RMS = 0.142 A

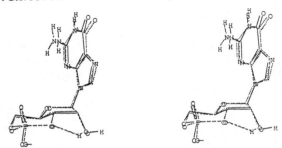

Figure 2. Stereoview of cAMP calculated (- - -) and X-ray (——) RMS = 0.337 A

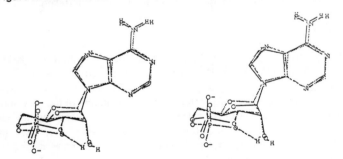

REFERENCES

1. E.W.Sutherland, Science, 1972, 177, 401, and references cited therein.
2. Obtained from Prof. K.Steliou, Department of Chemistry, University of Montreal, Montreal, QP, Canada H3C 3J7. We also would like to thank him for his suggestions and help with the calculations and use of the programs.
3. S. Topiol, T.K.Morgan, Jr, M.Sabio and W.C.Lumma, Jr, J. Am. Chem. Soc., 1990, 112, 1452.

NORMAL COORDINATE ANALYSIS OF 5'-GMP. APPLICATION TO THE ASSIGNMENT OF ITS VIBRATIONAL SPECTRA IN AQUEOUS SOLUTIONS.

R. Escribano, R. Navarro* and A. Hernanz*

*Instituto de Estructura de la Materia, CSIC
Serrano, 119, 28006-Madrid, Spain.*

**Departamento de Química Física, Facultad de Ciencias, UNED
Senda del Rey, s/n, 28040-Madrid, Spain.*

INTRODUCTION

A big effort has been dedicated to force constants calculation and normal coordinate analysis of nucleic acid bases[1,2]. These treatments have been applied to some mononucleotides[3]. Nevertheless, the assignment of the vibrational spectra of mononucleotides in solution cannot be established without difficulties[4].

In order to facilitate the interpretation of the behaviour of these compounds in aqueous solutions a normal coordinate analysis of 5'-GMP has been performed as a first step.

NORMAL COORDINATE ANALYSIS

The normal coordinate analysis has been carried out using the Wilson GF method with some particularities. The G matrix is written in internal coordinate representation, using an extended basis of internal coordinates, which includes redundancies. Through diagonalization of G we get rid of these redundancies whose corresponding eigenvalues are zero, plus the non-null eigenvalues G. The eigenvectors of G provide a "symmetry" coordinate basis, which is then applied to the F matrix in internal coordinates, to yield a "symmetrized" F matrix. The product GF gives rise to the usual secular equation, whose eigenvalues are the harmonic vibration wavenumbers. The analysis has been performed using computer programmes developed in our group[5] and modified to allow the handling of the large matrices involved.

The internal coordinates are based on those used for 5'-dGMP[3]. But in this case out-of-plane modes for the guanine residue are considered. The force constants (valence force field) are taken from the values given in the literature for 5'-dGMP[3], guanine residue[2] and tetrahydrofuran[6]. Geometrical data for 5'-GMP are obtained from X-ray patterns of disodium guanosine 5'-phosphate heptahydrate (molecules A and B)[7].

Some of the stronger bands observed in the infrared (ir.) spectra of 5'-GMP in solid phase (s), KBr pellets, and in the ir. and Raman spectra of 5'-GMP solutions in deuterium oxide, are tentatively assigned in the light of this analysis (see Table 1).

Calculated modes for molecules A and B of the asymmetric unit, inside the unit cell, are identical in the 4000-2700 cm^{-1} region. But modes whose fundamentals are expected at lower wavenumbers result mostly different for both types of molecules.

Table 1   Some of the stronger bands of the vibrational spectra of 5'-GMP.
(Deconvolved wavenumbers are enclosed between parenthesis)

v(obs.)/cm^{-1}			v(calc.)/cm^{-1}		Assignments
ir.(s)	ir.(^{2}H$_2$O)	Raman(^{2}H$_2$O)	mol.A	mol.B	(main contributions)
3380	3413		3304	3305	ν(N2H)-ν(N2H')
3130	3131	1681	[3149,3149]	[3151,3149]	[ν(O2'H),ν(O3'H)]
1685				1689	δ(HN2H') + ...
	1667			1667	δ(C2N2HH') out-of-plane
1650			1654		ν(C=O) + ...
	(1657)			1659	ν(C=O) + ...
	(1593)	1594		1593	δ(HN2H')+δ(N9C8H)+...
1576	1576		1571		δ(HC2'O2')-δ(C3'C2'H)...
1536	1538			1542	δ(HC2'O2')+δ(C3'C2'H)...
		1493		1491	δ(C3'C4'H)+δ(O4'C4'H)...
1487				1487	δ(O3'C3'H)+δ(C2'C3'H)...
	1462		1466	1456	δ(N7C8H)-δ(N9C1'H)...
1366		1371	1375		δ(C2'C1'H)-δ(O4'C1'H)...
		1339		1343	δ(O4'C1'H)-δ(C2'C1'H)...
1120				1121	δ(N7C8H)-δ(N9C8H)...
	1096		1095	1097	ν(PO11)-ν(PO12)
1076				1074	δ(O3'C3'H)+δ(C3'O3'H)...
978	975	976	996	993	ν(PO12)+ν(PO11)+ ...

The force field gives account of the main features of the vibrational spectra of 5'-GMP. Fundamentals of modes with CO stretching involved are expected at lower wavenumbers than the corresponding for the NH$_2$ group inside the 1670 cm^{-1} ir band profile. This result is in perfect agreement with our previous study on this band[8]. Refinement of the force field and its application to other mononucleotides would give a better insight into the nature of the vibrational spectra of this compounds.

REFERENCES

1.   M. Tsuboi, Y. Nishimura, A.Y. Hirakawa and W.L. Peticolas, in *Biological applications of Raman Spectroscopy*, T.G. Spiro (ed.), John Wiley, New York, 1987, vol. 2, pp. 109-179 and refs. therein.
2.   M. Majoube, *Biopolymers*, 1985, **24**, 1075; *J. Raman Spectrosc.*, 1985, **16**, 98; *J. Chim. Phys.*, 1984, **81**, 303.
3.   M. Ghomi and E. Taillandier, *Eur. Biophys. J.*, 1985, **12**, 153.
4.   P. Carmona and M. Molina, *J. Mol. Struct.*, 1990, **219**, 323.
5.   J.M. Orza, R. Escribano and R. Navarro, *J. Chem. Soc., Faraday Trans. 2*, 1985, **85**, 653.
6.   J.M. Eyster and E.W. Prohofsky, *Spectrochim. Acta*, 1974, **30A**, 2041.
7.   S.K. Katti, T.P. Seshadri and M.A. Viswamitra, *Acta Cryst.*, 1981, **B37**, 1825.
8.   R. Navarro and A. Hernanz, in *Chemistry and Properties of Biomolecular Systems*, E. Rizzarelli and T. Theophanides (eds.), Kluwer, Dordrecht, 1991.

INFLUENCE OF CHLORIDE IONS ON THE SERS SPECTRA OF CYTOSINE ON SILVER COLLOIDS

S.Sánchez-Cortés and J.V. García-Ramos

Instituto de Optica. CSIC.
Serrano 121.
28006 Madrid. Spain.

In a previous paper [1], we presented some preliminary results of SERS of 1,5-dimethylcytosine on silver and on copper colloids. In order to study more carefully this molecule, we shall compare its SERS spectrum with that of cytosine and of other related molecules.

In this work we present the effect of chloride ions on the SERS spectra of cytosine on silver colloids.

Figure 1 shows the SERS spectra of cytosine obtained before and after adding sodium chloride. In the first case (Fig. 1a), it appears a broad band at ca 913 cm^{-1} which is associated to some anionic products coming from borohydride reduction. This band disappears when NaCl is added to the solution (Fig. 1b). This suggests that the Cl$^-$ ions replace the reduction anions on the metal surface due to their larger affinity for silver.

Also, the intensity of some bands in the region 1100-1700 cm^{-1} which are assigned to in-plane ring vibration, increases when NaCl is added. One of these bands (the one which appears at 1306 cm^{-1}) may correspond to that which appears at 1290 cm^{-1} in the conventional Raman spectrum of cytosine in aqueous solution. This frequency, characteristic of cytosine residues in polynucleotides, is usually assigned to ν(C-N) vibrations of the bonds close to the carbonyl group (2,3).

The addition of Cl$^-$ ions also produces a progressive narrowing of the SERS bands in this region, probably due to the reorientation of the cytosine molecules on silver surface. In this process the molecules evolve from one state in which several positions on the surface are allowed to another in which the co-adsorption of Cl$^-$ reduces the number of positions and the available space on the metal.

Simultaneously, a shift of the ν(C=O) band to lower frequencies is observed as the amount of NaCl increases. This indicates that the reorientation mentioned above affects specially the carbonyl group. Therefore, the interaction of this group with the metal surface becomes stronger.

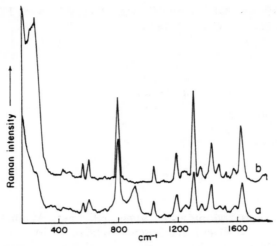

Figure 1. SERS spectra of cytosine (final concentration 10^{-5} M) on silver colloids before (a) and after (b) adding NaCl (final concentration $8 \cdot 10^{-3}$ M). $\lambda_{exc} = 514.5$ nm, 15 mW.

The existence of Cl$^-$ ions on the surface is suggested by the appearance of an intense ν(Ag-Cl) band at ~233 cm^{-1}.

The next step of our current research is the study of SERS spectra of cytosine on copper colloids in which these results have revealed to be very useful.

ACKNOWLEDGEMENTS

This work was financially supported by the Dirección General de Investigación Científica y Técnica (CICYT) under project PB87-0200.

REFERENCES

1. S. Sánchez-Cortés and J.V. García-Ramos. J. Raman Spectrosc. **21** (1990) 679.
2. M. Mathlouthi and A. M. Seuvre. Carbohydr. Res. **146** (1986) 1.
3. Y. Nishimura and M. Tsuboi. Chem. Phys. **98** (1985) 71.

LASER RAMAN AND IR SPECTRA OF 5-BROMOURACIL

V K Rastogi[1*], Sushma Arora[1], S L Gupta[2] and D K Sharma[3]
[1]Physics Department, L.R.College, Sahibabad-201005, India
[2]Physics Department M M College, Modinagar, India.
[3]Deptt. of Science and Industrial Research, Technology
Bhawan, New Mahrauli Road, Delhi India.

Recent spectroscopic studies of uracil and its
derivatives have been motivated by their biological
importance. Nucleic acids that contain 5- halogenated
uracil bases display unusual biological   properties
which may be a consequence of mis-pairing between the
uracil derivatives and guanine. DNA normally contains
uncommon nucleotides usually in very small amount. 5-
bromouracil is one of the well known uncommon nucleotide bases.Considering that the Raman spectroscopy
in conjunction with I R spectroscopy offers many more
possibilities to distinguish different symmetry species or to separate bands of close frequencies, we un-
dertook the study of the vibrational spectrum of 5-
bromouracil.

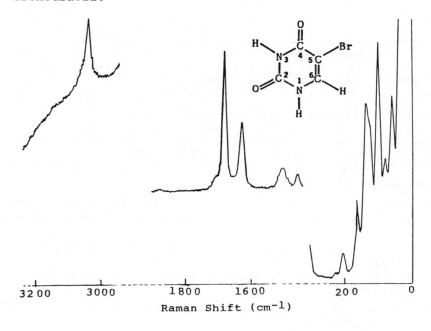

Raman Shift (cm^{-1})

TABLE 1. Assignment of fundamental frequencies (in cm^{-1}) of 5-bromouracil

Symmetry species	IR	Raman	Assignments.
a'	3205	–	N-H stretching
a'	3100	–	N-H stretching
a'	3045	3040	C-H stretching
a'	1700	–	C=O stretching
a'	1655	1670	C=O stretching
a'	1620	1620	C=C stretching
a'	1560	–	N-H in-plane-bending
a'	1438	1450	C-N stretching
a'	1420	1405	N-H in-plane bending
a'	1340	1340	C-N stretching
a'	1242	1240	C-H in-plane bending
a'	1224	1225	C-N stretching
a'	1058	–	C-N stretching
a'	1000	1010	Ring in-plane bending
a"	940	945	N-H out-of-plane bending
a"	860	–	N-H out-of-plane bending
a"	804	–	C-H out-of-plane bending
a"	778	785	Ring out-of-plane bending
a'	750	750	Ring stretching
a'	740	–	C=O in-plane bending
a'	640	635	Ring in-plane bending
a'	610	612	C=O in-plane bending
a'	540	550	C-Br stretching
a"	440	–	Ring out-of-plane bending
a"	418	425	C=O out-of-plane bending
a"	–	300	C=O out-of-plane bending
a"	–	205	C-N out-of-plane bending

The laser Raman and IR spectra of this compound in solid phase have been recorded. For recording Raman spectra, 4800 Å radiation from a 120 mW Argon ion laser was used. The molecule has been assumed to belong to C_s point group. The wavenumbers of the observed fundamental bands alongwith their proposed assignments are given in Table 1. The vibrations associated with C=O ... H-N hydrogen bonding have been observed at 160, 136, 100,80 62 cm-1 in Raman spectrum. These assignments are in agreement with those available for uracil derivatives [1-4].

REFERENCES

1. V.K. Rastogi, Y.C. Sharma and D.K. Sharma in Procd XII Int. Conf. on Raman spectroscopy, John Wiley & Sons (1990) p 178.

2. H. Susi and J. S. Ard, spectrochim Acta, 1971, 27A, 1549.

3. J. Bandekar and G. Zundel, Spectrochim Acta, 1983, 39A, 343.

4. V.K.Rastogi,H.P.Mital and S.N.Sharma,Indian J.Phys, 1990, 64B, 312.

VIBRATIONAL SPECTRA OF BIOMOLECULE : 6-AZATHYMINE

V.K.Rastogi[1*],H.P.Mital[2],Y.C.Sharma[3] and S.N.Sharma[4]

[1]Physics Department, L.R.College,Sahibabad-201005,India
[2]Physics Department,Meerut College,Meerut-250001,India
[3]Physics Department,N.A.S.College,Meerut-250002,India
[4]Indian Institute of Petroleum,Dehradun-248 005,India

N-heterocyclic molecules e.g. pyrimidine,cytosine, thymine, uracil and their derivatives are of considerable importance,because some of them are the basic constituents of DNA and RNA and play an important role in constitution and properties of nucleic acids.We have been investigating the vibrational features of biologically active N-heterocyclic molecules with a view to investigate the effect of different substituents at different positions[1-3]. A complete study of vibrational spectrum of 6-azathymine has not been made so far. Hence, the present investigation was undertaken to study the vib.spectrum of 6-azathymine and to identify the frequencies of different modes of vibration in this molecule.

The laser Raman and IR spectra of this compound in solid phase have been recorded and the observed fundamentals have been assigned on the basis of group frequency approach (Table-1). From the structural point of view,the molecule has been assumed to belong to C_s point group and the assignments for the fundamental modes of vibration as well as internal modes of vibration due to the substituent (CH_3) group have been proposed.

The N-H stretching vibrations associated with ring are expected to lie in the region 3000-3300 cm^{-1}.Therefore,the strong bands observed at 3265 and 3170 cm^{-1} in the present molecule have been assigned to this mode. An additional strong band at 3070 cm^{-1} may be due to the polymeric association with another molecule. Very strong bands observed at 1720 and 1665 cm^{-1} in the IR and at 1680 cm^{-1} in Raman spectra have been clearly identified

TABLE 1.Assignment of fundamental frequencies (in cm^{-1} of 6-azathymine.

IR	Laser Raman	Assignments
-	188 m	CH_3 torsion
310 m	327 w	Ring out-of-plane bending
350 m	-	$C-CH_3$out-of-plane bending
395 m	410 w	C=O out-of-plane bending
422 w	-	C=O out-of-plane bending
486 vs	485 w	C=O in-plane bending
529 vs	534 w	C=O in-plane bending
577 m	590 m	Ring in-plane bending
722 s	716 s	Ring out-of-plane bending
787 s	-	N-H out-of-plane bending
852 s	-	N-H out-of-plane bending
959 w	946 m	Ring in-plane bending
1012 m	1005 w	Ring stretching
1131 m	1135 s	Ring stretching
1196 vs	1200 w	$C-CH_3$stretching
1285 s	1280 s	Ring stretching
-	1355 m	N-H in-plane bending
1380 s	1385 s	CH_3sym. deformation
1430 s	-	CH_3asym. deformation
1478 m	-	N-H in-plane bending
-	1593 m	Ring stretching
1604 sh	-	2 x 802
1665 vs	1680 s	C=O stretching
1720 vs	-	C=O stretching
2806 m	-	1430 + 1380
2838 sh	-	CH_3sym.stretching
2900 w	2920 m	CH_3asym.stretching
2950 sh	-	1665 + 1285
3070 s	-	N-H stretching
3170 s	-	N-H stretching
3265 vs	3252 m	N-H stretching

as due to the C=O stretching mode. The vibrations associated with C=O...H-N hydrogen bonding have been observed at 136,130 and 80 cm^{-1}in Raman spectrum.

REFERENCES:

1. V.K.Rastogi,H.P.Mital and S.N.Sharma in Procd. XI
    Int.Conf.on Raman Spectroscopy,John Wiley & Sons 1988
2. V.K.Rastogi,H.P.Mital and S.N.Sharma,Indian J. Phys,
    1990, 64B,312.
3. V.K.Rastogi,Y.C.Sharma and D.K.Sharma in Procd.XII
    Int. Conf.on Raman spectroscopy,John.Wiley & Sons
    (1990)
4. H.Susi and J.S.Ard,Spectrochim Acta, 27A,1549,1971.
5. J.Bandekar & G.Zundel,spectrochim Acta,38A,815,1982.

Excited electronic states of the purine chromophore

Bo Albinsson and Bengt Nordén

Department of Physical Chemistry
Chalmers University of Technology
412 96 Göteborg
SWEDEN

Excited state properties of purine and different derivatives of purine (e.g adenine) have attracted considerable attention over the past three decades, primarily as a result of extensive applications of nucleic acid spectroscopy but also in order to understand potential photo reactions that may occur in the genome of living organisms. Among questions that remain to be answered one may note: why do not the DNA bases fluoresce? and, do $n\pi^*$ transitions contribute to any significant extent to the absorption, CD and emission spectra?

We have performed experiments with polarized light spectroscopy (IR and UV linear dichroism) together with quantum mechanical calculations in order to characterize the excited states of a number of substituted purines. More specifically, we have determined transition moment directions and intensity distributions (oscillator strengths) for the near-UV transitions ($\lambda > 200$ nm). In addition, emission spectra at low and ambient temperature were investigated in an attemt to correlate the energy splitting between the lowest $n\pi^*$ and $\pi\pi^*$ states to the fluorescence and phosphorescence quantum yields.

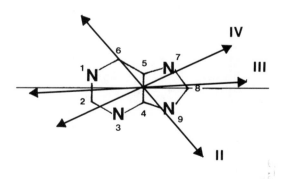

Figure 1 Tentative in-plane transition moment directions for the near-UV electronic transitions of the purine chromophore. Transition I is out-of-plane polarized.

The purines divide into three classes with respect to substitution effects. Unperturbed or weakly perturbed purines (e.g 9-methylpurine) show phosphorescence with about unit efficiency and have isolated $n\pi^*$ states as their lowest excited states (Table 1) whereas 2-aminopurine shows strong fluorescence at 77 K and has a $\pi\pi^*$ state lowest. Adenine is found in between these extremes, having low quantum yields for both fluorescence and phosphorescence. The lowest $\pi\pi^*$ transitions in the studied compounds (Fig 1.) seem to be possible to interpret within a common set of transitions with conserved intensities and directions although with varying energies depending on substitution (compare purine and 2-aminopurine in Table 1). This "optical stability" will most likely simplify the analysis of the luminescence properties (work in progress) since these can be expected to be determined effectively by the $\pi\pi^*$-$n\pi^*$ energy gap.

Table 1 Transition energies corresponding to transitions shown in Figure 1.

Transition	Character	Energy/10^3 cm^{-1} (wavelength/nm in parenthesis)				
		Purine	7-methyl purine	9-methyl purine	Adenine	2-amino purine
I	$n\pi^*$	35.6(281)	35.2(284)	34.4(291)	>35.6	?
II	$\pi\pi^*$	37.5(267)	36.7(272)	37.4(267)	37.7(265)	32.2(310)
III	$\pi\pi^*$	39.4(254)	38.7(258)	39.2(255)	40.8(245)	40.8(245)
IV	$\pi\pi^*$	48.4(207)	48.1(208)	48.4(207)	47.8(209)	45.5(220)

CHEMICAL REACTIVITY OF ACYL HALIDES AND THEIR CARCINOGENIC ACTIVITY MECHANISM BY VIBRATIONAL SPECTROSCOPY

G. Bottura, P. Filippetti and V. Tugnoli

Centro di Studio Interfacolta' sulla
Spettroscopia Raman, Dipartimento
di Biochimica, Sezione di Chimica e
Propedeutica Biochimica, Via Selmi 2
Universita' degli Studi di Bologna (Italy)

The evaluation, by instrumental methodology, of the potential human health risk and hazard caused by organic halogenated compounds has had in the last decades an explosive growth which is reflected by a large number of reviews and symposia on halocompounds in general and haloalkanes and haloalkenes in particular. The growing interest for these classes of compounds originates from their wide use in anthropogenic activities and their consequent presence in the environment due to losses during manufacturing processing, distribution, use and disposal.

The analytical detection of the relative amount of these pollutants cannot be the only parameter in the determination of toxic or carcinogenic activity, neither can the study of reactivity maps by ab initio methods.

It has been assessed that halogenated alkanes and alkenes require enzymatic oxidative activation to initiate their tumoral action. Such activation leads to highly reactive electrophilic species, the "ultimate carcinogens", which interacting with the nucleic acid bases or other biomacromolecules trigger the transformation of normal to neoplastic cells.

In a previous work (1) we studied the case of 1,2 dichloroethane, whose ultimate carcinogen is monochloroacetaldehyde which covalently interacts with the DNA bases adenosine and cytidine inducing molecular modifications associable to their tautomeric iminic forms as shown by Ft-ir spectra.

The present study is aimed at the vibrational spectroscopic characterization of the interaction between acetyl chloride (AC), the ultimate carcinogen yielded in the metabolic oxidation of the procarcinogen 1,1 dichloroethane, and the base residues of nucleic acids. As model molecules the nucleosides adenosine and cytidine were used.

In order to detect spectroscopically the molecular modifications occurring between the interacting species (AC and nucleoside) a suitable vacuum cell provided with KBr windows has been arranged. Such vacuum cell allowed the commercially available AC vapours to be adsorbed in the presence of air by a thin layer of the nucleoside deposited on the KBr window and to perform experiments in controlled atmosphere (presence or absence of reactants, air, etc.).

Fig. 1a shows the Ft-ir spectrum of a layer of untreated solid adenosine (ut Ado) and Fig. 1b the Ft-ir spectrum of the same layer (t Ado) recorded under 10^{-2} mmHg vacuum after a 12 hour contact with a controlled air-AC atmosphere.

After 12 hours contact with the atmosphere of AC-air, the spectrum of t Ado shows a weakening of strong 1666 cm^{-1} and a new band at 1695 cm^{-1}, whilst the strong one at about 1600 cm^{-1} remains unchanged. Other modifications can be noted as the intensity lowering of the 1570, 1470 and 1410 cm^{-1} bands together with other less evident. The trend is similar to that observed in the ir spectra of solid Ado obtained from a neutral aqueous solution and from an acidic solution (2).

Such an analogy suggests that AC acts as an acid towards the base residue of Ado.

407

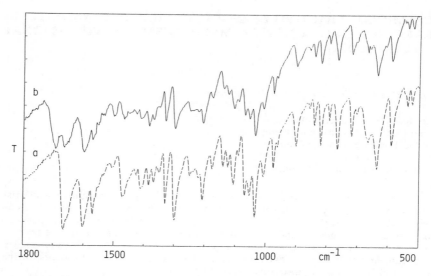

Fig. 1. Ft-ir spectra of solid adenosine: (a) untreated, (b) treated with AC

Previous ir spectroscopic results (3) showed AC can interact with surface Lewis centers ($AlCl_3$) giving the $CH_3(CO)^+$ (2290 cm^{-1}) or the CH_3COCl (1635 cm^{-1}) adsorbed species.

The absence of these bands in the spectrum of t Ado and the analogy with the spectrum of solid Ado from an acidic solution, suggests that AC interacts by its hydrolysis product, acetic acid which can be yielded in the experimental conditions of the present work simulating the real ones. This hypothesis is proved by the observations that the ir spectra recorded at different adsorption times show a gradual intensity decrease of the typical bands of AC and a gradual increase of the bands attributable to acetic acid.

In analogy with the deductions of the authors above mentioned and of X-ray diffraction studies (4), it can be reasonably argued that the N1 atom of the adenosine ring is the active center of main interest in the electrophilic attack, even if the ir bands of the residue of the acetic acid interacting with the base are not visible as they are masked with more intense ones.

Cytidine, which has as part of the aromatic ring the same N1C6N6 grouping, behaves in the same manner as adenosine.

Therefore, differently from chloroacethaldehyde which interacts with the adenosine base modifying the relative tautomery (1) and hence its reactivity localized on the base, AC modifies the H-bond interactions of the base causing miss-pairing of the canonical couplings. Both phenomena are involved in the molecular mechanism of chemical cancerogenesis.

REFERENCES
1. G. Bottura, P. Filippetti and A. Tinti, LALS '90, Moscow, in press.
2. M. Tsuboi, Y. Kyogoku, T. Shimanouchi, Biochim. Biophys. Acta, 1962, 55, 1.
3. A. Bertoluzza, Rend. Accad. Nat. XL, Serie IV, 1969, Vol. XX, 1.
4. W. Cockran, Acta Cryst., 1951, 4, 81; T.J. Kistenmacher and T. Shigematsu, Acta Crystallogr., Sect. B, 1974, 30, 166.

VIII

HAEM PROTEINS

AND

RELATED SYSTEMS

RESONANCE RAMAN AND TRANSIENT ABSORPTION STUDIES OF MYOGLOBIN DYNAMICS

Paul M. Champion

Department of Physics
Northeastern University
Boston, Ma. 02115

The electronic structure and covalent linkage of the heme prosthetic group clearly plays a central role in the biochemical activity and functional diversity of heme proteins. Nevertheless, the interaction between the heme and the protein motions is also of inherent interest, since the fluctuations and conformational changes of the surrounding protein matrix can lead to dynamic modulations of the heme geometry that affect its reactivity.

One prototypical example we have considered in some detail involves the binding of diatomic ligands to myoglobin (Mb), where the heme is covalently linked to the protein through a single amino acid (the proximal histidine). Flash photolysis experiments have revealed the existence of protein conformational substates that interconvert on time scales much slower than the ligand recombination times, at temperatures below 180K[1]. The strong electron-nuclear coupling, associated with the electronic spin-state change of the heme iron, leads to significant nuclear rearrangement or reorganization (localized at the heme-histidine site) when a diatomic ligand binds to, or dissociates from, the heme. The nuclear coordinate most profoundly affected by the iron electronic state change involves the doming of the heme and the subsequent movement of the iron-histidine moiety away from the porphyrin plane. This rearrangement can be quantified by the mean out-of-plane displacement or equilibrium position of the heme iron atom, $< a >$. For example, in the CO bound state $< a > \sim 0$ and in the "deoxy", or unbound, state $< a > \equiv a_0 \sim 0.45$Å. Because of this large change in the nuclear coordinates, the relatively weak force constants (K), expected for heme doming motions, can still lead to a substantial nuclear reorganization (or Stokes shift) energy that must be considered in the analysis of the ligand binding reaction. For example[2], if we take $K \sim 15$N/m ~ 0.15 mdyn/Å, we find that the heme localized reorganization energy is on the order of $\frac{1}{2}Ka_0^2 \sim 10$ kJ/mole (2.4 kcal/mole).

We have recently put forward a simple, quantitative, model[2] (SRC) that includes these heme doming forces and allows for the explicit coupling of the heme coordinate to a generalized protein coordinate. Thus, as the protein undergoes thermally driven conformational fluctuations (or, in the case of Hb, other structural rearrangements), the heme reactivity is modulated via the altered heme geometry. The coupling is most pronounced in the pentacoordinate "deoxy" state and leads to a spread in the iron equilibrium positions, σ_a, around a_0. The heme is taken to be uncoupled from the protein in the bound state, since the heme is "locked" into a planar configuration by the CO ligand. These ideas are consistent with spectroscopic observation[3].

A basic difference between the SRC model and the model developed by Agmon and Hopfield[4] (AH) can be traced to the explicit separation of the coordinates associated with the heme (Q) from those associated with the protein (x). In the SRC model, this allows the heme to relax in the low temperature (T<180K) photoproduct (Mb*, CO) even though the protein coordinates are held fixed at these temperatures (quenched disorder). The AH model describes the low temperature flash photolysis experiments depicted in Scheme I without any energetically significant heme relaxation or reorganization between MbCO and (Mb*, CO). Instead, all of the relaxation in the AH model occurs via "bounded diffusion" between (Mb*, CO) and Mb+CO. The two models of heme relaxation lead to significantly different extrapolations of the low temperature kinetics (which probes Mb*) into the high temperature regime, where the kinetic response arises from thermally averaged Mb states.

$$\text{Scheme I} \qquad \text{MbCO} \underset{\gamma}{\overset{\gamma}{\rightleftharpoons}} (\text{Mb}^*, \text{ CO}) \overset{\text{T>185K}}{\dashleftarrow\dashrightarrow} \text{Mb + CO}$$

Spectroscopic experiments that probe the optical transitions of Mb, Mb* and MbCO can help to resolve the amount of heme relaxation associated with the two steps depicted in Scheme I. In addition, they reveal the magnitude of the inhomogeneities associated with the absorption bands. Both the Soret band and band III (at ~ 760nm) shift by $\Delta\tilde{\nu}^*_{ABS} \sim 140$ cm^{-1}between Mb* and Mb (Fig. 1).

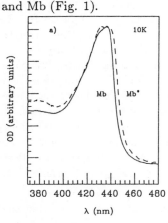

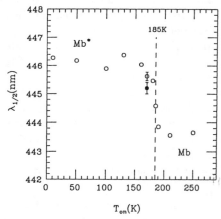

Figure 1

Figure 2

412

This shift is assigned to alterations in the mean iron out-of-plane equilibrium position (a_0^* vs. a_0) and can be used to independently calculate the optical coupling between this coordinate and the transition frequencies[2,5]. In addition, it can be used to probe the heme relaxation in the second step of scheme I. By varying the temperature, T_{on}, at which photolyzing cw laser light irradiates the sample (MbCO in 75% glycerol) as it is cooled, we can map out the transition temperature at which the heme relaxes from Mb* to Mb (see Fig. 2).

We can also utilize the SRC model[2] to self-consistently analyze the observed shifts, $\Delta \tilde{\nu}_{ABS}^*$, the kinetic holeburning (KHB) experiments, and the low temperature kinetics. These three experimental constraints lead to an unambiguous value for the iron out-of-plane equilibrium position in the photoproduct state (a_0^*=0.2±0.05Å) and to a well defined prediction of the room temperature Arrhenius barrier height for rebinding at the heme (E_A=16±2 kJ/mole). The SRC value for E_A should be contrasted with the much larger AH prediction, E_A=33 kJ/mole, which arises from the greater amount of relaxation between Mb* and Mb.

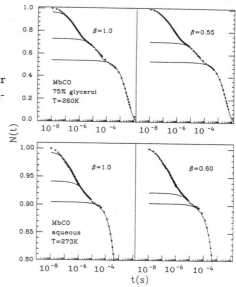

Experimental measurements of the geminate rebinding process at high temperature are extremely difficult for MbCO, because the amplitudes ($\sim$ 4%) are typically smaller than the transient absorption signal-to-noise. As a result, we developed a fast and stable signal averaging transient absorption spectrometer to monitor the full kinetic progress, subsequent to MbCO photolysis at room temperature. The results at 293K in aqueous phase are in good agreement with the work of Henry et al.[6], however careful analysis establishes that the rapid first phase is not exponential (Fig. 3). A second (exponential) geminate phase is also clearly revealed at lower temperatures ($\sim$273K) and in 75% glycerol solution (Figs. 4&5).

Figure 3

Analysis of the temperature dependence of the samples in 75% glycerol (Fig. 5) leads to a value of E_A=18kJ/mole which is in good agreement with the SRC prediction. The absolute magnitude of the geminate rebinding rates is somewhat smaller than predicted, unless the Arrhenius prefactor is reduced at high temperature. This is consistent with a larger distal pocket volume in the liquid phase than in the frozen solid. At 313K there is a distinct departure from the Arrhenius rate law (Fig. 5, insert), suggesting

413

that further protein conformational changes are taking place that increase the distal pocket volume at higher temperatures. The relative proportion, Φ, of the total geminate rebinding that is due to the second exponential phase is a distinct function of viscosity. Its physical significance will be discussed along with the possible origins of the "stretched" first phase.

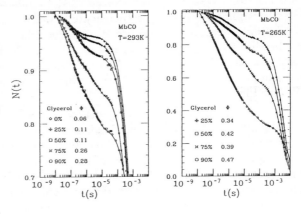

Figure 4

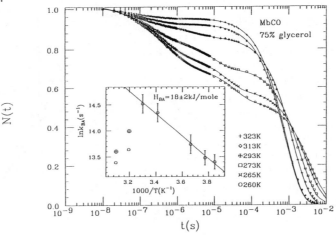

Figure 5

REFERENCES

1. R.H. Austin, K. Beeson, L. Eisenstein, H. Frauenfelder, I.C. Gunsalus (1975) *Biochemistry 14*, 5355.

2. V. Šrajer, L. Reinisch & P.M. Champion (1988) *J. Am. Chem. Soc. 110*, 6656.

3. V. Šrajer, K.T. Schomacker & P.M. Champion (1986) *Phys. Rev. Lett. 57*, 1267.

4. N. Agmon & J.J. Hopfield (1983) *J. Chem. Phys. 79*, 2042.

5. V. Šrajer & P.M. Champion (1991) *Biochemistry*, submitted.

6. E. Henry, J. Sommer, J. Hofrichter and W. Eaton (1983) *J. Mol. Biol. 110*, 443.

INTERACTIONS BETWEEN THE HEME POCKET AND THE AXIAL LIGANDS IN PEROXIDASES AS REVEALED BY RESONANCE RAMAN SPECTROSCOPY.

G. Smulevich[*1] and T.G. Spiro[2]

[1]Dipartimento di Chimica, Universita' di Firenze, Via G. Capponi 9, 50121 Firenze, Italy.

[2]Chemistry Department, Princeton University, Princeton, N.J. 08544, U.S.A.

Resonance Raman (RR) spectroscopy has been extensively applied to obtain information on heme structure, including the coordination and spin state, the status of the bond to the proximal ligand, and of the bonding of the carbon monoxide adduct of the reduced form of the protein.

Cytochrome c peroxidase (CCP), an enzyme which catalyzes the reduction of hydroperoxides via electrons from cytochrome c (cyt c)[1], has been particularly useful in probing the interaction between the prosthetic group and the protein residues in or near the heme pocket. The availability of site directed mutants of cloned protein[2,3] has provided specific structural information with regard to ligation and spin state of the heme[4-11] .

Figure 1 shows the heme crevice of ferric CCP[12] together with the amino acids replaced by site directed mutagenesis (Arg48-->Leu, Trp51-->Phe, His52-->Leu, His181-->Gly, Trp191-->Phe, Asp235-->Asn).

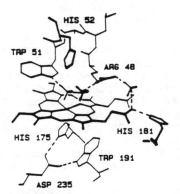

Figure 1 Structural diagram of the heme pocket of Bakers' yeast CCP[12]. The amino acid replaced by site directed mutagenesis and investigated by RR spectroscopy are included in the figure.

At neutral and low pH the reduced form of CCP and all the studied mutants is 5-c high spin, but at high pH it is 6-c low spin except when the distal His52 residue is replaced with Leu. It was shown[6] that the 6th ligand was very easy to photolyze, giving rise to a 5-c high spin Raman spectrum, while no photoeffect was observed for the Leu52 mutant[11]. These observation strongly support the view that the sixth ligand is the His52 imidazole.

Although the Leu52 mutant stays 5-c, the alkaline transition induces conformational changes which affects the vinyl group, as observed for CCP and all the other mutants.

The same behaviour is found for the Fe[III] forms. At alkaline pH the imidazole is again the sixth ligand, except for the Leu52 mutant which binds hydroxide giving a mixed-spin RR spectrum instead of 6-c low spin.

The Fe-imidazole stretching mode of Fe[II]CCP[13] and all the mutants[4] appears as a broad band with two components at ~245 and ~230 cm[-1] (Table I). These are interpreted as due to a double well potential for the imidazole proton between the proximal His175 imidazole and the Asp235 carboxylate group. The 230 cm[-1] component corresponds to the proton residing on the imidazole N atom and the 245 cm[-1] component to the proton residing on the carboxylate O atom of Asp235. The only exception to this behaviour is represented by the Asn235 mutant, which shows a narrow Fe-imidazole band at lowered frequency (205 cm[-1]) implying the absence of any H-bond.

At alkaline pH, since the sixth ligand is photolabile the status of the Fe[II]-imidazole bond can be determined in the resulting 5-c photoproduct. For CCP and CCP(MI) an increase of the intensity of the 230 cm[-1] band was observed[4,13], while in the Leu52-mutant a band at 205 cm[-1] appears together with the 230-245 doublet, indicating a partial loss of the proximal H-bond due to the distal alteration[11]. More dramatic is the effect of mutations which perturb the H-bonding network that extends from the distal to the proximal side of the heme. Leu48, Phe51 and Gly181 show only a band at 205 cm[-1], indicating a complete loss of the proximal Asp235-His175 H-bond. The conformational changes caused by the alkaline pH induce a stress on the Asp235-His175 H-bond which does not break only if the H-bond network connecting the distal with the proximal side is intact.

A coupling between the distal residues and the proximal H-bond is also revealed by the properties of the Fe[II]CO adducts. At acidic or neutral pH CCP and its mutants give two forms of CO adducts. Form I which is linear with a strong donor fifth ligand (imidazole-like) and form II, tilted and H-bonded by a distal group, with a weak donor fifth ligand (Table I). At alkaline pH there is only form I' which has no distal H-bond but the proximal imidazole donor strength is normal. Leu52 represents an exception, showing at both acidic and alkaline pH's a mixture of form II and I'. The proximal H-bond seems to be weakened even in the absence of distal H-bonding. On the other hand, Leu48 mutant gives rise to form I at acidic pH and to form I' at alkaline pH, but both forms appear to have a normal imidazole, not imidazolate. Thus, replacement of either His52 or Leu48 by Leu appears to weaken the proximal H-bond when the CO adduct is formed.

Table I   Vibrational frequencies (cm^{-1}) of the Fe-Imidazole and FeCO for CCP mutants.

Acidic pH

		Form I		Form II	
	Fe-Im	Fe-CO	CO	Fe-CO	CO
CCP(MI)[4,5]	246/233	503	1922		
Leu48[4,5]	242/227	500	1941		
Phe51[4,5]	247/229			528	1932
Leu52[11]	241/229			522	1928
Gly18[11]	241/229	496	1928	529	1928
Phe19[4,5]	246/225	505	1922	531	1922
Asn235[4,5]	205			531	1933

Alkaline pH

		Form I'	
	Fe-Im	FeCO	CO
CCP(MI)[4,5]	233	507	1948
Leu48[4,5]	206	500	1951
Phe51[4,5]	204	505	1948
Leu52[11]	205/229/248	508	1944
Gly18[11]	203	504	1944
Phe19[4,5]		505	1950
Asn235[4,5]	205	505	1951

REFERENCES

1. T. Yonetani, 'Enzymes' (3rd Ed.), 1976, 13, 345.
2. D.B. Goodin, A.G. Mauk and H. Smith, Proc. Natl. Acad. Sci. U.S.A., 1986, 83, 1295.
3. L.A. Fishel, J.E. Villafranca, J.M. Mauro and J. Kraut, Biochemistry, 1987, 26, 351.
4. G. Smulevich, J.M. Mauro, L.F. Fishel, A.M. English, J. Kraut and T.G. Spiro, Biochemistry, 1988, 27, 5477.
5. G. Smulevich, J.M. Mauro, L.F. Fishel, A.M. English, J. Kraut and T.G. Spiro, Biochemistry, 1988, 27, 5486.
6. G. Smulevich, M.A. Miller, D. Gosztola and T.G. Spiro, Biochemistry, 1989, 28, 3960.
7. G. Smulevich, A.R. Mantini, A.M. English and J.M. Mauro, Biochemistry, 1989, 28, 5058.
8. G. Smulevich, Y. Wang, S.L. Edwards, T.L. Poulos, A.M. English and T.G. Spiro, Biochemistry, 1990, 29, 2586.
9. G. Smulevich, Y. Wang, J.M. Mauro, J. Wang, L.A. Fishel, J. Kraut and T.G. Spiro, Biochemistry, 1990, 29, 7174.
10. T.G. Spiro, G. Smulevich and C. Su, Biochemistry, 1990, 29, 4497.
11. G. Smulevich, M.A. Miller, J. Kraut and T.G. Spiro, in preparation.
12. B.C. Finzel,T.L. Poulos and J. Kraut, J. Biol. Chem., 1984, 259, 13027.
13. S. Hashimoto, J. Teraoka, T. Inubishi, T. Yonetani and T. Kitagawa, J. Biol. Chem., 1986, 261, 11110.

RAPID MIXING RESONANCE RAMAN SPECTROSCOPY OF PEROXIDASE INTERMEDIATES

Vaithianathan Palaniappan, Ann M. Sullivan,
Melissa M. Fitzgerald, John R. Shifflett and James Terner

Department of Chemistry
Virginia Commonwealth University
Richmond, VA 23284-2006, U.S.A.

INTRODUCTION

Recent developments in our laboratory have opened up new possibilities for the investigation of peroxidase mechanisms by resonance Raman spectroscopy, in particular for the intermediate known as compound I. Our methodology utilizes excitation into near-ultraviolet absorptions of derivatives and intermediates of heme enzymes. In the past, near-ultraviolet laser excitation has been used only infrequently for resonance Raman studies of heme proteins, due to the weak resonance enhancements afforded below 370 nm, relative to the exceptionally strong enhancements obtained by excitation into the intense Soret absorption with violet laser wavelengths.

Our recent work has centered on horseradish peroxidase, a heme-containing glycoprotein which is one of the most well characterized of the peroxidases. The physiological intermediates of horseradish peroxidase are known as compounds I and II. Compound I is formed by a two-electron oxidation of the resting enzyme by peroxide and contains an Fe(IV) porphyrin π-radical cation heme. A one-electron reduction of compound I results in the compound II intermediate which contains an Fe(IV) (ferryl) heme.

Resonance Raman enhancement of derivatives and intermediates of horseradish peroxidase in the near-ultraviolet results in intensity and enhancement patterns that are different from those normally observed within the porphyrin Soret and α-β absorptions that occur in the visible. Signals are weak, nevertheless near-ultraviolet excitation allows the resolution of resonance Raman in-plane porphyrin skeletal modes of horseradish peroxidase compound I. Though previous attempts had been made to obtain resonance Raman spectra of compound I using visible excitation, the results had been unreliable and contradictory.

METHOD

Horseradish peroxidase compound I is generated by mixing a concentrated solution of the enzyme with H_2O_2 in a Ballou 4-jet mixer fed by two 100 mL syringes mounted on a syringe pump. The mixer produced a horizontal jet stream of sample passing through the open air. The sample jet is excited transversely by a focussed laser beam, with the Raman scattering collected at 90° by collection lenses. The sample jet is aimed into a small catch basin, and then pumped into a reservoir which contains sodium ascorbate to reduce the oxidized enzyme back to the ferric resting state. The regenerated enzyme is repurified by ion exchange chromatography. Further specifics are given in reference 1.

Compound I is susceptible to photolysis by the focussed laser excitation. This can be minimized by using high protein concentration, low laser power, and a rapidly flowing sample. A resonance Raman spectrum of the resting enzyme (800 μM) is shown in Figure 1c. Upon mixing with H_2O_2, the resonance Raman frequencies of the porphyrin π-radical cation (compound I) are observed if the laser power is kept sufficiently low (Figure 1a). If the laser power is not sufficiently low, significant photoreduction occurs, and signals from the resting enzyme dominate the spectrum (Figure 1b).

RESULTS

Compounds I and II are reputed to both contain six-coordinate low spin ferryl hemes, however compound I has an electron removed from the porphyrin ring forming the porphyrin π-radical cation. It is therefore instructive to compare the resonance Raman in-plane skeletal modes of compounds I and II. Resonance Raman spectra of horseradish peroxidase compound II in-plane skeletal modes have appeared previously in the recent literature [2].

The symmetry state of the porphyrin π-radical cation of horseradish peroxidase has been believed to be A_{2u} rather than A_{1u} However, the resonance Raman frequencies we have observed for horseradish peroxidase compound I are similar to those recently reported for A_{1u} metallo-octa-ethyl porphyrin cation radicals [3], in terms of frequency shifts from the non-radical six-coordinate low-spin ferryl heme of compound II.

Metallo-octaethyl porphyrin cation radicals have been proposed to be predominantly A_{1u} rather than A_{2u} [3]. The A_{2u} radicals have large spin density on the bridging meso carbons and pyrrole nitrogens, while the A_{1u} radicals have spin density concentrated at the pyrrole carbons. The Cb-Cb interactions are bonding in the A_{2u} orbital and antibonding in the A_{1u} orbital. Removal of an electron

Table I
(Frequencies in cm^{-1}, from ref. 1)

Mode	Cpd. I	Cpd. II	Cpd. I - Cpd. II	Mode Composition
ν_{10}	1636	1639	-3	Ca-Cm
ν_{37}	1614	1602	+12	Cb-Cb
ν_2	1606	1587	+19	Cb-Cb
ν_{11}	1570	1560	+10	Cb-Cb
ν_3	1504	1509	-5	Ca-Cm
ν_{28}	1458	1472	-14	Ca-Cm
ν_{29}	1402	1398	+4	Ca-Cb
ν_4	1359	1379	-20	Ca-Cn

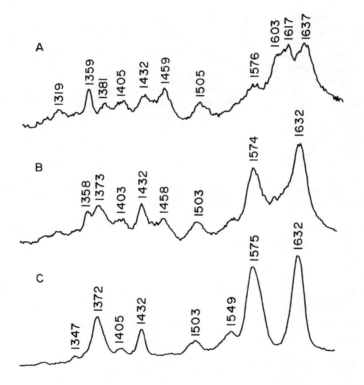

Figure 1. Resonance Raman spectra (3564 Å excitation) of (A) horseradish peroxidase (800 μM, pH 6.8) plus H_2O_2 (1.6mM) with 1 mW laser power; (B) same conditions except higher laser power (10 mW); and (C) resting horseradish peroxidase (800 μM, pH 6.8) without added oxidant (10 mW).

from the A_{2u} orbital is expected to result in decreases of frequency of the Cb-Cb modes, ν_2, ν_{11} and ν_{37}. However, removal of an electron from the A_{1u} orbital of model compounds has been reported to result in upshifts of Cb-Cb modes. Our data do, in fact, show substantial upshifts (10-20 cm^{-1}) of Cb-Cb modes ν_2, ν_{11} and ν_{37} (Table I). The 20 cm^{-1} downshift of ν_4 that we observe from compound II to compound I is consistent with ν_4 downshifts reported for octaethylporphyrin π-radical cations which range from 12 to 34 cm^{-1}. The resonance Raman frequency shifts for compound I, relative to compound II and the six-coordinate low-spin ferric derivatives, are consistent within normal mode compositions. In addition to the upshifts of the Cb-Cb modes noted above, downshifts are observed for the ν_3, ν_{10} and ν_{28} frequencies, whose mode compositions are primarily Ca-Cm.

In summary, near-ultraviolet excitation of the porphyrin π-radical cation absorption, high protein concentrations, and the use of rapidly flowing samples have been shown to be advantageous for resonance Raman investigation of the previous elusive compound I intermediate of horseradish peroxidase. The near-ultraviolet wavelengths allow one to avoid the strong Soret enhancements of the ferryl and ferric species which normally overwhelm the weaker signals of horseradish peroxidase compound I under visible excitation. The highly oxidized porphyrin π-cation radical is well known to occur during the catalytic cycles of many other important peroxidases such as lactoperoxidase and chloroperoxidase, and has been postulated to be a key intermediate in the mechanisms of many important heme enzymes such as cytochrome oxidase and cytochrome P-450.

Acknowledgements: This work is supported by Grant GM34443 from the National Institutes of Health.

References

1. V. Palaniappan and J. Terner (1989) J. Biol. Chem. 264, 16046-16053

2. J. Terner and D.E. Reed (1984) Biochim. Biophys. Acta 789, 80-86

3. R.S Czernuszewicz, K.A. Macor, X.-Y. Li, J.R. Kincaid and T.G. Spiro (1989) J. Amer. Chem. Soc. 111, 3860-3869

RESONANCE RAMAN SPECTRA OF PHYTOCHROME AND ITS MODEL COMPOUNDS

T. Kitagawa,[a] Y. Mizutani,[a] S. Tokutomi,[b] K. Aoyagi[c]

[a]Institute for Molecular Science and Graduate University for Advanced Studies, [b]National Institute for Basic Biology, Okazaki National Research Institutes, Myodaiji, Okazaki 444 Japan, and [c] National College of Technology, Taira, Iwaki, 970 Japan

Introduction

Phytochrome is a chromoprotein of green plants which acts as a photoreceptor for a variety of light-induced morphogenetic responses. The chromophore, 2,3-dihydrobiliverdin (Fig. 1a), is bound to a protein via thioether bonds at A-ring. This protein has two spectroscopically distinct and interconvertible forms denoted as P_r and P_{fr}; P_r is converted to P_{fr} by red light illumination and v.v by far-red light illumination. Only P_{fr} triggers physiological responses but structural differences between P_r and P_{fr} remain to be elucidated. Since the phytochrome is very fluorescent, RR spectroscopy had not been successfully applied to it until the far-red excited RR spectrum of oat P_r at 77K reported by Fodor et al.[1] Later SERRS of oat phytochrome on Ag colloids at 77 K were reported,[2,3] but in order to study the light-induced structural changes, it is desirable to observe the RR spectra for a natural state at an ambient temperature. Accordingly, we adopted blue excitation to circumvent the interference by fluorescence and the two color illumination to monitor either P_r or P_{fr} more favorably under a photo-steady state, and measured successfully the RR spectra of 'large' pea phytochrome, which lacks amino acid residues corresponding to 7 kDa from the 'intact' form.[4] Here we report the blue-excited RR spectra of P_r and P_{fr} of 'intact' pea phytochrome at 16°C which demonstrate that P_r and P_{fr} have different protonated struc-

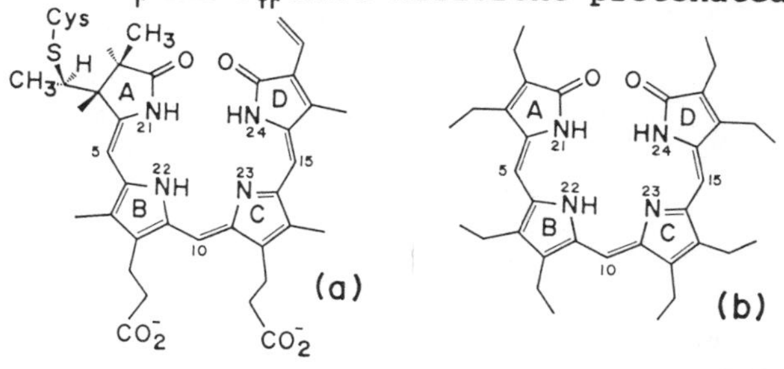

Fig. 1. Structure of phytochrome chromophore (a) and OEBV-h$_3$ (b)

tures of the chromophore.[5] In addition, we specify the protonation site from the studies on the isotope-labeled model compounds.

Materials and Methods

Intact pea phytochrome was isolated from 7 days-old etiolated seedlings of pea[6], precipitated by ammonium sulfate and resuspended in 50 mM HEPES and 1 mM EDTA, pH 7.8. Octaethylbiliverdin (OEBV-h_3, Fig. 1b) and its ^{15}N-derivative (OEBV-$^{15}N_4$-h_3) were synthesized from the corresponding Fe-porphyrins. The N-deuterated (OEBV-d_3) and protonated forms (OEBV-h_4^+) were also examined.

Raman spectra were observed at 16°C with a micro spinning cell, a double monochromator (Spex,1404) and an intensified diode array (PAR 1421HQ). Raman scattering was excited by the 407 nm line of a Kr^+ ion laser or the 364 nm line of an Ar^+ ion laser. The sample solution in the spinning cell was illuminated with either far-red (740 nm) or red (633 nm) light at another spot of the cell during the Raman measurements. The relative population of P_r and P_{fr} was determined with absorption spectra observed under the same two-color illumination conditions as used for Raman measurements. The 400 MHz ^{1}H- and 40.5 MHz proton-decoupled ^{15}N-NMR spectra were measured with a JEOL JNM GX400 FT NMR spectrometer. The chemical shifts were measured downfield in ppm relative to $Si(CH_3)_4$ for ^{1}H and to 2.9 M $^{15}NH_4Cl$ dissolved in 1N HCl for ^{15}N.

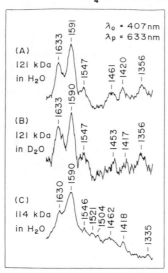
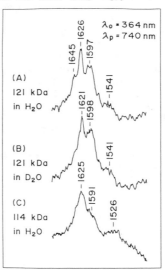

Fig. 2. The RR spectra of 'intact' (A and B) and 'large' (C) pea phytochromes excited at 364 nm under red light illumination (left) and at 407 nm under far-red light illumination (right); A and B were obtained for the H_2O solution and C for D_2O.

Results and Discussion

Figure 2 shows the RR spectra of 'intact' pea phytochrome excited at 363 nm under far-red illumination (left) and at 407 nm under red illumination (right) (A in H_2O and B in D_2O). The RR spectra of 'large' pea phytochrome in H_2O observed previously[4] under the same conditions as this measurement are also displayed by spectra (C). While the photo-steady state of 'large' phytochrome contains P_r, P_{fr}, and I_{bl}, in which I_{bl} is the bleached intermediate and is known to be accumulated under red light illumination,[7] that of 'intact' phytochrome little contains I_{bl} under the same condition. This may cause the relative intensity difference between spectra (A) and (C). Since P_r and P_{fr} of intact pea phytochrome have the second absorption band around 380 and 410 nm, respectively, the excitations of Raman scattering at 364 and 407 nm are expected to probe selectively the Raman bands of P_r and P_{fr}, respectively, due to the resonance effect. It is emphasized that the most prominent Raman band of P_r at 1626 cm^{-1} exhibits downshift to 1621 cm^{-1} in D_2O. In contrast, while none of double bond stretching RR bands of P_{fr} around 1550-1650 cm^{-1} shows a shift in D_2O, the RR band at 1461 cm^{-1} shifts to 1453 cm^{-1} and the main band around 1420 cm^{-1} disappears in D_2O. These observations indicate that the protonation structures of the chromophore of P_r and P_{fr} are different.

Figure 3 shows the RR spectra of OEBV-h_3 (A), OEBV-d_3 (B), OEBV-h_4^+ (C) and OEBV-d_4^+ (D) of ^{14}N- (left) and ^{15}N - compounds (right). These RR spectra did not change when spinning of the Raman cell was abolished, the laser power was raised, or the sample was cooled to $-50°C$. Upon N-deuteration of the A-, B-, and D-rings, the 1617 and 1467 cm^{-1} bands were shifted downward by -3 and -7 cm^{-1}, respectively. The RR spectrum of OEBV-h_3 changed definitely upon protonation of C-ring as shown by spectrum (C). The most prominent RR band of the protonated form is seen at 1616 cm^{-1} which exhibits

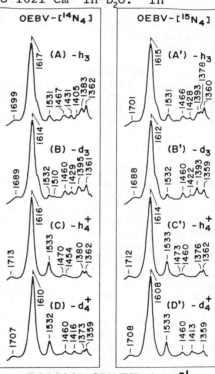

RAMAN SHIFT / cm^{-1}

Fig. 3. RR spectra of OEBV-h_3 (A) OEBV-d_3 (B), OEBV-h_4^+ (C), and OEBV-d_4^+ (D) in $CHCl_3$. left panel: ^{14}N compounds right panel: ^{15}N compounds

deuteration shift by -6 cm^{-1}. This well corresponds to the deuteration shift of the most prominent band of P_r at 1626 cm^{-1} by -5 cm^{-1}. The 1616 cm^{-1} band was shifted little by ^{15}N substitution. Therefore, it is most likely that the 1616 cm^{-1} band of OEBV-h$_4^+$ arises mainly from methine-bridge C=C stretching vibrations adjacent to C-ring. While most of RR bands show the ^{15}N frequency shifts smaller than 4 cm^{-1}, the 1405 cm^{-1} band of OEBV-h$_3$ and 1429 cm^{-1} band of OEBV-d$_3$ exhibit downshifts by 7-12 cm^{-1} upon ^{15}N substitution and seem to disappear upon protonation to C-ring. These may be associated mainly with the C=N stretching mode of C-ring.

In the ^{1}H NMR spectra (not shown), the methine proton showed a downfield shift and new signals appeared at 10.1 and 13.0 ppm upon protonation. For OEBV-d$_4^+$ these signals were weakened and with ^{15}N-enriched compound they were split into doublets. Therefore, the 13.0 and 10.1 ppm signals were assigned to B- and C-rings and A- and D-rings, respectively. The ^{15}N resonances of A- and D-rings and of B- and C-rings appeared at 107.1 and 184.3 ppm, respectively. Upon protonation, the latter signal showed a large upfield shift (51 ppm) while the former remained unshifted. This evidently indicates protonation to C-ring.

Recently Fodor et al.[8] pointed out that a RR band of P_{fr} around 1600 cm^{-1} was shifted in D$_2$O contrary to the present results. It is highly likely that the blue and far-red excitations probe different vibrational modes due to different resonant electronic transitions. If A-, B- and D-rings are primarily probed by far-red excitation, the RR spectrum of P_{fr} could be deuteration-sensitive.

The population ratio of constituents present in the photo-steady state of 'intact' phytochrome is scarcely affected by the bulk pH while that of 'large' phytochrome is affected.[7] This suggests that a proton is not transferred between bulk media and protein in the photoconversion between intact P_r and P_{fr}. On the other hand, the present experiment demonstrated that the protonation level of the chromophore is different between P_r and P_{fr}. Consequently, it is plausible that the extra 7kDa polypeptides of 'intact' phytochrome contain an acceptor residue for the proton dissociated from the chromophore.

1. S. P. A. Fodor, J. C. Lagarias, and R. A. Mathies, Photochem. Photobiol. 1988, 48, 129.

2. B. N. Rospendowski, D. L. Farrens, T. M. Cotton, and P.-S. Song, FEBS Lett, 1989, 258, 1.

3. D. L. Farrens, R. E. Holt, B. N. Rospendowski, P.-S. Song, and T. M. Cotton, J. Am. Chem. Soc. 1989, 111, 9162.

4. S. Tokutomi, Y. Mizutani, H. Anni, and T. Kitagawa, FEBS Lett, 1990, 269, 341.

5. Y. Mizutani, S. Tokutomi, K. Aoyagi, K. Horitsu and T. Kitagawa, to be published.

6. S. Tokutomi, M. Kataoka, J. Sakai, M. Nakasako, F. Tokunaga, M. Tasumi, and M. Furuya, Biochim. Biophys. Acta 1988, 953, 297.

7. S. Tokutomi, Y. Inoue, N. Sato, K. T. Yamamoto, and M. Furuya, Plant Cell Physiol., 1986, 27, 765.

8. S. P. A. Fodor, J. C. Lagarias, and R. A. Mathies, Biochemistry, 1990, 29, 11141.

THE "T=0 FS" SPECTRUM OF THE EXCITED STATE
OF THE REACTION CENTER OF *RHODOBACTER SPHAEROIDES* R-26
MEASURED AT 10K

Marten H. Vos[1,2], Jean-Christophe Lambry[1], M. Ashokkumar[1], Jacques Breton[2] and Jean-Louis Martin[1*]

1 Laboratoire d'Optique Appliquée, INSERM U275, Ecole Polytechnique ENSTA, 91120 Palaiseau, France

2 Service de Biophysique, Département de Biologie, CEN Saclay, 91191 Gif-sur-Yvette Cedex, France

The precursor of electron transport in reaction centers of photosynthetic organisms is the excited state of the primary donor P^*. Detailed knowledge of the spectrum of this state may be very valuable to understand the nature of the very fast rate (~1 ps at 10K[1]) of primary electron transport. In the reaction center of purple bacteria, P is a bacteriochlorophyll dimer. The reaction center contains four more pigments, two bacteriochlorophylls (B_L and B_M) and two bacteriopheophytins (H_L, which acts as electron acceptor, and H_M). The Q_Y transitions of the pigment complex of *Rhodobacter sphaeroides R-26* are located in the 750-850 nm region (Figure 1A), except the broad low exciton band of P, which is centered at 890 nm at 10K. Kinetic measurements have shown that formation of the P^* state, apart from bleaching of the latter band, is accompanied by spectral changes in the whole Q_Y region[1-3]. It is also known that the spectrum evolves on a timescale faster than $P^+H_L^-$ formation[3,4]. The spectrum of P^* has never been directly determined at high resolution. Here we present such a spectrum at both high temporal (50 fs) and spectral (1 nm) resolution; the measurements were performed at 10K.

Such a high resolution requires not only pump and probe pulses of short duration, but also the compensation of the group velocity dispersion within the broad band probe pulse over the entire spectral range (110 nm). In these experiments a 40 fs pump pulse at 870 nm and a near dispersion-free (50 fs) continuum probe pulse have been used.

The results are shown in Figure 1B. It can be seen that the t=0 fs P^*-P ΔA spectrum is very rich in structure. Remarkable are the two separate bands which appear near 797 and 806 nm, the broad shoulder in the 780 nm region, where the ground state does not absorb much, and the small but significant bleaching near 760 nm.

Some features can be readily assigned. The bleaching near 820 nm is presumably that of the higher exciton band of P (known to be at about 810 nm[5]), partly masked by the induced absorption band at 807 nm. The latter band can be attributed directly to the disruption of the dimeric coupling of P, as such bands can also be observed in the spectra of $P^+H_L^-$ [6] and P^T [7]. The broad feature at the blue side of the spectrum presumably reflects the $S_1 \rightarrow S_n$ transition of the excited state.

The presence of many more features than would be expected from the formation of an excited state of an isolated dimer indicates that the spectrum of the other (B and H) bands is seriously perturbed. A quantitative analysis of the transient spectrum shows that the induced bands in the 800 nm region are shifted, narrowed and increased in strength compared to the ground state spectrum. We suggest that these alterations reflect changes in the direct excitonic coupling of the pigments and/or in the pigment-protein interactions. More extensive analysis of the femtosecond spectral evolution of P* (in preparation) indicates that these interactions play an important role in the charge separation process.

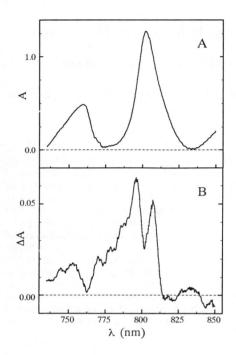

Figure 1

Ground state absorption spectrum (A) and 0±50 fs transient absorption spectrum due to excitation at 870 nm with a 40 fs pump pulse (B) of reaction centers of Rhodobacter sphaeroides R-26 at 10K. Q_A was kept reduced by the addition of 50 mM dithiothreitol. About 20% of the centers were excited at each shot. Pump and probe beam were polarized in parallel. The spectra were obtained with the same experimental setting.

ACKNOWLEDGMENT. MHV is recipient of an EEC Science Program grant.

REFERENCES

1. J. Breton, J.-L. Martin, G.R. Fleming, and J.-C. Lambry, Biochemistry, 1988, 27, 8267.
2. J.-L. Martin, J. Breton, A.J. Hoff, A. Migus and A. Antonetti, Proc. Natl. Acad. Sci. U.S.A., 1986, 83, 957.
3. W. Holzapfel, W. Finkele, W.M. Kaiser, D. Oesterhelt, H. Scheer, U. Stilz and W. Zinth, Chem. Phys. Lett., 1989, 160, 1.
4. C. Kirmaier and D. Holten Proc. Natl. Acad. Sci. U.S.A., 1990, 87, 3552.
5. J. Breton, Biochim. Biophys. Acta, 1985, 810, 235.
6. C. Kirmaier, D. Holten and W.W. Parson, Biochim. Biophys. Acta, 1985, 810, 49.
7. H.J. Den Blanken and A.J. Hoff, Biochim. Biophys. Acta, 1982, 681, 365.

PROTEIN STRUCTURE AND DYNAMICS FROM UV RESONANCE RAMAN SPECTROSCOPY

T. G. Spiro, Y. Wang, R. Purello, T. Jordan, C. Su, and K. Rodgers

Department of Chemistry
Princeton University
Princeton, New Jersey 08544, U.S.A.

INTRODUCTION

The advent of reliable pulsed lasers whose frequency can be shifted into the ultraviolet region has made possible the application of ultraviolet resonance Raman (UVRR) spectroscopy to biological molecules. For proteins, appropriate tuning of the laser yields selective enhancement of vibrational modes of the aromatic side chains phenylalanine, tyrosine and tryptophan, or of the amide bonds.[1-3] The amide UVRR bands are responsive to the protein secondary structure. Enhancements of the classic amide II and III bands decrease with increasing α-helical content,[4] while an additional UVRR band, amide S, is even more sensitive.[5-8] The aromatic side chain UVRR bands show frequency and/or intensity changes in response to changes in their local environment, particularly to changes in H-bonding.[9-12] They can therefore monitor protein conformation changes that alter the tertiary or quaternary structure.

The UVRR experiment can be applied in a time-resolved mode to monitor structural changes associated with transient processes in proteins. The R-T quaternary transition in hemoglobin is currently being studied in this manner, using a pair of nanosecond pulsed lasers, one to photolyze the CO adduct, and the other to generate UVRR spectra.[13-15] Changes in the tyrosine and tryptophan UVRR signals which develop 20 microseconds after photolysis are attributed to the alteration of subunit contacts during the R-T rearrangement. A preceding tryptophan signal is associated with interior tryptophan residues whose H-bonding is sensitive to helix-helix motions within the subunits.[15] Thus the aromatic residue signals provide indicators of protein motions associated with both tertiary and quaternary rearrangements on the R-T pathway.

Amide S and Helix Content

The classical amide I, II and III modes are seen in RR spectra of proteins obtained with excitation in the UV region. The amide II and III enhancements are about four times lower for α helical than for β sheet or loop polypeptide segments,[4] consistent with the two-fold reduction in the absorptivity associated with the first allowed electronic transition, due to the alignment of the transition dipoles in α helices. An additional band at ca. 1395 cm[-1] appears prominently in UVRR spectra of amides and proteins.[5-8] It has been labelled amide S[5] because of its sensitivity to secondary structure. Its intensity decreases linearly with increasing α helix content and is zero for pure α helices.[6] This band has been proposed to be the overtone of the amide V N-H out-of-plane bend,[7,8] but we assign it to a bending mode of the Cα–H bond, on the basis of its disappearance upon CαH/D substitution.[6] This substitution also produces an upshift of the amide III frequency, from ca. 1265 to ca. 1340 cm^{-1}, showing amide S and III to be strongly mixed. The UVRR enhancement of amide S is explained by this mixing. The disappearance of amide S in UVRR spectra of α helices is attributed to unmixing from amide III when the Cα-H bond is rotated into an orientation cis to the C=O bond, in α helices, from its usual trans orientation in β sheet and loop structures. This unmixing is believed also to account for the high frequency, ca. 1290 cm^{-1}, of the α helix amide III mode. The amide S/III mixing effects have been modelled with normal coordinate calculations.[6] Because of the high sensitivity of the amide S band to helix formation, and because it occurs in a relatively uncluttered region of protein UVRR spectra, it is a promising monitor of helix content in polypeptides and proteins. The correlation between amide S intensity and helix content has been shown to be quite reliable for a range of proteins.[6]

Probes of Aromatic Residue Environment

UVRR enhancements for the aromatic residues phenylalanine (Phe), tyrosine (Tyr) and tryptophan (Trp) are sensitive to the polarity of the environment due to shifts in the electronic transition frequencies and intensities.[9-12] In the case of Trp, H-bonding from the indole NH group has specific effects on the excitation profiles,[13,15] and, for some of the bands, there are perceptible frequency shifts as well.[11] The 880 cm^{-1} mode is particularly sensitive to H-bonding since it has a significant contribution from the N-H bending coordinate. The Trp W3 mode, at ca. 1555 cm^{-1}, is sensitive to variations of the dihedral angle, $\chi 1,2$, about the bond connecting the indole ring to the Cβ atom of the residue.[11] In the case of Tyr, the OH group can be either an H-bond donor or acceptor. These interactions produce opposite shifts in the excitation profiles, and also in some of the vibrational frequencies themselves.[9,15] The ν_{8a} and ν_{8b} ring modes, at ca. 1615 and 1600 cm^{-1}, are particularly useful as monitors of H-bonding, since they are strongly enhanced at 230 cm^{-1}, the optimum excitation wavelength for discriminating Tyr from Phe (whose ν_{8a} and ν_{8b} modes overlap those of Tyr).[9] The ν_{8b} frequency shifts down significantly when p-cresol, a model for Tyr, interacts with H-bond acceptors,[9] and increases when it interacts with H-bond donors.[15] Experiments with ovomucoid proteins containing a Tyr residue H-bonded to a carboxylate side chain showed a clear upshift in ν_{8b} when the H-bond was broken at low pH.[9]

Dynamics of the Hemoglobin R-T Switch

When the 230 nm-excited UVRR spectrum of the CO adduct of Hb, which is in the R state, is subtracted from that of deoxy-Hb, which is in the T state, well-defined difference signals are obtained for the ν_{8a} and ν_{8b} bands of Tyr and the W3 band of Trp, as seen in Fig.[13,15] The same difference signals, albeit with lower amplitude, are obtained when the spectrum of HbCO is subtracted from that of the HbCO photoproduct, provided that the delay between the photolysis and probe pulses is 20 μs (Fig.). If there is no delay, there is no difference spectrum, showing that the Tyr and Trp signals are unaffected by the ligation of the heme per se. 20 μs is also the time constant determined by Gibson and coworkers for the transition from fast to slow recombining Hb, and suggested by them to be the R-T quaternary transition. The Hb UVRR data provide direct structural support for this suggestion. Kitagawa and coworkers have likewise determined a 20 μs time constant for a perturbation of the 880 cm^{-1} Trp band in 218 nm-excited UVRR spectra of the HbCO photolysis product.[14]

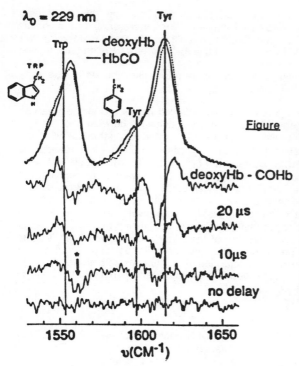

λ_0 = 229 nm

Trp — deoxyHb
— HbCO

TRP

Tyr

Tyr

deoxyHb - COHb

20 μs

10μs

no delay

1550   1600   1650
υ(CM⁻¹)

Figure 229 nm-excited UVRR spectra of deoxyHb and HbCO (top) and their difference, showing the changes associated with the W3 band of Trp and the ν_{8a} and ν_{8b} bands of Tyr. The bottom three traces are time-resolved difference UVRR spectra, obtained by subtracting the spectrum of HbCO from that of the photoproduct obtained at the indicated delays following a 532 nm photolysis pulse. The difference spectra all have a x3 scale expansion.

At a 10 μs delay, the Tyr difference signal is diminished, but the Trp difference signal alters its shape significantly, from a dispersive signal to a negative peak at slightly higher frequency (Fig.). At still shorter delays this signal also disappears. Thus the Trp W3 band monitors an additional process prior to the 20 μs R-T transition. The evolution of the W3 signal can be interpreted with the aid of the correlation between the W3 frequency and the indole $\chi_{1,2}$ dihedral angle discovered by Harada and coworkers.[11] When this correlation is applied to the $\chi_{1,2}$ angles calculated from the x-ray coordinates for the three inequivalent Trp residues of Hb, it is found that Trp β37, which is located at the interface between the α_1 and β_2 subunits, is predicted to have a frequency that coincides with a shoulder on the low-frequency side of the broad W3 band, while Trp α14 and Trp β15, which are located on the A helices in the subunit interiors, are predicted to have frequencies in the main part of the W3 band.[15] This assignment of the Trp β37 frequency was confirmed with a mutant protein, Hb Rothschild, in which Trp β37 is replaced by an arginine residue. The low-frequency shoulder was missing for this protein.

In comparing the R and T state crystal structures, one finds that the Trp β37 indole ring is H-bonded to the carboxylate group of Asp α94 in the T state but not in the R state. At 230 nm, H-bonding is expected to increase the W3 intensity, since the excitation profile, normally peaked at 220 nm, shifts to the red. Consequently the Trp β37 H-bond in the T state should produce a positive difference signal on the low-frequency side of the W3 band, and this is what is observed. The interior residues, Trp α14 and β15, are H-bonded to the OH groups of Thr and Ser residues, respectively, on the E helices . From the x-ray coordinates, the H-bond geometry is more favorable in the R than the T state, due to a slight realignment of the A and E helices. This change accounts for the negative difference signal in the main part of the W3 band, and the dispersive character of the overall difference signal. Moreover the altered shape of the difference signal at a 10 μs delay can also be understood

431

from these interactions, if we assume that the tertiary but not the quaternary movements have taken place at these earlier times. Then there should be a mainly negative difference signal in the main part of the W3 band, as is observed. We therefore infer that the realignment of the A and E helices is part of tertiary motion immediately preceding the R-T switch. It may be significant that the E helix forms the distal side of the heme binding pocket, and is therefore plausibly on the pathway from ligand dissociation to the R-T switch.

Hb has six inequivalent Tyr residues, all of which contribute to the $\nu 8a/\nu 8b$ bands. Only one of them changes its H-bond status significantly between the R and T states, however. Tyr $\alpha 42$, which is also at the $\alpha_1\beta_2$ interface, interacts with the side-chain of Asp $\beta 99$ in the T state, but not in the R state. It is therefore reasonable to associate the ν_{8a}/ν_{8b} difference signal with Tyr $\alpha 42$. Its interaction with Asp $\beta 99$, which is known to be important for cooperativity, has always been assumed to involve H-bond donation from the Tyr OH to the Asp carboxylate group. However, the UVRR difference signal does not show a downshift for ν_{8b}, as would be expected for H-bond donation, and as is seen in the ovomucoid proteins with carboxylate H-bond acceptors. Instead ν_{8b}, as well as ν_{8a}, shift up slightly, suggesting that the Tyr OH is acting as an H-bond acceptor. Examination of the x-ray coordinates of the T state show the Asp $\beta 99$ side chain to be surrounded by hydrophobic residues except for Tyr $\alpha 42$. Even though it is at the subunit interface, it is inaccessible to water. We therefore propose that the carboxylate group is protonated, and donates an H-bond to Tyr $\alpha 42$.

REFERENCES

1. I. Harada, H. Takeuchi in "Spectroscopy of Biological Systems," R.J.H. Clark and R. E. Hester, Ed., J. Wiley & Sons, New York, 1986, pp. 112-175.
2. B. Hudson and L. C. Mayne in "Biological Applications of Raman Spectroscopy," T. G. Spiro, Ed., J. Wiley & Sons, New York 1987, Vol. 2, pp. 181-210.
3. S. A. Asher, Ann. Rev. Phys. Chem., 1988, 39, 537-588.
4. R. A. Copeland and T. G. Spiro, Biochemistry, 1985, 26, 2134.
5. Y. Wang, R. Purrello, and T. G. Spiro, J. Am. Chem. Soc. 1989, 111, 8274-8276.
6. Wang, Y., R. Purrello, T. Jordan, and T. G. Spiro, J. Am. Chem. Soc. 1990, in press.
7a. S. Song, S. A. Asher, S. Krimm, and J. Vandekar, J. Am. Chem. Soc. 1988, 110, 8548-8550.
7b. S. Krimm, S. Song, and S. A. Asher, J. Am. Chem. Soc., 1988, 111, 4290-4294.
8. S. Song, S. A. Asher, and K. D. Shaw, J. Am. Chem. Soc. 1991, 113, 1155-1163.
9. P. Hildebrandt, R. A. Copeland, T. G. Spiro, J. Otlewski, M. Laskowski, and F. G. Prendergast, Biochemistry, 1988, 27, 5426.
10. H. Takeuchi, N. Watanabe, Y. Satoh, and I. Harada, J. Raman Spectrosc., 1989, 20, 233-237.
11. T. Miura, H. Takeuchi, and I. Harada, J. Raman Spectrosc., 1989, 20, 667.
12. G.-Y. Liu, C. A. Grygon, and T. G. Spiro, Biochemistry, 1989, 28, 5046.
13. C. Su, Y. D. Park, G.-Y. Liu, and T. G. Spiro, J. Am. Chem. Soc., 1989, 111, 3457.
14. S. Kaminaka, T. Ogura, and T. Kitagawa, J. Am. Chem. Soc., 1990, 112, 23.
15. C. Su, S. Subramaniam, and T. G. Spiro, J. Am. Chem. Soc. 1991, in press.
16. C. Sawicki and Q. H. Gibson, J. Biol. Chem., 1976, 251, 1533.

RESONANCE RAMAN INVESTIGATION OF CARBON MONOXIDE BINDING TO MUTANT MYOGLOBINS.

D. Biram, C.J Garratt and R.E Hester.

Department of Chemistry
University of York
Heslington
York YO1 5DD

Background.

The binding of carbon monoxide to myoglobin can be studied by resonance Raman spectroscopy since the CO vibrational modes are enhanced using Soret excitation. The Raman band at 508 cm^{-1} has been assigned to the υ (Fe-CO) stretching mode, a band at 577 cm^{-1} to the δ (Fe-C-O) bending mode and 1944 cm^{-1} to the υ (C-O) stretching mode.[1,2] The question of the exact geometry of the bound CO ligand and the role of the distal pocket amino acid residues in destabilising this ligand in relation to bound oxygen, has been the subject of much crystallographic and spectroscopic research.[2-4] The role of the distal histidine residue has been thoroughly investigated using site-directed mutagenesis and resonance Raman techniques,[5] but less attention has been given to the role of the valine E11 in myoglobin.[6]

Mutations.

Valine E11 has two methyl groups in close proximity to the bound ligand (see fig. 1). This has been changed to threonine which is isosteric with valine, but introduces another polar group into the pocket. This was expected to have the effect of further destabilising the CO ligand but stabilising bound oxygen, by providing the possibility of forming a second hydrogen bond in addition to that which exists between the ligand and the distal histidine. Kinetic experiments on this mutation[7] indicated an overall reduction in both the O_2 and CO affinity. It was thought that the hydroxyl group of the threonine side chain may be interacting with the distal histidine residue, perhaps disrupting the already existing hydrogen bond to the oxygen molecule. To investigate this possibility, further mutations were made: a double mutation involving the E11 valine to threonine and also the replacement of the E7 distal histidine with a valine residue. This removes the possibility of the interaction of threonine with the distal histidine side chain. The effect of introducing this additional mutation was investigated by producing the single mutation of E7 histidine to valine.

All plasmids containing mutant or wild-type genes for porcine myoglobin were constructed by S.J Smerdon.

Results.

Fig 2 shows the υ(Fe-CO) stretching mode for the mutant and wild-type myoglobins. This band has five components indicating five possible configurations of the CO ligand. The distal mutations show large variations in band shape, the major component of the band occuring at progressively lower wavenumber in the order : wild-type > val-thr > his-val > double mutation.

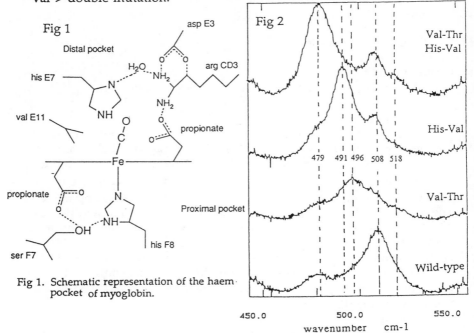

Fig 1. Schematic representation of the haem pocket of myoglobin.

Fig 2. Resonance Raman spectra, (450-550cm^{-1}), of wild-type and mutant, carbonmonoxy myoglobins in 50mM potassium phosphate buffer pH 8, excitation 413 nm, 3 cm^{-1} res

We attribute these effects to a combination of polar (hydrophilic) interactions involving the ligand O-atom and non-polar (hydrophobic) interactions associated with the ligand C-atom. Systematic changes in the relative stability of the ligand within the residues which define the binding site, can be linked to the mutations and these correlate well with the observed shift in the major υ(Fe-C) component. More subtle effects associated with mutation-induced changes in the general protein conformation must be involved to account for the presence of five component bands of varying intensity in all of the spectra.

References.
1. M Tsubaki, R.B Srivastava, N.T Yu. Biochemistry (1982) 21 1132
2. E.A Kerr, N.T Yu, D.E Bartnicki, H Mizukami. J.Biol.Chem (1985) 260 (14) 8360
3. J Kuriyan, S Witz, M Karplus, G.A Petsko. J.Mol.Biol (1986) 192 133
4. B.A Springer, K.D Egeberg, S.G Sligar, R.J Rohlfs, A.J Mathews J.S Olson. J.Biol.Chem (1989) 264 3057
5. D Morikis, P.M Champion, B.A Springer, S.G Sligar. Biochemistry (1989) 28 4791
6. K.D Egeberg, B.A Springer, S.G Sligar, T.E Carver, R.J Rohlfs, J.S Olson. J.Biol.Chem (1990) 265 (20) 11788
7. S.J Smerdon, D.Phil Thesis. (1990) University of York.

ESR SPECTROSCOPIC EVIDENCE FOR SITE-SPECIFIC RADICAL DAMAGE TO BSA AND METHAEMOGLOBIN

Michael J. Davies, Bruce C. Gilbert and Rachel M. Haywood

Department of Chemistry, University of York, Heslington, York YO1 5DD

Introduction

The modification of biological macromolecules by the reactions of oxygen-centred free radicals - formed in metal-catalysed reactions or by radiation - may have serious implications for the development of disease; whilst oxidative degradation of DNA induced by ·OH has been thoroughly investigated, related damage to proteins has been less well explored.

We report here the results of ESR investigations designed to establish that metal-peroxide reactions are responsible for radical attack (by ·OH) on a variety of proteins and to gain further information about the sites and consequences of damage. Incubation with suitable nitroso or nitrone spin-traps gives ESR signals from protein-derived radicals (labelled with nitroxides) which can be interpreted in terms of sites of attack (e.g. in terms of carbon-, sulphur- or oxygen-centred protein-derived radicals) and the extent of immobilization of the label. Emphasis is placed here on proteins which contain sulphur.

Radical damage to BSA

Reaction of Bovine Serum Albumin with ·OH (from Fe^{II}/H_2O_2) in the presence of the spin-trap 3,5-dibromo-4-nitrosobenzenesulphonic acid (DBNBS) gives a strong ESR spectrum characteristic of an immobilized spin adduct (evidently typical of a bulky carbon-centred precursor) as shown in the Figure. Enzymatic cleavage (protease or chymotrypsin) gives, after ca. 30 min, a mixture of sharper signals from mobile nitroxides which characterize sites of damage as *tertiary* radicals (probably back- bone derived) and *secondary* radicals (side-chain derived), as shown in the Figure. Complementary experiments with the cyclic spin-trap 5,5-dimethyl-pyrroline-N-oxide (DMPO) confirm that carbon-centred radicals are formed from BSA, though reduction in signal intensities from both traps in experiments with sulphur-blocked BSA (with N-ethylmaleimide; NEM) suggests that a proportion of initial attack occurs at thiol groups.

Reaction of BSA with cerium(IV) in the presence of DBNBS, which would be expected to oxidize thiol and phenoxyl groups, leads to the detection of a similar broad spectrum (removed when thiol groups are blocked) together with a relatively mobile spectrum (which is dependent on the presence of oxygen); with DMPO, both sulphur- and oxygen-centred radicals are trapped and recognized via their low values of $a_{\beta\text{-H}}$: cf. Figure.

Radical reactions of Methaemoglobin (MetHb)

The reaction of MetHb with hydrogen peroxide (which gives radicals even in the absence of exogenous iron) and DBNBS gives a broad ESR signal which is cleaved rapidly with enzymes (ca. 10 min) to give a nitroxide from a tertiary carbon-centred radical (and which is reduced in intensity when the protein is blocked with NEM). Reaction of MetHb with H_2O_2 in the presence of DMPO yields signals from sulphur- and oxygen-centred radicals (see Figure), though there is no evidence for the formation of carbon-centred radicals. Similar behaviour is observed with Ce^{4+}.

Conclusion

The hydroxyl radical reacts with BSA to give back-bone and side-chain radicals, via transfer of damage to some extent from thiol and, probably, phenoxyl-derived radicals. In contrast, initial radical damage to MetHb induced by hydrogen peroxide appears to be more selective; this involves thiyl radicals and also oxygen radicals, which may be obtained by electron transfer from cysteine and tyrosine residues to the oxidised haem centre, and subsequent reaction of the latter phenoxyl radicals with oxygen.

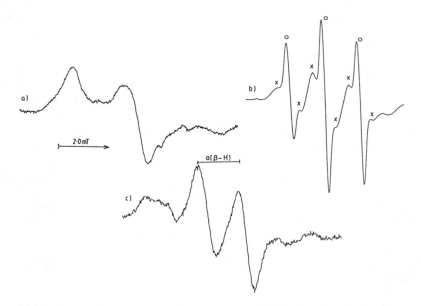

Figure: E.s.r. spectra from (a) BSA/Fe^{II}/H$_2$O$_2$ and DBNBS; (b) As (a), with enzymic cleavage; (c) MetHb, H$_2$O$_2$, DMPO, showing sulphur-adduct

PROBING THE ELECTROSTATIC ARRAY OF MYOGLOBIN WITH UV RESONANCE RAMAN AND HIGH PRESSURE FTIR SPECTROSCOPIES

O. SIRE[1], V. LE TILLY[1], B. ALPERT[1], P.-Y. TURPIN[2], L. CHINSKY[2] and P.T.T. WONG[3]

[1]Laboratoire de Biologie Physico-Chimique, Université de Paris VII, Paris, France.
[2]Laboratoire de Physique et Chimie Biomoléculaire, CNRS (U.A. 98), Institut Curie et Université Paris VI, Paris, France.
[3]Steacie Institute Molecular Science, NRC, Ottawa, Canada.

INTRODUCTION

To monitor the variations of polar interactions in the myoglobin matrix, far ultraviolet (223 nm) resonance Raman (UVRR) spectroscopy of tryptophan (Trp) and tyrosine (Tyr) residues was used. The subsequent H bond modifications involved in the secondary and tertiary structures were probed through a high pressure Fourier transform infrared study of the amide I' band.

This study on the monomeric myoglobin molecule was performed since it has been previously observed that the tetrameric hemoglobin molecule undergoes simultaneous solvent-induced tertiary strains and inter-subunit contact loosening[1].

METHODS

The UVRR spectra of myoglobin excited at 223 nm are dominated by the vibrational modes of the Trp and Tyr residues[2]. Owing to their polarisability properties, aromatic aminoacids are sensitive to any variation of the protein electrostatic field, and, consequently, their vibrational modes can be used to monitor such variations.

High pressure Fourier transform infrared spectra were obtained with myoglobin in the frequency region of the conformational sensitive amide I' band. Experiments were performed from 1 bar to 12 kbar in a diamond anvil cell. With a data treatment through Fourier self-deconvolution and derivation[3,4], it has been possible to probe the evolution of the protein conformation as a function of the pressure. Moreover, the high pressure technique also reveals the strength of hydrogen bonding[5,6]: an increase of the pressure results in a H bond strengthening, which is featured by a frequency downshift of the amide I' α-helix component. Since this downshift effect is all the more important than the H bond energy is higher[7], this allows to estimate the relative strengths of the H bonds in α-helices at atmospheric pressure in the protein.

RESULTS AND DISCUSSION

In the present work, the effects of pH (5.5, 7.0 and 8.5), ionic strength (0.05 and 0.1 M) and of iron-ligand nature (Mb⁺N₃⁻ vs Mb⁺H₂O) on polar interactions and H bonds strength were studied. In the pH range investigated, only His groups and the protein N terminal group undergo ionization state modifications. However, these alterations are not expected to directly modify the

437

vibrational modes of the Trp and Tyr (Trp 7 and 14; Tyr 103 and 146) residues, since no direct molecular contact can occur between the aromatic side chains and these histidine.

The main findings are:

i) pH drastically affects the aromatic amino-acid vibrational features of both azido and aquo metmyoglobin derivatives, this effect being largely dependent on ionic strength. Though the acid-base titration curve of $Mb^+N_3^-$ presents no sharp transition, the evolution of the corresponding UVRR spectra with pH is not monotonous.

ii) The strength of the α-helical H bonds is modulated by pH in a manner depending on ionic strengh: at low ionic strength (0.05 M) the H bonds strength increases with pH, whereas at higher ionic strength (0.1 M) the H bonds are stronger at pH 6.0 and 8.5 than at pH 7.0.

iii) Each aquo or azido derivative exhibits a specific evolution with pH of their UVRR spectra, indicating thereby that the protein matrix interactions are highly pH dependent whatever the nature of the Fe^{3+} ligand. Thus, Trp and Tyr vibrational modes depend on the iron-ligand nature though their remote location from the heme macrocycle.

iv) The α-helix H bonds are stronger in $Mb^+N_3^-$ than in Mb^+H_2O. From the IR amide I' band analysis it arises that the azido derivative exhibits a more regular conformation than the aquo derivative.

It is concluded that both the charge spatial distribution and the evolution of this distribution with pH depend on the iron-ligand nature. Each of the three parameters, ionic strength, pH and iron-ligand, modulates in a specific manner the electrostatic interactions inside the protein matrix. As a consequence, the strength of the H bonds in the molecule α-helices is highly dependent on these three parameters. Any variation in the electrostatic network - originating either in the protein matrix (His deprotonation) or at the active site (azide binding on the heme which results in a charge cancellation at the heme-iron) - induces a global reorganization of the interactions inside the protein matrix. It arises, hence, that the binding site properties of a globular monomeric protein such as myoglobin may be modulated by the solvent without significant protein conformational changes. This effect mainly results from the modulation of polar interactions inside the protein matrix.

REFERENCES
1. S. El Antri, O.Sire, and B. Alpert, Eur. J. Biochem. 1990, 191, 163.
2. S. El Antri, O.Sire, B. Alpert, P.Y. Turpin and L. Chinsky, Chem. Phys. Lett., 1989, 164, 45.
3. J.K. Kauppinen, D.J. Moffat, H.H. Mantsch and D.G. Cameron, Appl. Spectrosc., 1981, 35, 271.
4. D.G. Cameron and D.J. Moffat, Appl. Spectrosc., 1987, 4, 539.
5. P.T.T. Wong in Current Perspectives in High Pressure Biology, 1987 (H.W. Jannasch, R.E. Marquis, A.M. Zimmerman Eds.) 287, A cademic press, New York.
6. P.T.T. Wong and K. Heremans, Biochim. Biophys. Acta, 1988, 956, 1.
7. E. Fishman and H.G. Drickamer, J. Chem. Phys., 1956, 24, 548.

INVESTIGATION OF HEME DISTORTIONS IN ISOLATED SUBUNITS OF HUMAN HEMOGLOBIN BY RESONANCE RAMAN DISPERSION SPECTROSCOPY

R. Schweitzer-Stenner*, U. Dannemann and W. Dreybrodt

Institute of Experimental Physics
University of Bremen
2800 Bremen 33, Germany

INTRODUCTION

To probe the distortions of the heme groups in the isolated subunits of HbA, we measured the dispersion of the depolarisation ratio (DPR) of the Raman lines ν_4 (1375cm^{-1}) and ν_{10} (1638cm^{-1}) of α^{SH}-oxyHbA and β^{SH}-oxyHbA at various pH. They were analysed in terms of vibronic coupling parameters c^Γ_{es} given by[1]:

$$c^\Gamma_{es} = <e|dH_{e1}/dQ_R^{\Gamma(R)}\delta_{\Gamma(R)\Gamma(J)} \tag{1}$$

$$+ \sum_J (d^2 H_{e1}/dQ_R^{\Gamma(R)}dQ_J^{\Gamma(J)})\delta Q_J^{\Gamma(J)}|s>Q_R^{01}$$

where $|e>$, $|s> = |Q_x>$, $|Q_y>$, $|B_x>$, $|B_y>$ are the excited states of the porphyrin molecule. The operator $dH_{e1}/dQ_R^{\Gamma(R)}$ is the vibronic coupling operator of the Raman mode of symmetry $\Gamma(R)$ in ideal D_{4h}, whereas $(d^2 H_{e1}/dQ_R^{\Gamma(R)}dQ_J^{\Gamma(J)})$ relates to changes resulting from the symmetry classified perturbations $\delta Q_J^{\Gamma(J)}$ caused by the heme moiety and the peripheral side chains of the heme. Q_R^{01} is the transition matrix element of the Raman vibration. $\delta_{\Gamma(R)\Gamma(J)}$ is the Kronecker symbol.

RESULTS

I. α^{SH}-oxyHbA. While the DPR of the ν_4-mode shows dispersion, the DPR of the ν_{10}-mode does not (Fig.1). This clearly shows that the ν_{10}-mode is not subject to any significant B_{1g} and B_{2g}-distortions. The ν_4-mode, however, is affected by asymmetric B_{1g} and A_{2g}-distortions. The coupling parameters are listed in Table 1.

II. β^{SH}-oxyHbA. In contrast to α^{SH}-oxyHbA both modes display a significant dispersion of their DPR which varies with pH (Fig. 1). Especially the parameters c^{B1g}_{es} of the ν_4-mode reflecting asymmetric B_{1g}-distortions are significantly larger than those derived from the corresponding α^{SH}-oxyHbA-data. Moreover the ν_{10}-

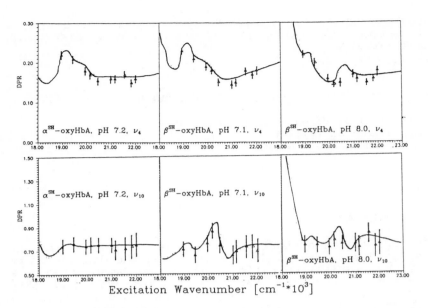

Excitation Wavenumber $[cm^{-1}*10^3]$

Fig. 1: Depolarization dispersion of the ν_4– and ν_{10}–modes of α^{SH}–oxyHbA and β^{SH}–oxyHbA.

mode is subject to strong B_{1g} and B_{2g} – distortions reflected by the coupling parameters c^{A1g}_{es} and c^{A2g}_{es}, respectively (Table 1).

DISCUSSION

Our results clearly show that in the ligated α–chains heme–protein coupling is significantly lower than in the ligated β–chains.

These findings are in accordance to low temperature kinetic and optical measurements by DeIorio et al.[3]. They found that the distribution of activation enthalpies for CO–rebinding is considerably broader for β^{SH}–HbA than for the α–chains. This indicates to differences of structural fluctuations.

Table 1: Vibronic coupling parameters of the ν_4 and ν_{10}–mode in cm^{-1} (α^{SH}–oxyHb: pH=7.2, β^{SH}–oxyHb: pH=8.0)

mode	C^{A1g}_{QQ}	C^{A1g}_{BB}	C^{A1g}_{BB}	C^{B1g}_{QQ}	C^{B1g}_{QB}	C^{B1g}_{BB}	C^{A2g}_{QB}
ν_4 (α)	46	99	59	−21	35	21	−24
ν_4 (β)	82	84	70	−66	72	−30	−41
ν_{10} (α)	−	−	−	−41	163	41	−
ν_{10} (β)	75	76	1	54	130	24	64

REFERENCES

1. R. Schweitzer-Stenner, Q. Rev. Biophys. 1989,22,381
2. E.DeIorio et al., Biophys.J., 1991, 59, 1–13

NEAR INFRARED SPECTRA OF LOW SPIN FERRIC HEMOGLOBIN AND MYOGLOBIN AT CRYOGENIC TEMPERATURES: AN EXAMPLE OF STRONG VIBRONIC COUPLING IN HEMEPROTEINS.

M. Leone*, A. Cupane, E. Vitrano and L. Cordone

Istituto di Fisica dell'Università and GNSM-CISM
90123 - Palermo - ITALY

The effect of vibrational coupling on the optical absorption spectra of metal proteins and other biomolecules has recently received increasing attention[1-6]. The interest of such studies arises from the fact that they provide information on the electronic structure of the active site and on the dynamic properties of the protein, in the proximity of the chromophore. Recent methodological advances in the analysis (transform and time-correlator techniques) permit to relate the absorption bands to the resonance Raman excitation profiles and therefore to determine normal mode frequencies and vibronic coupling constants and also to investigate various sources of spectral broadening in biomolecules. For metal proteins investigated to date, however, only weak coupling (linear coupling constants $S < 1$) with several vibrational modes has been observed, so that the optical spectra remain deceivingly smooth and without any clearly resolved vibronic structure even at cryogenic temperatures. We report here the temperature dependence of the near infrared absorption spectra of the two low spin ferric heme proteins cyanomet-hemoglobin (Hb-CN) and -myoglobin (Mb-CN), in two different solvents. At low temperature these spectra exhibit a well resolved fine structure; a suitable analysis, within the framework of (harmonic) Franck-Condon approximations taking into account vibronic coupling, shows that they are compatible with four electronic transitions strongly coupled to a single vibrational mode at 365 cm^{-1}. To our knowledge this is the first report of strong vibronic coupling in the spectra of metallo-proteins. Near infrared absorption spectra (1800-800 nm) of Hb-CN and Mb-CN in D_2O solutions at room temperature were already reported by Eaton and coworkers[7], who attribute the observed bands to charge transfer transitions from the top four filled porphyrin π orbitals to the d_{yz} iron orbitals. This assignment is also supported from magnetic circular dichroism and single crystal polarized absorption spectroscopy. The observed spectra can be made consistent with the suggested assignment by taking into account a strong vibronic coupling. Assuming that the electronic transitions are strongly coupled with a single vibrational mode and that the functional form of the absorption lineshape is Gaussian, the following expression is used to fit the spectra:

$$A(v)/v = \sum_{k=1}^{4} I_k \, \exp(-S_k) \sum_{m_1=0}^{10} S_k^{m_1}/m_1! \, \exp[-(v - v_{0k} - m_1 \Omega_1)^2 / 2\sigma_k^2] \tag{1}$$

where Ω_1 and m_1 are the frequency and the occupation number of the vibrational mode considered, I_k, v_{0k}, σ_k and S_k are the intensity, fundamental frequency, halfwidth and linear coupling constant of the k-th electronic transition, respectively. In Fig.1 are reported the nir spectra of Mb-CN at 25 K and 295 K, together with the fitting in terms of Eq.1; as can be seen fine structures are clearly resolved. These structures

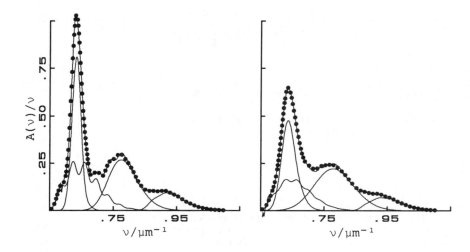

Figure 1 Nir spectra of Mb-CN in 65% glycerol $(OD)_3/D_2O$ at 25 K (left panel) and 295 K (right panel). The continuous lines represent the experimental spectra and the component structure relative to each electronic transition; dots represent the lineshape synthesized according to Eq.1.

remain unaltered even if the spectra of carbonmonoxy derivatives are used as baseline. The quality of the fitting indicates that the mentioned assumptions can be considered correct. The value of the parameter Ω_1 is 365 cm^{-1} for both Hb-CN and Mb-CN and does not depend upon temperature and solvent composition. The value of 365 cm^{-1} is typical of out-of-plane deformation modes involving pyrrole and/or peripheral substituents tilting. In contrast the values of the other parameters that describe the electron-vibration coupling depend upon the particular protein, solvent composition and temperature. We think that a finding worth of note is that the coupling between the charge transfer transitions and the out-of-plane vibrational modes is not a local property of the heme group but rather is sensitive also to heme-globin-solvent interactions, as evidenced by the differences observed between hemoglobin and myoglobin in different solvent conditions.

REFERENCES

1. K. T. Schomacker and P. Champion, J. Chem. Phys., 1986, 84 5314.
2. A.R. Mantini, M.P. Marzocchi and G. Smulevich, J. Chem. Phys.,1989, 91, 85.
3. L. Cordone, A. Cupane, M. Leone and E. Vitrano, Biophys. Chem., 1986, 24, 259.
4. M. Leone, A. Cupane, E. Vitrano and L. Cordone, Biopolymers, 1987, 26, 1769.
5. L. Cordone, A. Cupane, M. Leone, E. Vitrano and D. Bulone, J. Mol. Biol., 1988, 199, 213.
6. A. Cupane, M. Leone, E. Vitrano and L. Cordone, Biopolymers, 1988, 27, 1977.
7. A. Schejter and W.A. Eaton, Biochemistry, 1984, 23, 1081.

MIGRATION OF SMALL MOLECULES THROUGH PROTEINS: A MEASURE OF THEIR STRUCTURAL DYNAMICS

Jehuda Feitelson

Department of Physical Chemistry
The Hebrew University of Jerusalem
Jerusalem 91904, Israel

A variety of experiments[1,2] as well as calculations[3] indicate that proteins are dynamic entities which undergo structural fluctuations. It is these movements within the protein which are thought to enable small molecules to migrate through the otherwise compact protein structure.

We propose that the quenching of an optical probe within the protein can be used as a measure for the dynamic structural changes in the latter, provided that the probe is: (a) at a well defined position within the protein and (b) the actual quenching step is very rapid. Under condition (b) the overall quenching rate is diffusion controlled.

Triplet state quenching by small molecules of Zn protoporphyrin (ZnPP), which was substituted for the native Fe porphyrin in myoglobin, Mb was studied.[4] The solution was excited by a laser flash and the ^{3}ZnPP decay measured as a function of quencher concentration and of temperature. Zn hematoporphyrin, ZnHP, was used to measure the quenching rate of the chromophore in solution. It can be shown that as long as the quenching rate of ^{3}ZnHP in solution is significantly larger than in the protein, the observed quenching rate is determined by the migration rate of the quencher through the protein matrix.

The results in Table 1 show that the room temperature quenching rates in the proteins are by an order of magnitude smaller than in solution and that the activation energies, E_a, for all three quenchers O_2, anthraquinone sulfonate (AQS) and methylviologen (MV) are very similar. These results can be interpreted in terms of Northrup and McCammon's gated reaction theory[5] in which the process in a restricted medium is described by a reaction coordinate proper and by an auxiliary coordinate which is, in our case, related to changes in the protein conformation. If the direct passage between two locations in the protein A and B, is blocked by a high potential barrier, there might exist a different protein conformation where the free energy of activation of A $\rightarrow$ B is lower.

Table 1: ZnMb, ZnHb and ZnHP Quenching by O_2, AQS and MV

	$\alpha Zn\beta FeHb$	Zn_4Hb	ZnMb	ZnH
$k_q(AQS)$	2.1×10^8 M^{-1}s^{-1}	1.7×10^8	2.9×10^8	22×10^8
$k_q(MV)$	8.2×10^7	8.3×10^7	4.5×10^7	36×10^8
$k_q(O_2)$	9×10^7	1.5×10^8	1×10^8	11×10^8
$E_a(AQS)$	5.4 Kcal/mol	6.0	5.8	3.1
$E_a(MV)$	5.6	6.21	7.4	3.9
$E_a(O_2)$	6.3	6.9	6.0	3.2

The theory shows that our data can be explained if the observed activation energy is assumed to measure the energetics of the protein structural change i.e., the opening of a "gate". Once this conformational change has taken place, small and large quencher molecules alike can pass through the previously constricted region. This means that a rather wide gate of about 6Å is required to accommodate the AQS and MV molecules and that the activation energy of the gating is $E_a \sim 6$ kcal/mol.

We might ask whether the quaternary structure of hemoglobin (Hb) with its four Mb-like subunits and its R-T state transitions affects the above small molecule migration rate. To this end, ZnPP was substituted into the α subunits of Hb while the β subunits were oxidized to the Fe^{3+} state. This $\alpha Zn\beta FeHb$ served as a model for the R state Hb. Tetra ZnPP substituted Hb, Zn_4Hb, represents the T state of the protein. Again, as for Mb, the quenching rate of the ^3ZnPP in the Hb was measured.[6]

Table I shows that the E_a values in both R and T Hb are remarkably similar to those of ZnMb. This means that the migration of ligands into the binding site and hence the subunit conformational dynamics and its energetics is not affected by the Hb quaternary structure. Of course, the binding of ligands depends very much upon the (R or T) state of Hb.

REFERENCES

1. H. Frauenfelder, G.A. Petsko and D. Tsernoglou, Nature, 1979, 558.
2. J.R. Lakowicz and G. Weber, Biochem., 1973, 12, 4171.
3. M. Karplus and J.A. McCammon, CRS Crit. Revs. Biochem., 1981, 9, 293.
4. N. Barboy and J. Feitelson, Biochem., 1989, 28, 5450.
5. S.H. Northrup and J.A. McCommon, J. Am. Chem. Soc., 1984, 106, 930.
6. J. Feitelson and G. McLendon, Biochem., 1991, in press.

INTERACTIONS BETWEEN CYTOCHROME C AND PHOSPHOLIPID VESICLES STUDIED BY RESONANCE RAMAN SPECTROSCOPY

P. Hildebrandt[*], T. Heimburg[¶], and D. Marsh[¶]

Max–Planck–Institut für Strahlenchemie, Stiftstr. 34–36, D–4330 Mülheim, F.R.G.
[¶]Max–Planck–Institut für biophysikalische Chemie, Postfach 2841, D–3400 Göttingen, F.R.G.

INTRODUCTION

Binding of cytochrome c to negatively charged surfaces such as metal electrodes, polyanions, and phospholipid vesicles can produce significant conformational changes in the heme pocket[1-3]. Detailed resonance Raman (RR) studies have revealed that in such complexes cytochrome c exist in a conformational equilibrium between two states (I, II). While in state I the structure of the heme pocket is unchanged compared to the unbound protein, in state II the heme crevice opens, weakening the coordinative bond of the heme iron with the methionine–80 ligand. Thus, a thermal coordination equilibrium between a six–coordinated low–spin (6cLS) and a five–coordinated high–spin configuration (5cHS) is established in state II. These structural changes were also detected in complexes with the physiological redox partner cytochrome c oxidase[4], suggesting that they are of functional significance for the biological electron transfer.

In the present work we have continued these studies by systematically analyzing the interactions of ferri–cytochrome c with negatively charged phospholipid vesicles. RR spectroscopy was employed to determine the conformational equilibria in cytochrome c formed in complexes with different lipid systems. The relative contributions of the various species of cytochrome c were determined from the relative intensities of those RR bands which have been found to be characteristic markers for the individual conformational states[3]. The effect on protein binding on the lipid bilayer structure was analyzed by ^{31}P–NMR spectroscopy. The main goal of this study was to find correlations between the structural changes in the protein and in the lipid upon complex formation. A detailed description of the experimental methods, data analysis, and sample preparation is given elsewhere[4,5].

RESULTS AND DISCUSSION

Fig. 1 shows the so obtained concentration ratio of state II (including both the 5cHS– and the 6cLS–configuration) and state I of cytochrome c bound to dioleoyl phosphatidylglycerol (DOPG) – dioleoyl glycerol (DOG). The admixture of uncharged lipids (DOG) significantly lowers the fraction of state II. At 10% DOG, the equilibrium constant has decreased by a factor of ~ 2, but a further increase of DOG up to 30% lowers the equilibrium constant only from ~ 1.6 to ~ 1.2. On the other hand, the binding energy is approximately the same (– 39 kJ/M) in pure DOPG and DOPG–DOG (70:30). Since the main contribution to the binding energy originates from the electrostatic interactions between the anionic phopholipid headgroups and the positively charged lysine residues around the heme crevice of cytochrome c, these findings imply that complex formation with cytochrome c induces an asymmetric distribution of charged and uncharged lipids at the protein binding site. This would in turn affect the bilayer structure of the phospholipid vesicles which is, in fact, reflected by the ^{31}P–NMR spectra. While in the absence of cytochrome c the ^{31}P–NMR spectra reveal the characteristic signature of a regular bilayer structure, admixture of DOG favors the the formation of an inverted hexagonal phase which is prevailing at 50% DOG. In the presence of cytochrome c, however, all the spectra are dominated by an isotropic peak which may result from local curvatures on the bilayer structure. Such curvatures may facilitate the electrostatic interactions with the lysine residues on the surface of the spherically–shaped cytochrome c.

The state II/state I ratio of cytochrome c in the pure DOPG system is constant in the temperature range between 15 and 30°C (Fig. 2). On the other hand, in the dimyristoyl phosphatidylglycerol (DMPG) system there is an abrupt increase of the concentration ratio at ~ 25°C. This temperature corresponds to the phase transition gel→fluid of DMPG, which in the case of DOPG is about 20° lower. Below and above this temperature, the concentration ratio is largely constant and in both phases of DMPG clearly larger than in the DOPG system.

It is not possible to correlate the equilibrium constants of the conformational equilibrium with the surface charge density of the various lipid systems since in the absence of specific ion binding effects, the lipid surface charge density is expected to be greater in the gel phase than in the fluid phase of DMPG and yet smaller in the fluid phase of DOPG. Hence, there must be some other effect contributing to the electrostatic control of the conformational equilibrium. Based on previous studies of cytochrome c bound to charged interfaces it was argued that depending on the individual lysine residues which are involved in the complex formation either the conformational state II or I are stabilized. Thus, not only the number but also the specific spatial arrangement of the phospholipid headgroups at the protein binding site may govern the conformational equilibrium. The latter parameter should strongly depend on the surface flexibility of the lipid bilayer which is substantially different in the fluid and in the gel phase as well as in saturated and unsaturated lipid systems. Thus, surface charge density and the structural flexibility may exert opposite effects on the conformational distribution of the bound cytochrome c.

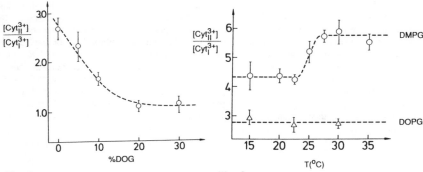

Fig. 1:
Concentration ratio of state II and state I in cytochrome c/DOPG–DOG, as a function of DOG content.

Fig. 2:
Temperature–dependence of the conformational equilibrium state II/state I for cytochrome c bound to DMPG and DOPG.

REFERENCES

1.   P. Hildebrandt and M. Stockburger, Biochemistry, 1989, 28, 6710.
2.   P. Hildebrandt, Biochim. Biophys. Acta, 1990, 1040, 175.
3.   P. Hildebrandt, T. Heimburg, and D. Marsh, Eur. Biophys. J., 1990, 18, 193.
4.   P. Hildebrandt, T. Heimburg, D. Marsh, and G. L. Powell, Biochemistry, 1990, 29, 1661.
5.   T. Heimburg, P. Hildebrandt, and D. Marsh, submitted to Biochemistry.

INFLUENCE OF VIBRATIONAL COUPLING ON RESONANCE RAMAN SPECTRA OF DIOXYGEN ADDUCTS OF COBALT PORPHYRINS

Leonard M. Proniewicz[1*] and James R. Kincaid[2]

[1] *Regional Laboratory of Physicochemical Analyses and Structural Research of the Jagiellonian University. Jagiellonian University, ul. Karasia 3, 30-060 Cracow, Poland.*
[2] *Chemistry Department, Marquette University, Milwaukee, Wisconsin 53233, U.S.A.*

The oxygen transport proteins, hemoglobin (Hb) and myoglobin (Mb), are probably the most thoroughly studied of all biomolecules. The essential goal of these studies remains the elucidation of the structural and electronic factors that control varied functional properties of their active sites. Vibrational spectroscopy has long served as a powerful probe of the structure and bonding of exogeneous ligands of heme proteins[1]. The major obstacle in studies of dioxygen adducts of Hb and Mb by resonance Raman (RR) is the inability to observe enhancement of $\nu(O-O)$ in these proteins. Fortunately, this vibration is efficiently enhanced in cobalt-substituted heme proteins[2]. However, RR spectra of these proteins yield complex vibrational patterns that were originally interpreted in terms of multiple dioxygen adduct structures. Recently, we[3-6] offered an alternative interpretation which invoked vibrational coupling of the bound dioxygen with the internal modes of the trans nitrogeneous base. This well documented model does not require the existence of multiple dioxygen adducts.

In the present work we focus on the issue of axial ligand disposition using cobalt(II) "unprotected" (CoTPP) as well as bis-diamidostrapped porphyrins having short (9 carbon atoms), CoAz2, and long (12 carbon atoms), CoDe2, strap lengths in an attempt to induce distortion of the trans-axial ligand. Spectral measurements were carried out by using the "minibulb" technique[3]. The RR spectra were recorded on a Spex Model 1403 double monochromator equipped with a Hamamatsu R928 PMT and a Spex DM1B computer. Excitation at 406.7 nm (10 mW at the sample) was accomplished with a Coherent Model I100-K3 krypton ion laser.

The RR spectrum of the $^{18}O_2$ adduct of the CoTPP-d8 complex with 3,5-lutidine in CH2Cl2 shows a strong isolated band at 1081 cm^{-1} that is due to the $\nu(^{18}O-^{18}O)$. Thus, a simple harmonic oscillator approximation yields an expected $\nu(^{16}O-^{18}O)$ at 1147 cm^{-1}.

However, the $^{16}O_2$ adduct exhibits two relatively strong bands at 1160 and 1138 cm^{-1}. The appearance of this doublet is clearly ascribable to coupling of the $\nu(O-O)$ (inherent frequency of 1147 cm^{-1}) with the 1150 cm^{-1} internal mode of coordinated base. Thus, the stronger 1138 cm^{-1} component which contains a larger contribution from $\nu(O-O)$ is down-shifted by 9 cm^{-1} from its inherent value of 1147 cm^{-1}, while the weaker 1160 cm^{-1} component is up-shifted 9-10 cm^{-1} from its inherent frequency of 1150 cm^{-1}. This doublet is replaced by a single band at 1153 cm^{-1} due to $\nu(O-O)$ when CoAz2 is used under the same conditions. Obviously, the presence of the short strap imposes an encumbrance to trans-axial ligand binding. Inspecting molecular models indicates, that in the case of the 9 carbon atom polymethylene strap, 3,5-lutidine is too large to fit into the cavity. While this base ligand can assume either a parallel or perpendicular orientation relative to the strap, in either case the rotation of the 3,5-lutidine relative to the porphyrin mean plane is inhibited. In addition, both the strap and the axial-base deviate from the plane perpendicular to the mean porphyrin plane. Thus, the distortion experienced by the bulky 3,5-lutidine must be sufficient to effectively eliminate coupling of the internal modes. It is interesting to note that this apparent ligand distortion is relaxed in the case of the dioxygen adducts with CoDe2, which possesses a longer, more flexible strap. Thus, in this case, the RR spectrum looks identical with that of CoTPP-d8.

It has to be mentioned that similar behavior as reported above is observed also in the RR spectra of dioxygen adducts in the presence of other nitrogeneous trans-axial ligands such as pyridine or imidazole.

REFERENCES

1. see for example T.G. Spiro (Ed.), *Biological Applications of Raman Spectroscopy*, Wiley and Sons, New York 1987, vol. 3.
2. M. Tsubaki, N.-T. Yu, Proc. Natl. Acad. Sci. U.S.A., 1981, 78, 3581.
3. L.M. Proniewicz, K. Nakamoto, J.R. Kincaid, J. Am. Chem. Soc., 1988, 110, 4541.
4. A. Bruha, J.R. Kincaid, *ibid.*, 1988, 110, 6006.
5. L.M. Proniewicz, A. Bruha, K. Nakamoto, E. Kyuno, J.R. Kincaid, *ibid.*, 1989, 111, 7050.
6. L.M. Proniewicz, J.R. Kincaid, *ibid.*, 1990, 112, 675.

LOW TEMPERATURE OPTICAL SPECTROSCOPY OF COBALT-SUBSTITUTED HEMOCYANIN FROM *Carcinus maenas.*

E. Vitrano, A. Cupane*, M. Leone, V. Militello and L. Cordone
Istituto di Fisica, Università di Palermo, 90123 - Palermo - Italy.

B. Salvato, M. Beltramini, L. Bubacco and P. Rocco
Dipartimento di Biologia, Università di Padova, 35100 - Padova - Italy.

Studies on cobalt-substituted hemocyanins are stimulated by the fact that substitution of the native metal with Co^{2+} is a useful tool to probe the structure of the active site, owing to the peculiar spectroscopic properties of the cobalt ion which are dependent on its coordination number, geometry and microenvironment[1]. In particular, since the binuclear $(Co^{2+})_2$-hemocyanin does not bind oxygen even at high pressure, this derivative has been taken as a good model for the structure of the active site of deoxy-hemocyanin[1]. In this work we report the temperature dependence of the optical absorption spectra of the binuclear cobalt substituted hemocyanin from *Carcinus maenas*; in fact, optical spectroscopy in wide temperature ranges has proved to be a useful experimental tool to study the structural and dynamic properties of the active site of metallo-proteins[2,3]. In the presently studied hemocyanin derivative the two Co^{2+} ions are thought to be in a symmetrical distorted tetrahedral geometry, coordinated by three histidyl residues from the protein and by a bridging exogenous ligand (probably a water molecule or an OH^- ion)[4]. The room temperature absorption spectrum exhibits broad bands in the wavelength range 500-650 nm that are generally attributed to cobalt d-d transitions[1].

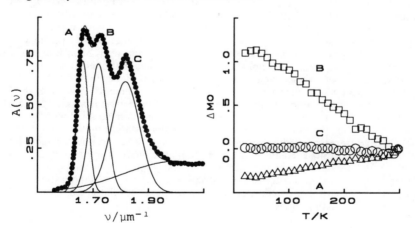

Figure 1 Left panel: deconvolution of $(Co^{2+})_2$-Hc spectrum at 20 K. Dots are the experimental points; the continuous lines are the gaussian components and the synthesized band profile. Right panel: temperature dependence of the integrated intensities (M_0). ΔM_0 is defined as $M_0(T)/M_0(295\ K)-1$.

449

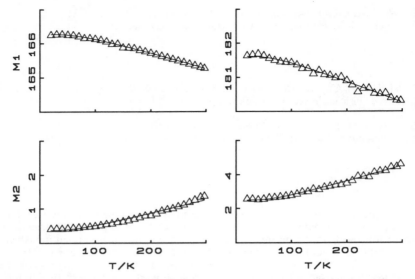

Figure 2 M_1 and M_2 as a function of temperature. Left: band A; right: band C. The M_1 scale is cm^{-1} 10^{-2}; the M_2 scale is cm^{-2} 10^{-5}.

Fig.1 shows the optical spectrum of $(Co^{2+})_2$-hemocyanin at 20 K; three bands are clearly resolved and are called A, B and C in order of increasing frequency. The fact that only three bands are resolved even at cryogenic temperatures is consistent with the hypothesis that the two cobalt ions have symmetrical environments in the active site. The right panel of Fig.1 shows the temperature dependence of the integrated intensities; by lowering the temperature bands A and B display large intensity variations whereas the intensity of band C remains almost unchanged. Fig.2 shows the temperature dependence of the first moment (M_1, peak frequency) and of the second moment (M_2, halfwidth) of bands A and C. As can be seen they exhibit a marked temperature dependence that is indicative of their coupling with low frequency vibrational modes of the nearby nuclei; the mean effective frequency of these modes is about 140 cm^{-1}. In contrast, M_1 and M_2 relative to band B are almost temperature independent (data not shown); this indicates that band B is coupled to high frequency modes whose mean effective frequency is higher than 800 cm^{-1}. The reported results enable to suggest the following band assignments. In view of the large intensity variation with temperature bands A and B should arise from charge transfer transitions and in particular band A from a N_{his}-Co transition, coupled with low frequency vibrational modes (probably N_{his}-Co-N_{his} deformation modes), and band B from a charge transfer transition from the π orbitals of the exogenous bridging ligand into the cobalt d orbitals, coupled with high frequency modes (probably ligand-Co stretches). We also tentatively attribute band C, whose intensity is temperature independent, to a cobalt d-d transition; moreover its M_1 and M_2 temperature dependence indicates that cobalt d orbitals are coupled with the above mentioned N_{his}-Co-N_{his} deformation modes.

REFERENCES

1. B. Salvato and M. Beltramini, Life Chem. Rep. , 1990, 8, 1.
2. L. Cordone, A. Cupane, M. Leone and E. Vitrano, Biophys. Chem., 1986, 24, 259.
3. A. Cupane, M. Leone, E.Vitrano and L. Cordone, Biophys. Chem., 1990, 38, 213.
4. J. Lorosch and W. Haase, Biochemistry, 1986, 25, 5850.

STRUCTURAL CHANGES OF THE CHROMOPHORE OF PHYTOCHROME PROBED BY INFRARED SPECTROSCOPY

J. SAKAI,[*],[**] E. H. MORITA,[*] H. HAYASHI,[*] M. FURUYA[**]
and M. TASUMI[*]

[*] Department of Chemistry, Faculty of Science,
     The University of Tokyo, Bunkyo-ku, Tokyo 113, Japan
[**] Plant Biological Regulation Laboratory,
     Frontier Research Programs, RIKEN Institute,
     Wako, Saitama 351-01, Japan

Phytochrome is a chromoprotein which controls a red and far-red photoreversible reaction in green plants. It consists of two equivalent subunits, each of which has a molecular weight of 124000 and contains a verdin type tetrapyrrole chromophore. The most prominent feature of phytochrome is its reversible phototransformation between the red light absorbing form (Pr) and the far-red light absorbing form (Pfr).

Although there are many studies on the phototransformation of phytochrome, few observations of structural changes of the chromophore in situ are reported, except for those using circular dichroism[1] and resonance Raman scattering[2-4] measurements. Recently we reported red-light induced infrared spectral changes for the so-called 'large' phytochrome.[5] An advantage of using infrared spectroscopy is that infrared light in the region below 2000 cm^{-1} induces neither phototransformation nor any other photoreactions, and therefore can be used safely for spectral measurements. To assign the observed spectral changes between Pr and Pfr, we have used in the present study biliverdin dimethylester and its geometric isomers as model compounds of the chromophore. The results enable us to discuss the photoconversion mechanism and the local environment of the chromophore.

Biliverdin dimethylester (BVDME) was synthesized from biliverdin (Sigma), treated with methanol-BF$_3$, and purified with alumina column chromatography. Geometric isomers of BVDME were obtained as follows. BVDME was adsorbed on alumina powder and irradiated with white actinic light in n-hexane. Then the pigment mixture was extracted and three isomers (ZZZ, EZZ, and ZZE) were separated from each other with alumina column chromatography.

Infrared measurements were made on an FT-IR spectrophotometer (JEOL JIR-5500). Infrared spectra of BVDME were measured in KBr disc and in chloroform solution. A CaF$_2$ cell (thickness 50 μm) was used for measuring solution spectra.

In the carbonyl stretching region, the spectrum in KBr disc shows two peaks at 1732 and 1697 cm^{-1}, while that in chloroform solution shows three peaks at 1738, 1701 and 1678 cm^{-1}. According to an X-ray analysis, BVDME exists as a dimer in crystal and one of the two lactam CO groups is hydrogen-bonded. Therefore we assign the bands at 1732 and 1738 cm^{-1} to the ester CO, the bands at 1697 and 1701 cm^{-1} to the free lactam CO, and the band at 1678 cm^{-1} to the hydrogen-bonded lactam CO.

Difference spectra in the carbonyl stretching region obtained for BVDME between ZZE and ZZZ and between EZZ and ZZZ are compared with the difference spectra between P_r and P_{fr}. The results show that the ZZE - ZZZ difference spectrum is similar to the P_{fr} - P_r difference spectrum, but the EZZ - ZZZ difference spectrum is not. This suggests that the structural change of the chromophore occurring in the P_r to P_{fr} phototransformation corresponds to the ZZE to ZZZ isomerization.

Infrared spectra were measured for ZZE and ZZZ in mixtures of chloroform and methanol with various mixing ratios. With increasing methanol concentration, the bands due to the CO stretches shift to lower wavenumbers without changing band shapes.

The ZZE - ZZZ difference spectra in the mixed solvents were compared with the P_{fr} - P_r difference spectrum. The ZZE - ZZZ difference spectrum obtained from a 50 : 50 mixture of chloroform and methanol shows similarities with the P_{fr} - P_r difference spectrum in the lactam CO region, whereas the ZZE - ZZZ difference spectrum obtained from neat chloroform solution is similar to the P_{fr} - P_r difference spectrum in the ester CO region. The results indicate that in phytochrome the lactam CO groups are hydrogen-bonded, but the ester CO groups are not.

REFERENCES

1. K. Schaffner, S.E. Braslavsky and A.R. Holzwarth, 'Advances in Photochemistry', D.H. Volman, G.S. Hammond and K. Gollnick, eds., John Wiley & Sons, Inc., New York, 1990, Vol. 15, p. 229.
2. S.P.A. Fodor, J.C. Lagarias and R.A. Mathies, Photochem.Photobiol., 1988, 48, 129.
3. D.L. Farrens, R.E. Holt, B.N. Rospendowski, P.-S. Song and T.M. Cotton, J.Am.Chem.Soc., 1989, 111, 9162.
4. S. Tokutomi, Y. Mizutani, H. Anni and T. Kitagawa, FEBS Lett., 1990, 269, 341.
5. J. Sakai, E.H. Morita, H. Hayashi, M. Furuya and M. Tasumi, Chem.Lett., 1990, 1925.

ASSIGNMENT OF RESONANCE RAMAN SPECTRA OF PHYCOBILIPROTEINS. EXPLANATION FOR CONFORMATION SENSITIVITY OF CERTAIN REGIONS.

[1]Szalontai, B., [2]Gombos, Z. and [3]Lutz, M.

[1]Institute of Biophysics, [2]Institute of Plant Physiology, Biological Research Center, Hungarian Academy of Sciences, H-6701 Szeged, P.O.B. 521, Hungary
[3]Département de Biologie Cellulaire et Moleculaire, C.E.N. Saclay, 91191 Gif-sur-Yvette, Cédex, France

In cyanobacteria and red algae there are special chromoproteins for light harvesting. They are called phycobiliproteins due to their covalently bound open-chain tetrapyrrole bilin chromophores. For optimal light-harvesting phycobiliproteins are organized into macromolecular complexes named phycobilisomes[1]. There are three major types of phycobiliproteins, phycoerythrin (PE), phycocyanin (PC) and allophycocyanin (APC).

PC and APC contain identical chromophores called phycocyanobilin (PCB). PCB is almost the same as the chromophore of phytochrome, the light sensory pigment of higher plants[2]. Therefore any information obtained on PCB conformation, PCB-apoprotein interaction or on the assignment of PCB resonance Raman (RR) spectra in phycobiliproteins can also be applied for phytochromes which are much more difficult to prepare and handle as compared to phycobiliproteins.

We have studied RR spectra of PC, APC both with visible and UV excitation in the aggregated, monomeric and denatured states of the proteins[3,4,5]. Both monomerisation and denaturing was achieved by decreasing the pH of the protein solution by dialyzing against a buffer of the required pH. From these studies we could conclude that while chromophore conformations were similar in the native proteins there were differences between PC and APC in the process of folding of the chromophores upon denaturing these proteins but chromophore conformations in the denatured proteins were again very close to each other[5]. We could also find evidence for excited state interactions between chromophores in APC, thus supporting the excitonic origin of the characteristic 652 nm band of its absorption spectrum[5].

Here we present the analysis of ^{14}N-^{15}N isotope substitution experiments which meant the exchange of all nitrogen atoms of the proteins and the chromophores. Observing isotopic shifts in several RR bands we could assign them as it is shown in Fig. 1. The behavior of the marker band region[5] (1590-1650 cm^{-1}) is rather strange, showing large isotope sensitivity for the 1643 cm^{-1} band in native proteins which disappears in its downshifted position around 1625 cm^{-1} in denatured protein.

By using computer-aided decomposition of this complex region we have revealed that (i) the downshift of the 1642 cm^{-1} band in the UV-excited RR spectra is the result of the disappearance of a component around 1650 cm^{-1} and the appearance of a new band around 1622 cm^{-1} upon folding of the chromophore. The 1650 cm^{-1} component we assign to C=C methin bridge vibration between pyrrole rings C and D when the methin bridge has an *anti* conformation in the extended chromophore. The 1622 cm^{-1} component should correspond to the vibration of the same methin bridge having *syn* conformation in the denatured protein. (ii) Another important finding is that the downshift (5-7 cm^{-1}) of the 1642 cm^{-1} band of the UV-excited RR spectra upon ^{15}N substitution is not the result of a direct isotope effect on any of the component bands, but it reflects an overall change of the chromophore conformation due to perturbations caused by the isotope substitution in the structure of the apoprotein. This fact shows on one hand, what a delicate equilibrium is maintained the biologically optimal conformation in phycobiliproteins, and on the other hand, how sensitive method RR spectroscopy, is being able to reveal these minor changes. Our results offer opportunities to discuss assignments made on RR and SERR spectra of phytochrome as well.

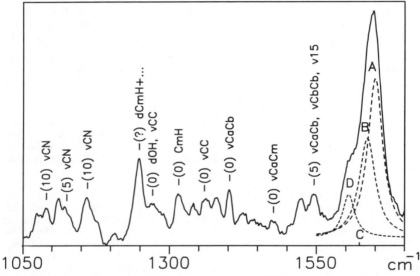

Fig. 1. Allophycocyanin resonance Raman spectrum (exc.:363.8nm) Numbers in brackets indicate downshifts observed upon ^{14}N-^{15}N substitution. Above the bands are the proposed assignments. Dashed curves indicate the components of the complex 1642 cm^{-1} band. Their assignments: A and C - νC=C at the methin bridge between rings C and D of the chromophore in *anti* and *syn* conformation, respectively (C has zero intensity in native proteins where the chromophores are in an extended conformation; B - νC=C in the pyrrole ring; D - νC=C in a methin bridge.

REFERENCES

1. E. Gantt, Ann.Rev.Plant.Physiol., 1981, 32, 327
2. H. Scheer, Angew.Chem.Int.Edn.Engl., 1981, 20, 241
3. B. Szalontai, Z. Gombos, V. Csizmadia,
   Biochem.Biophys.Res. Commun., 1985, 130, 358
4. B. Szalontai, Z. Gombos, V. Csizmadia and M. Lutz,
   Biochim. Biophys.Acta, 1987, 893, 296
5. B. Szalontai, Z. Gombos, V. Csizmadia, K. Csatorday and
   M. Lutz, Biochemistry, 1989, 28, 6467

SURFACE ENHANCED RESONANCE RAMAN SPECTRA OF PHYCOCYANIN AND ALLOPHYCOCYANIN

[1]Debreczeny, M., [2]Gombos, Z., and [1]Szalontai, B.

[1]Institute of Biophysics, [2]Institute of Plant Physiology,
Biological Research Center, Hungarian Academy of Sciences,
H-6701 Szeged, P.O.B. 521, Hungary

In the case of application of Surface-Enhanced Resonance Raman (SERR) spectroscopy for biological compounds the question frequently arisen whether the SERR spectra reflect native molecules or the biological compound is degraded due to the close interaction with the metal surface. It is especially of interest in the case of proteins. Although several investigation has shown (e.g. ref 1) that the native protein structure can be preserved on the surface of metal colloids, no SERR study was done on the basis of a complete set of resonance Raman (RR) data (i.e. both UV- and visible-excited RR spectra).

Here we present the results of such a study carried out on C-phycocyanin (CPC) and on allophycocyanin (APC), the light-harvesting proteins of cyanobacteria[2]. Our previous RR studies[3-5] have shown that visible- and UV-excited RR spectra of these proteins are not the same. There are bands which are in resonance with only one of the electronic transitions. The chromophore structures in the two proteins were very close to each other; however, the stability of the apoprotein on pH changes was higher in CPC monomers as compared to that of APC. RR data also gave evidence for the excitonic origin of the 652 nm strong absorption band of APC. By the use of nitrogen isotope substitution we have assigned several bands of the phycobiliprotein RR spectra as well. There were two regions, the 1230-1300 cm^{-1} and the 1590-1650 cm^{-1}, which were sensitive to minor alterations of the chromophore conformation as well. These informations were useful in deciding whether the SERR spectra of these proteins reflect native states or not. In the case of altered states on the metal surfaces the RR data offered a possibility to understand the nature of the changes which took place upon the metal-protein complex formation.

This approach in the study of SERR spectra of biliproteins might have more general implications as well, since similar problems ought to be resolved in the vibrational spectroscopy of an other open-chain pyrrole-containing protein, the light sensory pigment of higher plants, phytochrome. Hence, our results on phycobiliproteins may help to interpret differences between the RR and SERR spectra of phytochrome as well.

It seems that the crucial point in the SERR spectroscopy of proteins is to find the optimal conditions for the 'activation' of the silver sol. In our case 65 µl 1% ascorbic acid was added to 2 ml silver sol which lowered the pH of the sol (prepared according to ref 6) to around pH 4.3. After this treatment we obtained high quality reproducible SERR spectra from the added (~20µl) proteins being previously dialysed at different pH values between 7.0 and 1.6.

Fig.1. shows the visible-excited RR and SERR spectra of phycocyanin. It can be seen that (i) the native pH 7.0 spectra are very similar, indicating that surface interaction did not change significantly the chromophore conformation, (ii) identical changes take place in the two types of Raman spectra upon decreasing the pH of the protein solution. Decreasing the pH causes folding of the chromophores from their extended conformation to a cyclo-helical one. This is best indicated by the decreasing intensity of the 1597 cm^{-1} band and the downshift of the band around 1650 cm^{-1}. An other important conclusion of the experiments is thatthe structure of the protein which was adjusted with previous pH equilibrium is not altered either during the adsorption or after. The SERR spectra of protein solutions of different pH added to the pH 4.3 sol show the same differences as the RR spectra do.

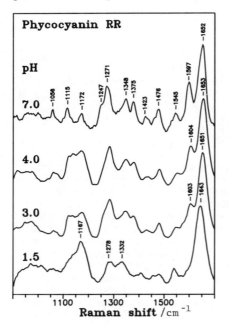 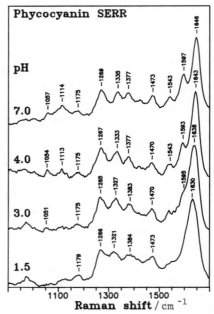

Fig. 1. Phycocyanin resonance Raman (RR) and surface- enhanced resonance Raman (SERR) spectra (exc.:488.0 nm) at differed pH values of the protein solutions. The silver sol was 'activated' with ascorbic acid and its pH was always around pH 4.3.

REFERENCES

1. P. Hildebrandt, T.G. Spiro, J.Chem.Phys., 1988, 92, 3355
2. E. Gantt, Ann.Rev.Plant.Physiol., 1981, 32, 327
3. B. Szalontai, Z. Gombos, V. Csizmadia,
    Biochem.Biophys.Res. Commun., 1985, 130, 358
4. B. Szalontai, Z. Gombos, V. Csizmadia and M. Lutz,
    Biochim. Biophys.Acta, 1987, 893, 296
5. B. Szalontai, Z. Gombos, V. Csizmadia, K. Csatorday and
    M. Lutz, Biochemistry, 1989, 28, 6467
6. P. Hildebrandt, M. Stockburger, J.Phys.Chem., 1984, 88, 5935

PHOTOINDUCED ELECTRON TRANSFER IN THE SYSTEM MODELLING PHOTOSYNTHESYS : PROTOPORPHYRIN IX DICATION - INORGANIC ANION, STUDIED BY PHOTOCHEMICAL HOLE BURNING.

M. A. Drobizhev and M. N. Sapozhnikov

P. N. Lebedev Physics Institute of the Academy of Sciences of the USSR, Leninsky prospect 53, 117924 Moscow, USSR

Under irradiation of frozen glassy solution of molecules by monochromatic laser light some molecules absorb light resonantly via narrow zero-phonon lines (ZPL) (linewidth $\propto 10^{-3} - 1$ cm^{-1} at T = 1 - 20K). Being in the electronic excited state such molecules can be transformed into the product having shifted absorption spectrum. This process results in an appearence of the narrow zero-phonon persistent hole in the broad inhomogeneous absorption band of the solute molecules. The spectral hole parameters (depth, width, shape) depend on the absorbed dose. From the kinetics of hole burning rate constant k, quantum yield of phototransformatin φ and true ZPL width γ are usually found under conditions when the laser line is much narrower then the ZPL [1].

We have proposed the method of determination of k, φ and γ at any relations between laser linewidth and γ.

In acid ethanolic solution of the protoporphyrin IX the dication is formed, when two addition protons link with nitrogen atoms in the center of the tetrapyrrol ring. Two counter ions (acid anions) are connected with the center by hydrogen bonds and are located under and over the molecule plane on the axis of symmetry. The absorption spectrum and photochemical properties of dications and metalloporphyrins are similar. It is known that under broad band irradiation porphyrin dications can accept electrons from counter ions in frozen ethanol at 77K. Hence, we supposed that the photochemical hole results from the same mechanism.

The rates of electron tunneling from anion to excited protoporphyrin dication have been measured using photochemical hole burning kinetics at 5K. We have studied the dependence of the electron transfer rate on the reaction enthalpy, which is determinated by ionization potential of investigated anions (Cl$^-$, Br$^-$,

I^-, NO_3^-, SO_4^{2-}) in frozen solution at 5 K. This dependence is described within the framework of Jortner's model[2] by Poisson distribution: $W \propto e^{-S} S^p/p!$, where $S = E_R/\langle h\nu \rangle$, $p = \Delta E/\langle h\nu \rangle$, E_R is the reorganization energy, $-\Delta E$ is the reaction enthalpy, $\langle h\nu \rangle$ is the average vibronic quantum, participating in the reaction heat exchange. A fitting of experimental curve to the model calculations gives the following parameters: $S = 20$, $\langle h\nu \rangle = 0.2$ eV, $\Delta E = 0.8$ $(Cl^-) - 2.5(I^-)$ eV. The value of S coincides with that found for chlorophyll – cytochrome pair in Chromatium (J. Jortner, 1976). The value of $\langle h\nu \rangle = 0.2$ eV corresponds to the intense vibration (1600 cm^{-1}) of porphyrin skeleton. We have also estimated the ΔE for Cl^-, according to the literature data, equal to 1 eV, which is near the experimental value.

This presented system may be a good model for photoreduction of chlorophyll in Chromatium.

REFERENCES

1. 'Persistent Spectral Hole Burning : Science and Applications', Ed. W. E. Moerner, Springer-Verlag, Berlin Heidelberg N.Y. London Paris Tokyo, 1988.
2. J. Jortner, J. Chem. Phys., 1976, 64, 4860.